Sub-contracting under the JCT 2005 Forms

To our wives, Carol and Karen, and our families, with our grateful thanks for all of the help and support given to us whilst writing this book, without which help and support we would not have been able to do so.

Sub-contracting under the JCT 2005 Forms

Peter Barnes

and

Matthew Davies

WILEY-BLACKWELL

A John Wiley & Sons, Ltd., Publication

Library of Congress Cataloging-in-Publication Data

Barnes, Peter, 1954–
Sub-contracting under the JCT 2005 forms/Peter Barnes and Matthew Davies.
      p. cm.
Includes bibliographical references and index.
ISBN 978-1-4051-7788-7 (hardback: alk. paper)   1. Construction contracts–England.
2. Construction industry – Sub-contracting–England.   I. Davies, Matthew.   II. Title.
KD1641.B376 2008
343.42′078624–dc22
2008017934

A catalogue record for this book is available from the British Library.

Set in 9.5/11.5 pt Palatino by SNP Best-set Typesetter Ltd., Hong Kong
Printed in Great Britain by TJ International Ltd, Padstow, Cornwall

1   2008

Contents

Preface

The construction industry is almost entirely dependent upon sub-contractors, and on nearly all construction projects a vast majority of the work is carried out by sub-contractors.

Despite this, traditionally, the contract terms relating to sub-contractors have not been given the same consideration as the contract terms in the employer/main contractor relationship. Possibly because of this, there is a clear shortage of books that deal with contract law in the context of sub-contracts.

However, the modern construction industry is fully aware of the importance of sub-contractors, and understands that sub-contract terms must be given as much consideration as any other terms. Often the obligations and liabilities of sub-contractors are a vital part of the overall contractual chain leading from the employer/purchaser passing though the contractor and passing down to the sub-contractors and the sub-subcontractors.

It is clear that the JCT recognises the importance of sub-contract terms, and, in its 2005 suite of contracts and sub-contracts, it has published a large range of sub-contract forms to be used with many of the current standard forms.

One of the co-authors of this book (Peter Barnes) has already written a book on the JCT 05 Standard Building Sub-contract, and this latest book now deals with all of the other JCT 05 sub-contracts that have been published up to the date of publication of this book.

The 'other' JCT 05 sub-contracts dealt with within this book are:

- the Design and Build Sub-contract (Revision 1, 2007);
- the Major Project Sub-contract (Revision 1, 2007);
- the Intermediate Sub-contract (Revision 1, 2007);
- the Intermediate Sub-contract with sub-contractor's design (Revision 1, 2007);
- the Intermediate Named Sub-contract (Revision 1, 2007);
- the Minor Works Sub-contract with sub-contractor's design;
- the Short Form of Sub-contract (Revision 1, 2007); and
- the Sub-subcontract (Revision 1, 2007).

Given the wide range of this book, some matters dealt with in considerable detail in the JCT 05 Standard Building Sub-contract book mentioned above have, by necessity, been dealt with more superficially in this book. Therefore, in that regard, this book may be considered as being complementary to that earlier book and the reader is therefore encouraged to refer to that earlier book when a more detailed analysis of legal or contractual matters that are common to all of the JCT 05 sub-contract forms is required.

It is expected that this book will principally be used by sub-contractors and main contractors. However, it will also be of particular interest to other construction professionals and lawyers who need to have an understanding of the contractual relationship and the allocation of risk between main contractors and sub-contractors under the JCT sub-contract forms.

Both of the authors of this book have had wide and direct experience in contractor/sub-contractor relationships, and have also had extensive experience in avoiding and resolving disputes in those relationships. Consequently, between us, we have encountered nearly every type of problem that can occur in a contractor/sub-contractor relationship, and we have attempted to interweave some of that knowledge and experience into the text of this book.

This book obviously cannot take into consideration amendments to the standard sub-contract forms that may be made by main contractors. However, it is considered that this book will provide a very useful guide as to the allocation of risk between the parties that exists in the unamended form, and also will help both contractors and sub-contractors to understand the possible effects that amendments made to the text of the standard sub-contracts will have on the parties' respective rights and obligations.

Peter Barnes and Matthew Davies
March 2008

Abbreviations

ADR	Alternative Dispute Resolution
CIMAR	Construction Industry Model Arbitration Rules
CIS	Construction Industry Scheme
CPR	Civil Procedure Rules
CSCS	Construction Skills Certification Scheme
DBSub	Design and Build Sub-contract
DBSub/A	Design and Build Sub-contract agreement
DBSub/C	Design and Build Sub-contract conditions
HGCRA 1996	Housing Grants, Construction and Regeneration Act 1996
ICSub	Intermediate Sub-contract
ICSub/A	Intermediate Sub-contract agreement
ICSub/C	Intermediate Sub-contract conditions
ICSub/D	Intermediate Sub-contract with sub-contractor's design
ICSub/D/A	Intermediate Sub-contract with sub-contractor's design agreement
ICSub/D/C	Intermediate Sub-contract with sub-contractor's design conditions
ICSub/NAM	Intermediate Named Sub-contract
ICSub/NAM/A	Intermediate Named Sub-contract agreement
ICSub/NAM/C	Intermediate Named Sub-contract conditions
ICSub/NAM/E	Intermediate Named Sub-contractor/employer agreement
ICSub/NAM/IT	Intermediate Named Sub-contract invitation to tender
ICSub/NAM/T	Intermediate Named Sub-contract tender
JCT	The Joint Contracts Tribunal
MPSub	Major Project Sub-contract
MWSub/D	Minor Works Sub-contract with sub-contractor's design
SCDP	Sub-contractor Designed Portion
SCWa/F	Sub-contractor Collateral Warranty for a Funder
SCWa/P&T	Sub-contractor Collateral Warranty for a Purchaser or Tenant
ShortSub	Short Form of Sub-contract
SubSub	Sub-subcontract

Chapter 1
Background and Introduction

The purpose and use of JCT sub-contracts

The Joint Contracts Tribunal (JCT) was established in 1931 and for 75 years has produced standard forms of contracts, guidance notes and other standard documentation for use in the construction industry. In 1998, the JCT became incorporated as a company limited by guarantee and commenced operation as such in May 1998.

Currently, JCT forms require the agreement of eight constituent bodies before they are issued by the JCT. Those bodies are:

- The Association of Consulting Engineers
- The British Property Federation
- The Construction Confederation
- The Local Government Association
- The National Specialist Contractors Council
- The Royal Institute of British Architects
- The Royal Institution of Chartered Surveyors
- The Scottish Building Contract Committee Limited

The above listed bodies are intended to be reasonably representative of the interests across the construction industry, namely, the employers, the consultants, the contractors and the sub-contractors; and the JCT forms are naturally a reflection of these competing interests.

In 2005, the JCT authorised the publication of an entirely new suite of contracts and sub-contracts, etc.

Amongst the sub-contracts that have been issued are the following sub-contracts which are the subject matter of this book:

- Design and Build Sub-contract
- Major Project Sub-contract
- Intermediate Sub-contract
- Intermediate Sub-contract with sub-contractor's design
- Intermediate Named Sub-contract
- Minor Works Sub-contract with sub-contractor's design
- Short Form of Sub-contract
- Sub-subcontract

The only JCT sub-contract that has currently been issued that is not covered by this book is the Standard Building Sub-contract which is the subject matter of a separate book[1] by one of the authors of this book.

In January 2008, the JCT Works Contracts under the JCT Management Contract were published. These works contracts are for individual works packages and are not sub-contracts as such and are therefore outside the scope of this book.

The Design and Build Sub-contract

The Design and Build Sub-contract is only suitable for sub-contracts where the main contract is the 2005 edition of the JCT Design and Build Contract. This sub-contract is in effect a modern-day version of the well known DOM/2 form of sub-contract.

Within this book the JCT Design and Build sub-contract will be referred to as DBSub (the designation given to it by the JCT).

This sub-contract can be used even where the sub-contractor is not to carry out any design work at all, and can be used for either lump sum or remeasurement contracts.

The sub-contract can be used when the main contract works are to be carried out in Sections as detailed in the main contract.

The sub-contract comprises two documents:

- Agreement (denoted by the JCT and in this book as DBSub/A).
- The conditions (denoted by the JCT and in this book as DBSub/C).

DBSub/A incorporates by reference (under article 1 of DBSub/A) the DBSub/C sub-contract conditions; therefore it is not necessary in formulating a sub-contract to include the said conditions in the documents to be exchanged. Despite this, and particularly where the parties do not regularly use this form of sub-contract, it is probably sensible that the DBSub/C is included in the documents to be exchanged so that each party has a complete set of the documentation applicable to the sub-contract.

The Major Project Sub-contract

The Major Project Sub-contract is only suitable for sub-contracts where the main contract is the JCT Major Project construction contract. Within this book the JCT Major Project Sub-contract will be referred to as MPSub (the designation given to it by the JCT).

This sub-contract can be used whether or not the sub-contractor is responsible for design, and can be used for either lump sum or remeasurement contracts.

The sub-contract can be used when the main contract works are to be carried out in Sections as detailed in the main contract.

[1] *The JCT 05 Standard Building Sub-Contract* written by Peter Barnes and published by Blackwell Publishing.

This sub-contract does not have a separate Agreement document, but relies on a section within the sub-contract called the Sub-contract Particulars to provide the information that would normally be included in the appendix to a pre-2005 JCT form.

The Intermediate Sub-contract

This sub-contract is suitable for sub-contracts where the main contract is the JCT Intermediate Building Contract, *or* where the main contract is the JCT Intermediate Building Contract with contractor's design, but, even where the main contract form with contractor's design is used, the Intermediate Sub-contract is only for use when the sub-contractor is not liable for design.

Within this book the Intermediate Sub-contract will be referred to as ICSub (the designations given to it by the JCT).

This sub-contract can be used when the main contract works are to be carried out in Sections as detailed in the main contract.

This sub-contract cannot be used where the sub-contractor is to be named or where the sub-contractor is to carry out any design work, but can be used for either lump sum or remeasurement contracts.

The sub-contract comprises two documents:

- The agreement (denoted by the JCT and in this book as ICSub/A).
- The conditions (denoted by the JCT and in this book as ICSub/C).

ICSub/A incorporates by reference (under article 1 of ICSub/A) the ICSub/C sub-contract conditions; therefore it is not necessary in formulating a Sub-contract to include the said conditions in the documents to be exchanged. Despite this, and particularly where the parties do not regularly use this form of sub-contract, it is probably sensible that the ICSub/C is included in the documents to be exchanged so that each party has a complete set of the documentation applicable to the sub-contract.

The Intermediate Sub-contract with sub-contractor's design

This sub-contract is suitable for sub-contracts where the main contract is the JCT Intermediate Building Contract with contractor's design.

Within this book, the Intermediate Sub-contract with sub-contractor's design will be referred to as ICSub/D (the designations given to it by the JCT).

This sub-contract can be used when the main contract works are to be carried out in Sections as detailed in the main contract.

This sub-contract is to be used where the sub-contractor is to design all or part of the sub-contract works, and can be used for either lump sum or remeasurement contracts.

This sub-contract is not to be used where the sub-contractor is to be named.

The sub-contract comprises two documents:

- The agreement (denoted by the JCT and in this book as ICSub/D/A).
- The conditions (denoted by the JCT and in this book as ICSub/D/C).

ICSub/D/A incorporates by reference (under article 1 of ICSub/D/A) the ICSub/D/C sub-contract conditions; therefore it is not necessary in formulating a sub-contract to include the said conditions in the documents to be exchanged. Despite this, and particularly where the parties do not regularly use this form of sub-contract, it is probably sensible that the ICSub/D/C is included in the documents to be exchanged so that each party has a complete set of the documentation applicable to the sub-contract.

The Intermediate Named Sub-contract

This sub-contract is suitable for sub-contracts where the main contract is the JCT Intermediate Building Contract *or* where the main contract is the JCT Intermediate Building Contract with contractor's design.

Within this book the Intermediate Named Sub-contract will be referred to as ICSub/NAM (the designations given to it by the JCT).

As its name implies, this sub-contract is to be used where a sub-contractor is named (by the employer) to carry out the sub-contract works.

The Intermediate Building Contract and the Intermediate Building Contract with contractor's design are the only contracts in the JCT suite of contracts that allow for the naming of a sub-contractor. It should be noted that a named sub-contractor does not have the status that a nominated sub-contractor had under the JCT 98 Main Contract Form[2], and after it is appointed a named sub-contractor effectively simply becomes a domestic sub-contractor. The procedure for appointing a named sub-contractor and the status of a named sub-contractor is dealt with later within this chapter.

This sub-contract can be used when the main contract works are to be carried out in Sections as detailed in the main contract.

It is not suitable for any sub-contract work that forms a part of the contractor's designed portion, but can be used where the sub-contractor is required to design all or part of the sub-contract works. In other words, any design carried out by a named sub-contractor cannot be design that forms part of the contractor's designed portion.

This sub-contract can be used for either lump sum or remeasurement contracts.

The Sub-contract comprises four documents:

- the invitation to tender (denoted by the JCT and in this book as ICSub/NAM/IT);
- the tender (denoted by the JCT and in this book as ICSub/NAM/T);
- the agreement (denoted by the JCT and in this book as ICSub/NAM/A); and
- the conditions (denoted by the JCT and in this book as ICSub/NAM/C).

As noted above, a named sub-contractor only applies where the main contract is the JCT Intermediate Building Contract or the JCT Intermediate Building

---

[2]It is not possible to nominate sub-contractors under the JCT 05 Standard Building Contract, nor under any other JCT 05 main contract form.

Contract with contractor's design, and is a person (or company) named by the employer (or architect/contract administrator) who is to be employed by the main contractor[3].

As noted earlier within this chapter, the named sub-contract procedure must not be used in relation to work (or its design) falling within the contractor's designed portion.

The process of a named sub-contractor being appointed starts with an invitation to tender being issued.

Invitation to tender (ICSub/NAM/IT)

The ICSub/NAM/IT is completed and sent to the proposed named sub-contractor by the employer or the architect/contract administrator, *not by the main contractor.*

The ICSub/NAM/IT sets out the information that is required to enable the prospective named sub-contractor to provide a tender, and that standard form advises the prospective named sub-contractor that he should not in any way seek to vary the basis upon which he is being asked to tender.

The sub-contractor then provides a tender.

Tender (ICSub/NAM/T)

Upon receipt of the invitation to tender form ICSub/NAM/IT, the proposed named sub-contractor provides his tender using form ICSub/NAM/T.

The sub-contractor's tender is to provide, amongst other things, the sub-contractor's price and the basis upon which that price is supplied, the required programme details, and any *additional* attendance items (the word 'additional' is used because a basic list of attendances that will be provided by the main contractor is given with the invitation to tender (ICSub/NAM/IT)).

Form ICSub/NAM/T is returned by the proposed named sub-contractor to the employer (or the architect/contract administrator); *it is not sent to the main contractor.*

It is only after this stage that the main contractor becomes involved with the named sub-contractor.

Main contractor's involvement with, the appointment of, and the status of the named sub-contractor

The main contractor first becomes aware of a prospective named sub-contractor when the sub-contractor's tender (and the invitation to tender) is included in:

- the original tender enquiry to the main contractor;
- an architect's/contract administrator's instruction for the expenditure of a provisional sum under the main contract; or
- in an architect's/contract administrator's instruction for the expenditure of a provisional sum naming a replacement named sub-contractor.

---

[3] There is the facility under the Major Project Form for there to be a named specialist, but no particular 'named specialist' procurement or appointment procedure is specified.

The Intermediate Building Contract and the Intermediate Building Contract with contractor's design set out the various procedures that must be followed by a contractor in respect of the appointment of a named sub-contractor (and those procedures deal with the situation where a contractor objects to the proposed named sub-contractor or the proposed terms of that sub-contractor).

It is for the main contractor and the prospective named sub-contractor to agree upon the actual terms of the sub-contract. Those terms will of course be based upon the ICSub/NAM/IT and the ICSub/NAM/T, and the main contractor therefore in effect accepts the named sub-contractor's tender, but there may be amendments and/or additions agreed to the terms contained in the above documents. Any such amendments and/or additions are to be noted in the numbered documents referred to under the second recital of ICSub/NAM/A, and listed under the numbered documents section of the contract particulars of ICSub/NAM/A.

When a sub-contract is entered into between a main contractor and a named sub-contractor, the sub-contract form used is the ICSub/NAM. When this occurs, the named sub-contractor becomes, for all intents and purposes, a normal domestic sub-contractor of the main contractor, and the main contractor has the same liability in respect of the named sub-contract works as it would for the works of a normal domestic sub-contractor. The only exception to this position is if the contractor becomes aware of events that may lead to the termination of a named sub-contractor's employment, the contractor must notify the architect/contract administrator accordingly. If termination subsequently takes place because of the default or insolvency of a named sub-contractor, then, with the architect/contract administrator's consent, the contractor is entitled to some relief from the financial consequences of that event.

Named sub-contractor's liability to the employer

Use of the named sub-contractor procedure is intended primarily for work involving a design input by the named sub-contractor, and this may be the case even where the installation of the specialist work is not of a complex nature. Under paragraph 11.1 of schedule 2 of the main contract, the contractor is in such cases expressly relieved of responsibility to the employer for defects in the named sub-contractor's design of the sub-contract works. This relief, however, does not affect the contractor's obligation in regard to the supply of goods, materials and workmanship.

However, as noted above, any sub-contract that is placed is formed between the main contractor and the named sub-contractor. Consequently, there is no contractual link between the employer and the named sub-contractor.

Because there is no contractual link between the employer and the named sub-contractor, and as the contractor is relieved of responsibility for the named sub-contractor's design, the JCT suggests that an employer may consider using an intermediate named sub-contractor/employer agreement (ICSub/NAM/E) which provides a direct contractual link between the employer and the named sub-contractor for use where the named sub-contractor is to carry out design work or is to procure or fabricate materials or goods prior to the letting of the main contract. Also, the use of ICSub/NAM/E may be considered by the employer where the employer requires:

- undertakings from the named sub-contractor in respect of the sub-contract works and any related design work that the named sub-contractor is to carry out; and/or
- the named sub-contractor to give collateral warranties to purchasers/tenants/ funders of the main contract works.

The Minor Works Sub-contract with sub-contractor's design

This sub-contract is suitable for sub-contracts where the main contract is the JCT Minor Works Building Contract with contractor's design.

Within this book the Minor Works Sub-contract with sub-contractor's design will be referred to as MWSub/D (the designations given to it by the JCT).

This sub-contract is for use with small sub-contract packages with a straightforward content and a low risk involved, and is only to be used where the sub-contractor is required to design all or part of the sub-contract works.

This sub-contract does not have a separate Agreement document, but relies on the Recitals and the Articles to provide the information that would normally be included in an appendix to a pre-2005 JCT form.

The Short Form of Sub-contract

This sub-contract is suitable for any sub-contracts (other than for named sub-contractors) where the main contract is a JCT contract. However, it is not suitable where provisions which are fully back to back with a particular main contract are required.

Within this book the Short Form of Sub-contract will be referred to as ShortSub (the designations given to it by the JCT).

This sub-contract can be used when the main contract works and/or the sub-contract works are to be carried out in Sections, and can be used for either lump sum or remeasurement contracts.

This sub-contract does not have a separate Agreement document, but relies on the Recitals and the Articles to provide the information that would normally be included in an appendix to a pre-2005 JCT form.

The ShortSub is intended for use in respect of small sub-contract packages of straightforward content and with low risk involved. It is not suitable where the sub-contractor is to design any part of the sub-contract works.

The Sub-subcontract

This sub-contract is an entirely new concept for the JCT in that it is a tertiary contract. It is intended for use when a sub-contractor wishes to place a sub-contract with his own sub-contractor(s), i.e. the sub-subcontractor(s).

It is suitable for sub-subcontracts where the main contract is a JCT contract, and can be used with any sub-contract. However, it is not suitable where provisions which are fully back to back with a particular sub-contract are required.

Within this book the Sub-subcontract will be referred to as SubSub (the designation given to it by the JCT).

This sub-subcontract can be used when the main contract works and/or the sub-contract works are to be carried out in Sections, and can be used for either lump sum or remeasurement contracts.

This sub-subcontract does not have a separate Agreement document, but relies on the Recitals and the Articles to provide the information that would normally be included in an appendix to a pre-2005 JCT form.

Sub-contracting generally

All of the above contracts are sub-contracts or sub-subcontracts.

The relationship between the various contract forms referred to above is:

- The employer *contracts* with the contractor using, for example, the JCT Design and Build Contract, the JCT Major Project construction contract, the JCT Intermediate Building Contract, the JCT Intermediate Building Contract with contractor's design, or the JCT Minor Works Contract.
- The contractor *contracts* with each of his sub-contractors using either the Design and Build Sub-contract, the Major Project Sub-contract, the Intermediate Sub-contract, the Intermediate Sub-contract with sub-contractor's design, the Intermediate Named Sub-contract, the Minor Works Sub-contract with sub-contractor's design, the Short Form of Sub-contract, or alternatively (in most cases) the sub-contract form of his choice.
- The sub-contractor *contracts* with each of his sub-subcontractors using the Sub-subcontract, or alternatively the sub-subcontract form of his choice.

The key point in respect of the above relationships is that, although the term sub-contract or sub-subcontract is used under the second and third bullet points, so that they can be distinguished from the main contract, and so that it is apparent that a sub-contractor or sub-subcontractor is operating under the general *umbrella* of a main contract or sub-contract (as appropriate), in *all* of the above cases a *contract* is formed between two parties only.

With that in mind, it would be useful, therefore, to understand some basic principles of contract law.

Most aspects of the law of contract are set down in case law; however, there are some notable exceptions where provision is made in statute (for example the Sale of Goods Act 1979, the Unfair Contract Terms Act 1977, and the Supply of Goods and Services Act 1982).

Because of the nature of this book, the basic principles of contract law as provided below can naturally be dealt with in outline only.

Contract formation

What is a contract?

There are many definitions of a contract, but in simple terms it can be considered as being: 'an agreement which gives rise to obligations which are enforced or recognised by law'. Under English law, only the actual parties to a contract can acquire rights and liabilities under the contract. This is known as 'privity of contract'.

In respect of a main contract situation, the practical consequences of the doctrine of privity of contract are two-fold:

- the main contractor carries responsibility for a sub-contractor's work, etc., so far as the employer is concerned; and
- the employer cannot take direct action in contract against the sub-contractor, unless there is a separate contract between the employer and the sub-contractor.

The effect that the Contracts (Rights of Third Parties) Act 1999 has upon this position in respect of the JCT sub-contracts considered in this book is dealt with later within this book.

How is a contract formed?

The essence of any contract is *agreement*.

In deciding whether there has been an agreement, and what its terms are, the court looks for an *offer* to do or to forbear from doing something by one party, and an unconditional *acceptance* of that offer by the other party, turning the offer into a promise.

In addition, the law requires that a party suing on a promise must show that he has given *consideration* for the promise, unless the promise was given by *deed*.

Further, it must be the intention of both parties to be legally bound by the agreement, and the parties must have the capacity to make a contract, and any formalities required by law must be complied with.

Both the consideration and the objects of the contract must not be illegal.

If there is *fraud or misrepresentation* the contract may be *voidable*, while if there is a *mutual mistake* about some serious fundamental matter of fact this may have the effect of making the contract *void*.

Finally, there must be sufficient *certainty of terms*.

The other matter that needs to be considered is the significance, or otherwise, of *letters of intent*.

These concepts are expanded upon below.

Offer and acceptance (and 'invitations to treat')

Background – the 'mirror image' rule

As stated above, for there to be an agreement the very starting point is an offer being made by one party and an acceptance of that offer by the other party. This is not always as simple as it sounds, particularly when there have been prolonged negotiations between the parties. The courts have developed their own rules for looking at such matters.

In terms of offer and acceptance, the courts adopt what is sometimes known as the 'mirror image' rule of contract formation. That is to say that there must be a clear and unequivocal offer which is matched by an equally clear and unequivocal acceptance of that offer.

What is an 'invitation to treat'?

Before looking at what an offer is, it is important to be aware of the fundamental difference in legal status between an 'invitation to treat' and an 'offer'.

An invitation to treat is simply an expression of willingness to enter into negotiations which may lead to a contract at a later date.

Therefore, a tender enquiry would normally be taken to be an invitation to treat, that is, it is not an offer, but an invitation to another party to make an offer.

Therefore, when a contractor sends a tender enquiry to a sub-contractor, the tender enquiry would in effect be an invitation to the sub-contractor to provide an offer (or, in other words, submit a tender) to execute the required sub-contract works.

The question of whether a statement is an offer or an invitation to treat depends primarily on the intention with which the statement was made. The actual wording of that statement is not necessarily conclusive.

Thus, a statement may be construed as being an invitation to treat even if it contains the word offer[4]; whilst on the other hand a statement may be an offer although it is expressed to be an acceptance[5], or where it requests the person to whom it is addressed to make an offer[6].

Of course, if the invitation to treat expressly provides that the issuing party is not to be bound merely by the receiving party's notification of acceptance, but only when he has signed the document in which the statement is contained, then it cannot be construed as being an offer.

What is an 'offer'?

An offer is a statement by one party of a willingness to contract on definite stated terms and intended to be binding provided that these terms are, in turn, unequivocally accepted by the party or parties to whom the offer is addressed.

There is generally no requirement that the offer be made in any particular form; it may be made orally, in writing, or by conduct. Of course, if a dispute arose in the future then it would be beneficial for the offer to be in writing.

In whichever form an offer is made it must be sufficiently definite to be capable of resulting in a contract if accepted. Its terms and conditions must be clear and unequivocal, and it must be made with the intention that it is to become binding as soon as it is accepted by the person to whom it is addressed. In this context a person includes a corporation because, in law, a corporation is a legal person; that is to say, a corporation is regarded by the law as a legal entity quite distinct from the person or persons who may for the time being be the member or members of the corporation.

Putting the above into context, it is generally the case that when the sub-contractor submits his estimate (i.e. his tender) this is an offer which the contractor can either accept or reject.

---

[4] *Spencer* v. *Harding* (1870) LR 5 CP 561.
[5] *Bigg* v. *Boyd Gibbins Ltd* [1971] 1 WLR 913.
[6] *Harvis Investments Ltd* v. *Royal Trust Co. of Canada* (1789) 3 TR 148; *British Car Auctions Ltd* v. *Wright* [1972] 1 WLR 1519.

With the above in mind, it must be noted by sub-contractors that the submission of a tender does not (normally) conclude a contract. Therefore, the preparation of a tender in response to a tender enquiry (which would, in the normal course of events, become an offer when submitted) may involve the sub-contractor in (sometimes considerable) expense, but the cost of tender preparation is not normally recoverable as a discrete cost. Obviously the cost of tender preparations is included within the head office overhead percentage that is added by sub-contractors onto their tenders, and the tender preparation costs so incorporated are therefore recovered by sub-contractors when their tenders are successful.

Must the sub-contractor's tender match the contractor's tender enquiry letter to be a valid offer?

A sub-contractor's tender can still be a valid legal offer even where it does not comply with the contractor's stipulated tender enquiry requirements (assuming the offer is made in definite terms capable of being accepted).

For a sub-contractor's tender to fulfil the legal requirements of a valid offer it must simply be something which invites, and is intended by the person making the offer (i.e. the sub-contractor) to invite, acceptance. It simply falls to the contractor to either accept it or reject it.

How is the offer accepted?

For agreement to be reached there must be a clear and unequivocal acceptance of a clear and unequivocal offer. The acceptance must be unqualified, i.e. as noted earlier it must 'mirror' the offer.

Therefore, if in a tender enquiry a sub-contractor was required to use Welsh slates but submitted his tender on the basis of using Spanish slates, the contractor, upon receiving the sub-contractor's tender, accepted the sub-contractor's tender without qualification, then the contract, when formed, would be on the basis of the terms and conditions which formed part of the tender (i.e. based on using Spanish slates rather than Welsh slates).

In such a situation, if a future dispute arose, the contractor would not be able to rely on the terms and conditions forming part of the tender enquiry (i.e. that Welsh slates were required) because those terms and conditions would not form part of the contract between the parties.

Must acceptance be communicated to be effective?

As a general rule silence does not constitute acceptance[7]; neither does inactivity.

Given this, the general rule is that an acceptance has no effect until it is communicated (either in writing or orally) to the party making the offer. The main reason for this being that it could cause hardship to the party making the offer if he were bound without knowing that his offer had been accepted.

---

[7] *Felthouse* v. *Bindley* (1862) aff'd (1863).

What if the sub-contractor's tender is rejected by the contractor?

Another rule is that refusal of an offer puts an end to that offer[8]. Hence, if a contractor rejects a sub-contractor's offer (i.e. his submitted tender) it is, in legal parlance, extinguished and is no longer legally capable of being accepted.

Can an offer be withdrawn, and if so when?

An offer can be withdrawn at any time before it is accepted[9]. This rule even applies where an offer is stated to be open for a fixed time. Therefore, if a sub-contractor submitted a tender (or a quotation) and stated that it remained open for acceptance within 30 days, there would be nothing to prevent the sub-contractor from withdrawing that offer in a period less than 30 days (unless the express terms of the sub-contract stated otherwise).

Can an offer be accepted after the fixed time for acceptance has lapsed?

If an offer is stated to be open for a fixed time, then it cannot (without the agreement of the party making the offer) be accepted after that time.

However, if no time is stated in the offer, then the offer is taken to lapse after a reasonable time. The word 'reasonable' is, of course, open to interpretation, but is based on the facts of the particular case.

Can an offer by one party be accepted by the other party's conduct?

In certain circumstances, depending on the facts of each particular case, acceptance may be made by conduct[10] (e.g. allowing possession, making payments in line with the agreed terms, etc.) or by performance[11] (e.g. by carrying out and completing the sub-contract works, etc.).

What is a counter-offer?

If the acceptance does not clearly and unequivocally accept the offer (e.g. it seeks to add to or vary the terms contained in the offer) then it is, simply, a counter-offer (not an acceptance) and this simply destroys the original offer.

A counter-offer has the same status as an offer in the formation of a contract, and consequently, a counter-offer must be clearly and unequivocally accepted before agreement has been reached.

Counter-offers should be distinguished from requests for information, which will not necessarily amount to a counter-offer[12]. Care must be taken when requesting further information to ensure this is not construed, in fact, as being a counter-offer.

---

[8] *Hyde* v. *Wrench* (1840) 3 Beav 334.
[9] *Dickinson* v. *Dodds* (1876) 2 Ch D 463.
[10] *G Percy Trentham* v. *Archital Luxfer Ltd* (1992) 63 BLR 44, CA.
[11] *Brogden* v. *Metropolitan Railway Co.* (1877) 2 App Cas 666.
[12] *Stevenson, Jaques & Co.* v. *McLean* (1880).

In respect of construction works, in particular, there are frequently a whole series of counter-offers made before an acceptance is made. This may be because the parties are negotiating about the terms or because they are each trying to impose their own terms.

This latter situation is often called the 'battle of the forms', and variants of this battle of the forms are at the base of many disputes between contractors and sub-contractors.

What is the 'battle of the forms'?

The expression 'battle of the forms' refers to an offer followed by a series of counter-offers where each party successively seeks to stipulate different terms, often based on their own standard printed terms. It is sometimes extremely difficult to determine whether or when a concluded contract came into existence, particularly where no formal contract is ever signed, and it is then left to the courts to determine the matter.

Clearly this is far from satisfactory, and it makes sense to ensure that the sub-contract terms are agreed and recorded in writing to avoid the need to rely on a court interpreting the intention of the parties from an analysis of the various exchanges of communication.

Agreement

As noted previously, the essence of any contract is agreement.

The test for the existence of an agreement is objective rather than subjective. In other words, the existence is tested on the facts, rather than on what may have been perceived to be the intention of the parties.

The principal justification for the adoption of this test is the need to promote certainty.

An agreement is reached either:

- when a statement of agreement is signed; or
- when one party makes an unambiguous offer capable of being accepted, and the other party accepts it unequivocally.

Consideration

Other than where a contract is executed as a deed (see further commentary later in this chapter) an agreement requires consideration to be exchanged between the contracting parties before it becomes binding.

The classic definition of consideration was expressed in *Currie v. Misa*[13] in the following terms:

---

[13] *Currie* v. *Misa* (1875) LR 10 Ex 153.

'a valuable consideration, in the sense of the law, may consist either in some right, interest, profit or benefit accruing to the one party, or some forbearance, detriment, loss or responsibility given, suffered or undertaken by the other.'

In the ordinary building sub-contract situation the consideration given by the contractor is the price paid or the promise to pay, and the consideration given by the sub-contractor is the carrying out of the works or promise to carry them out.

What are the rules regarding consideration?

The rules which make up the doctrine of consideration may be divided into three categories:

(1) The first is that consideration must be sufficient but it need not be adequate.
(2) The second is that past consideration is not good consideration (i.e. the general rule is that the consideration must relate to future works. Any acts already performed that are the subject matter of the agreement do not, generally, constitute good consideration and are unenforceable).
(3) The third is that the consideration must move from the party to whom the promise has been made.

Executed as a deed

As noted above, consideration is not required in the case where a contract is executed as a deed. Contracts, other than made by deed, are termed simple contracts, whether made orally or in writing.

Fraud or misrepresentation

What is a misrepresentation?

A misrepresentation may be defined as an unambiguous, false statement of fact which is addressed to a party and which misleads that party and which materially induces that party to enter into a contract.

Whether made fraudulently or innocently, the general effect of such a misrepresentation is to render the contract voidable. As a general rule, simple silence is not an example of misrepresentation[14].

The particular statement in question must be one of existing fact and the statement must be addressed to the party misled. This may either be by way of direct communication to that party, or it may be communicated to that party by a third party acting under the instruction of the party making the representation.

Note also that the statement must be a material inducement to make a party enter into the contract. A material inducement is one that would affect the judgment of a reasonable man in deciding whether to enter into the contract on those

---

[14] *Keates v. Cadogan* (1851) 10 CB 591.

terms (although a person who has been fraudulent cannot succeed with an argument that the representation was immaterial).

Finally, the ability of a party to exclude liability for misrepresentation is controlled by section 3 of the Misrepresentation Act 1967, which subjects any term which purports to exclude or restrict liability or a remedy for misrepresentation to a reasonableness test.

Are there any different types of misrepresentation?

Yes. There are four different types of misrepresentation.

(1) Fraudulent misrepresentation
    Fraudulent misrepresentation is made when it is proved that a false representation has been made (a) knowingly or (b) without belief in its truth or (c) recklessly or carelessly whether it is true or false.

(2) Negligent misrepresentation (actionable at common law)
    Negligent misrepresentation is actionable at common law where there is a *Hedley Byrne*[15] relationship (i.e. some special relationship of proximity) between the parties. The existence of such a relationship depends upon a number of factors, including the knowledge of the party making the representation, the purpose for which the statement was made and the reasonableness of the reliance by the party receiving the representation.

(3) Negligent misrepresentation (arising under section 2(1) of the Misrepresentation Act 1967)
    Section 2(1) of the Misrepresentation Act 1967 states that where a misrepresentation has been made by one contracting party to another, the party making the representation is liable to the other in damages unless he can prove that he had reasonable grounds to believe and did believe up to the time that the contract was made that his statement was true.

(4) Innocent misrepresentation
    An innocent misrepresentation is a misrepresentation which is neither fraudulent, nor negligent.

What are the legal remedies for misrepresentation?

The principal remedies for misrepresentation are rescission (i.e. revoking the contract) and damages.

Rescission is in principle available for all types of misrepresentation. The effect of rescission is to put the parties as far as possible into the position which they would have been in if the contract had not been concluded.

Damages can be claimed for all types of misrepresentation. However, in respect of innocent misrepresentation only, the court has discretion to award damages in lieu of rescission under section 2(2) of the Misrepresentation Act 1967.

---

[15] *Hedley Byrne & Co. v. Heller & Partners* [1963] 3 WLR 101, HL.

Mistake

What is mistake?

This is the position where the parties to the contract are mutually mistaken about some serious material fact fundamental to the contract which may render the contract void.

This is to be distinguished from circumstances where one party thought he was getting a better bargain than he actually did; this does not affect the contract's validity.

Mistake operates within very narrow confines.

A mistake can be mutual, unilateral or common.

A mutual mistake occurs when both parties make a mistake (i.e. A is offering one thing whilst B is accepting something different); a unilateral mistake occurs where only one party makes a mistake; and a common mistake occurs when both parties make the same mistake.

If both parties, or even one party, intended something different from that which the documents record, the courts have power to rectify or alter the document so as to give effect to the true agreement.

Void or voidable

A contract that is 'void' is a contract that has no legal effect (i.e. it is a nullity) as distinct from a contract that is voidable, which is simply a contract that is capable of being set aside (or rescinded) by one of the parties.

Certainty of terms

Even if there is clearly agreement through offer and acceptance, a contract may fail to come into existence because of uncertainty as to what has been agreed[16].

Although it has been found that 'the parties are to be regarded as masters of their contractual fate in determining what terms are essential'[17]; and 'it is for the parties to decide whether they wish to be bound and, if so, by what terms, whether important or unimportant'[18]; it is generally considered that agreement as to parties, price, time and description of works is normally the minimum necessary to make the contract commercially workable.

Silence by the parties as to either price or time may not alone prevent a contract coming into existence, for if the other essential terms are agreed a reasonable charge or time for completion may be implied by the Supply of Goods and Services Act 1982.

Letters of intent

All too frequently, the desire or pressure to commence works quickly (either on or off site) is given priority by the parties over having a formal contract in place

---

[16] *Scammell* v. *Ouston* (1941) 1 All ER 14.
[17] *Pagnan* v. *Feed Products* [1987] 2 Lloyds Rep 601, CA.
[18] *Mitsui Babcock Energy Ltd* v. *John Brown Engineering Ltd* (1996) CILL 1189.

beforehand, and in such circumstances it is not uncommon for the work to start on the basis of a letter of intent issued by the contractor to the sub-contractor, sometimes whilst sub-contract negotiations continue over the sub-contract terms. There are, however, risks involved for both parties by proceeding on this basis, particularly if no formal sub-contract is ever concluded.

What is a letter of intent?

A letter of intent arises out of a situation where parties are in the course of negotiating a contract and they express in writing their intention to enter into a contract in the future.

Although there is often much debate regarding this matter, it is in fact a misconception that letters of intent are or can ever be contracts. A true letter of intent does no more than set out the parties' intention to enter into a contract at a later date.

Despite this, there have been circumstances when the courts have found that what one of the parties intended to be a letter of intent was actually a contract in its own right.

In reality, as identified in the *Monk* v. *Norwich Union*[19] case, the question of the legal effect of letters of intent has only three possible outcomes. These are:

- that a simple contract may be formed under which each party assumes reciprocal obligations to each other;
- that there may be an 'if' contract, under which party A requests performance from party B and, if party B performs, he will receive remuneration (this is sometimes referred to as a 'quasi-contract'); or
- there is no contract formed at all.

How can a letter of intent be an ordinary contract?

The main consideration is whether the terms set out in the letter of intent are sufficiently certain for a contract to be formed, and to what extent the matters that still need to be resolved are contained within the letter of intent.

The intention of the parties from the surrounding circumstances is also relevant.

In the *British Steel* v. *Cleveland Bridge*[20] case, Judge Goff said:

'the question whether . . . [e] any contract has come into existence must depend on a true construction of the relevant communications which have passed between the parties and the effect (if any) of their actions pursuant to those communications.'

Even if some matters remain outstanding but all essential terms have been agreed, a contract may be formed. Authority for this is found in the *Mitsui* v. *John Brown*[21] case, where it was stated:

---

[19] *Monk Construction Ltd* v. *Norwich Union Life Assurance Society* (1992) 62 BLR 107.
[20] *British Steel Corporation* v. *Cleveland Bridge* [1984] 1 All ER 504.
[21] *Mitsui Babcock Energy Ltd* v. *John Brown Engineering Ltd* (1996) CILL 1189.

'My review of the authorities leads me to the conclusion that there is no reason in principle why two parties should not enter into a binding agreement, even though they have agreed that some proposed terms should be the subject of further discussions and later agreement.'

From the above, it would appear that for a letter of intent to be enforceable in law as a contract the following elements must be present:

- The parties must have demonstrated their intention to contract – either from the content of the letter of intent or from their conduct.
- The parties must have agreed the essential terms (e.g. parties, description of works, price and time) with sufficient certainty so as to allow the contract to become commercially workable.
- Acceptance of the offer must be clear either by words or by conduct.

What if the letter of intent has a capped price or capped time period?

It should be noted that there are many instances where a letter of intent has a capped price or a cap on time. In such a situation, the courts have generally enforced spending limits in letters of intent strictly (see *Monk Construction* v. *Norwich Union*[22], where the letter of intent authorised mobilisation and ordering of materials up to a maximum expenditure of £100,000, and the court held that Monk Construction could not recover proven costs for works performed in excess of that sum).

Does a letter of intent provide for adjudication?

This is a grey area and has caused much debate in the construction industry, particularly given the widespread use of letters of intent.

The question is does the letter of intent set out all the terms of the contract for the purposes of complying with section 107 of the Housing Grants, Construction and Regeneration Act 1996 (HGCRA 1996) (in particular section 107(2)(c))?

Section 107(2) of the HGCRA 1996 provides that an agreement is in writing if:

(1) it is made in writing;
(2) it is made by an exchange of communications in writing; or
(3) the agreement is evidenced in writing.

In interpreting the above, and at the date of writing, the surrounding case law[23] states that for an agreement to satisfy s107 of the HGCRA 1996 all of the terms of the contract must be evidenced in writing, not just the material ones. The court's view appears to be that the speed of adjudication demands certainty in respect of the construction contract's terms; it is not a forum for the parties to be arguing what the unwritten or obscure terms of the contract were.

---

[22] *Monk Construction Ltd* v. *Norwich Union Life Assurance Society* (1992) 62 BLR 107.
[23] *RJT Consulting Engineers Ltd* v. *DM Engineering (Northern Ireland) Ltd* [2002] EWCA Civ 270.

From the foregoing, whether or not a letter of intent satisfies s107 of the HGCRA 1996 will depend entirely on the wording of the letter of intent in each case. Certainly, a pure letter of intent (which is not a contract in any event) will not satisfy this requirement. Even when a letter of intent goes further than this, it will only permit statutory adjudication to be used if it is a contract and if all of the material terms of that contract are recorded in writing[24].

Standard forms of contract and sub-contract

There are obvious benefits in using standard forms of contract and sub-contract. These include:

- There is no need to produce (and incur the legal costs of producing) ad hoc contracts and sub-contracts for every project.
- There is a degree of certainty regarding the interpretation of the clauses of the contract (particularly those standard forms that have been in existence for some time and where some of the more important clauses may have been tested in the courts).
- The parties know (with reasonable certainty) the consequences of various possible courses of action.

Standard forms of contract (perhaps less so of standard forms of sub-contract) can be traced back to the nineteenth century (if not earlier). However, it appears to be a fairly recent phenomenon, largely emanating from the Latham Report (*Constructing the Team*, 1994), that standard forms and sub-contracts are now seen as having a dual purpose. These are:

- To set out the rights and obligations of the parties.
- To ensure that the risk of the project is allocated to the party that can best manage that risk.

In this regard there have been great steps taken, particularly in respect of main contracts, but it is questionable whether that same progress has been made in respect of sub-contracts.

It is still prevalent that main contractors produce their own sub-contract forms, or, more commonly, issue their own amendments to the standard sub-contract forms.

Perhaps unfortunately, under the JCT main contract forms, there is nothing to prevent this from happening in the future, and because of this it is considered by some that main contractors under JCT contracts should be obliged to sub-let works using JCT sub-contracts.

For instance, the JCT Design and Build Contract simply states, under clauses 3.3.1 and 3.3.2, that:

---

[24] The position in respect of contractual adjudications was considered in the case of *Treasure & Son Ltd* v. *Martin Dawes* [2007] EWHC 2429 (TCC).

Clause 3.3.1 'The Contractor shall not without the written consent of the Employer sub-let the whole or any part of the Works. Such consent shall not be unreasonably delayed or withheld but the Contractor shall remain wholly responsible for carrying out and completing the Works in all respects in accordance with clause 2.1 notwithstanding any such sub-letting.'

Clause 3.3.2 'The Contractor shall not without the written consent of the Employer sub-let the design for the Works. Such consent shall not be unreasonably delayed or withheld but shall not in any way affect the obligations of the Contractor under clause 2.17 or any other provision of this Contract.'

Clause 3.9 then notes that it shall be a condition of any sub-letting to which clause 3.3 applies that:

'3.9.1 the employment of the sub-contractor under the sub-contract shall terminate immediately upon the termination (for any reason) of the Contractor's employment under this Contract.

3.9.2 the sub-contract shall provide:

3.9.2.1 that no unfixed materials and goods delivered to, placed on or adjacent to the Works by the sub-contractor and intended for them shall be removed, except for use on the Works, unless the Contractor has consented in writing to such removal (such consent not to be unreasonably delayed or withheld) and that:

3.9.2.1.1 where, in accordance with clauses 4.8 and 4.13 or 4.14 of these Conditions, the value of any such material or goods has been included in any Interim Payment and that Interim Payment has been paid by the Employer to the Contractor, those materials or goods shall be and become the property of the Employer and the sub-contractor shall not deny that they are and have become the property of the Employer;

3.9.2.1.2 if the Contractor pays the sub-contractor for any such materials or goods before their value is included in any Interim Payment, such materials or goods shall upon such payment by the Contractor be and become the property of the Contractor.

3.9.2.2 for the grant by the sub-contractor of the rights of access to workshops or other premises referred to in clause 3.1 of these Conditions;

3.9.2.3 that if by the final date for payment stated in the sub-contract the Contractor fails properly to pay any amount, or any part of it, due to the sub-contractor, the Contractor shall in addition to the amount not properly paid pay simple interest at the Interest Rate for the period until such payment is made; such payment of interest to be on and subject to terms equivalent to those of clauses 4.10.6 and 4.10.12 of these Conditions;

3.9.2.4 where applicable, for the execution and delivery by the sub-contractor, in each case within 14 days of receipt of a written request by the Contractor, of such collateral warranties as comply with the Contract Documents;

3.9.2.5 that neither of the provisions referred to in clauses 3.4.2.1.1 and 3.4.2.1.2 shall operate so as to affect any vesting in the Contractor of any property in any Listed item required for the purposes of clause 4.15.2.1 of these Conditions;

3.9.3 the Contractor shall not give such consent as is referred to in clause 3.4.2.1 without the prior consent of the Employer under clause 2.21 of these Conditions.'

Therefore, in respect of the JCT Design and Build Contract, for example, and apart from the provisions under clause 3.9 above, the actual terms on which a contractor contracts with a sub-contractor are a matter for agreement between the contractor and the sub-contractor.

Very similar provisions operate in respect of the JCT Intermediate Building Contract, although under that main contract form the consent that the contractor is required to obtain is sometimes to be obtained from the architect/contract administrator and sometimes from the employer.

Therefore, in respect of both of the above main contract forms, other than for the few points specifically noted in the main contract conditions, the terms on which a contractor contracts with a sub-contractor are a matter for agreement between the contractor and the sub-contractor.

For example, the said main contract conditions do not require any particular conditions to apply in the sub-contracts in respect of, amongst other things:

- the consequences if an act or default of the sub-contractor causes the contractor to be in breach of the main contract;
- regular payments to the sub-contractor for work properly carried out;
- the recovery of any loss or expense caused to the sub-contractor by any act or default of the contractor or of the employer;
- the adjustment of the agreed period or periods for carrying out the sub-contract works where such adjustment is needed due to the occurrence of events outside the sub-contractor's control.

In respect of the Major Project construction contract and the ShortSub, there are no restrictions as to sub-contracting at all, apart from that under the Major Project construction contract the contractor is to appoint a named specialist to prepare any designs or execute any works that are identified by the requirements (within the main contract) as having to be undertaken by a named specialist. In such a situation the contractor is to notify the employer of the identity of any named specialist upon their appointment and shall supply to the employer a copy of the contract entered into by the contractor with the named specialist (other than for the financial details contained in it). In addition, the contractor is not to amend his contract with any named specialist or waive strict compliance by the named specialist with the performance of his obligations under such contract or estop himself from enforcing such obligations, without obtaining the prior written consent of the employer (consent that is not to be unreasonably withheld or delayed).

Chapter 2
The Sub-contract Agreement

In this chapter we will look at the following topics:

- the structure of the agreement;
- the recitals;
- the articles of agreement;
- the sub-contract particulars;
- the summary of interim and final payment provisions;
- the attestation forms;
- the schedule of information.

Terms and conditions – generally

People frequently refer to the terms and conditions of a contract.

Using this terminology, the terms of a sub-contract would normally constitute the particular insertions made by the parties to the sub-contract into the sub-contract, whereas the conditions would already be part of the standard published form.

In respect of some of the sub-contracts under consideration in this book, there is a separate agreement document where the terms are inserted by the parties (e.g. DBSub/A, ICSub/A, etc.) whilst for others (e.g. the MPSub) there is only one sub-contract document and the terms are inserted in the same way as would have been the case for an old style JCT appendix. In the cases where there are two separate sub-contract documents, there is a clause in the sub-contract (e.g. clause 1.3 of DBSub/C) which makes it clear that the sub-contract is to be read as a whole.

Whatever is inserted as a term into the executed version of the sub-contract will be an express term agreed between the parties.

The law will understand these express terms to be freely agreed between the parties, and any argument put forward at a later date that the express terms were not freely agreed is unlikely to meet with any success.

Also, the terms will be the only express terms applicable to the sub-contract. Whatever may have been discussed and/or written about before the execution of the sub-contract will not form part of the express terms of the sub-contract unless the matter so discussed and/or written about is specifically listed within the sub-contract.

Therefore, it is important for both the contractor and the sub-contractor that everything that should be referred to and listed within the sub-contract *is* referred to and listed within the sub-contract.

The corollary to the above is, of course, that anything that should not be referred to and listed within the sub-contract particulars should *not* be referred to and listed within the sub-contract particulars.

If a matter was discussed and/or written about prior to the execution of the sub-contract – for example that the sub-contractor was to clear his waste materials and rubbish to skips to be provided by the contractor – but this matter was not listed within the sub-contract; then, if this matter became an issue at a later date, the sub-contractor would not have the contractual right to insist that the contractor provide skips for the removal of the sub-contractor's waste materials and rubbish.

Conversely, if a matter was not discussed and/or written about prior to the execution of the sub-contract – for example that the sub-contractor would be charged for the use of mess rooms – but this matter was listed within the sub-contract; then, if this matter became an issue at a later date, the contractor would have the contractual right to charge the sub-contractor when the sub-contractor used the mess rooms.

The parties need to be aware that the meaning of a contract will be ascertained by a court from the meaning of the words actually used in that written contract, and a court may not hear evidence from the parties as to what they may have actually meant.

This 'rule' is sometimes known as the 'parole evidence rule'. Although the courts are moving away from the use of formal 'rules' it would still be generally futile (if a dispute arose) for a party to rely on any negotiations leading up to a written contract (if those negotiations are not recorded in some way within the written contract), in an attempt to change in some way the plain wording of the contract. This point can represent a trap for contractors and sub-contractors who may wish to refer to the sub-contract negotiations in an attempt to show that the alleged intention of the parties was different to the actual wording of the sub-contract.

It is extraordinary the number of times that contractors and sub-contractors take great care over agreeing terms that they both feel comfortable with, only to fail to properly record those agreements in the sub-contract. Experience shows that when disputes arise, people tend to have very selective memories; therefore if a term has been agreed, it must be recorded in the sub-contract if a party wishes to rely upon that term in future.

How are terms entered into the sub-contracts?

DBSub

DBSub is designed to be completed and executed by the parties to create a formal contract. DBSub has a separate agreement document (DBSub/A) which, when completed, set outs the terms of the sub-contract.

DBSub/A, as a blank document, does not (and cannot) set out the terms of the sub-contract; it is for the contractor and the sub-contractor to agree upon those terms and for those terms to be entered into DBSub/A.

MPSub

Unlike DBSub, MPSub does not have a separate agreement document, and there are no separate articles of agreement.

The sub-contract conditions must be executed by the parties, and the sub-contract particulars (within MPSub) must be carefully completed by the parties, together with other relevant schedules and documents (where required) which form part of the sub-contract. The sub-contract is comprised of the following:

- the sub-contract conditions;
- the sub-contract particulars;
- schedule 1 (relevant particulars of the main contract);
- schedule 2 (pricing document) plus its annexures;
- the proposals (if applicable).

ICSub

ICSub is designed to be completed and executed by the parties to create a formal contract. ICSub has a separate agreement document (ICSub/A) which, when completed, sets out the terms of the sub-contract.

ICSub/A, as a blank document, does not (and cannot) set out the terms of the sub-contract; it is for the contractor and the sub-contractor to agree upon those terms and for those terms to be entered into ICSub/A.

ICSub/D

ICSub/D is designed to be completed and executed by the parties to create a formal contract. ICSub/D has a separate agreement document (ICSub/D/A) which, when completed, sets out the terms of the sub-contract.

ICSub/D/A, as a blank document, does not (and cannot) set out the terms of the sub-contract; it is for the contractor and the sub-contractor to agree upon those terms and for those terms to be entered into ICSub/D/A.

ICSub/NAM

ICSub/NAM is designed to be completed and executed by the parties to create a formal contract. ICSub/NAM has a separate agreement document (ICSub/NAM/A) which, when completed, sets out the terms of the sub-contract.

ICSub/NAM/A, as a blank document, does not (and cannot) set out the terms of the sub-contract; it is for the contractor and the sub-contractor to agree upon those terms and for those terms to be entered into ICSub/NAM/A.

MWSub/D

The MWSub/D does not have a separate agreement document, and there are no separate articles of agreement. The MWSub/D is contained in one document. The parties fill in the specific sub-contract information in the recitals and article of agreement and execute the attestation page. The conditions form an integral part of the document.

ShortSub

All as MWSub/D above.

SubSub

All as MWSub/D above.

What is the structure of the agreement, i.e. the part that has to be completed by the parties?

DBSub

DBSub/A consists of:

- The date that the agreement is made (normally the date that the second party executes the agreement).
- The names and addresses of the parties (i.e. the contractor and the sub-contractor). Normally, the parties' company number should be inserted, and their registered office address should be used. If the contractor or the sub-contractor is a company incorporated outside England and Wales, particulars of its place of incorporation should be inserted immediately before the company number. If the contractor or the sub-contractor is neither a company incorporated under the Companies Act nor a company registered under the laws of another country, the references to a company number and to a registered office address should be deleted.
- Details (by identification only) of the sub-contract works, and the main contract works of which they are a part.

This is then followed by:

- first to eighth recitals;
- articles 1 to 6;
- sub-contract particulars items 1 to 16;
- a summary of interim and final payment provisions;
- attestation forms;
- a schedule of information;
- notes (to annexures to the agreement and to the schedule of information);
- supplementary particulars (containing information needed for formula adjustment, if applicable).

MPSub

The sub-contract conditions call for the completion on page 1 of:

- The date that the agreement is made (normally the date that the second party executes the agreement).
- The names and addresses of the parties (i.e. the contractor and the sub-contractor. Normally, the parties' company number should be inserted, and their registered office address should be used (see comments at DBSub where a party is not an incorporated company, or is a company incorporated outside England and Wales).

Page 25 onwards should also be completed:

- The sub-contract particulars (pages 25 to 29).
- Attestation form (page 31).

- Schedule 1: relevant particulars of the main contract (page 33).
- Schedule 2: pricing document (page 34).

ICSub

ICSub/A's structure is virtually identical to DBSub/A above, except as noted below:

- ICSUB/A has four recitals, whereas DBSub has eight.
- Sub-contract particulars items 1 to 14 (DBSub has 16 items).
- No supplementary particulars.

ICSub/D

ICSub/D/A's structure is virtually identical to DBSub/A above, except as noted below:

- Sub-contract particulars items 1 to 14 (DBSub has 16 items).
- No supplementary particulars.

ICSub/NAM

ICSub/NAM/A's structure is far shorter, as follows:

- ICSub/NAM/A has two recitals.
- Articles 1 to 6.
- Sub-contract particulars items 1 to 2. However, footnote 35 records the level of information (including the tender documents) and other relevant documentation which comprise the numbered documents (item 1) under this sub-contract.
- Attestation forms.

MWSub/D

The MWSub/D consists of:

- The date that the agreement is made (normally the date that the second party executes the agreement).
- The names and addresses of the parties (i.e. the contractor and the sub-contractor). Normally, the parties' company number should be inserted, and their registered office address should be used (see comments at DBSub where a party is not an incorporated company, or is a company incorporated outside England and Wales).

This is then followed by:

- First, second and third recitals.
- Articles 1 to 6.
- Attestation form.

ShortSub

The ShortSub consists of:

- The date that the agreement is made (normally the date that the second party executes the agreement).
- The names and addresses of the parties (i.e. the contractor and the sub-contractor). Normally, the parties' company number should be inserted, and their registered office address should be used (see comments at DBSub where a party is not an incorporated company, or is a company incorporated outside England and Wales).

This is then followed by:

- First and second recitals.
- Articles 1 to 6.
- Attestation form.

SubSub

As MWSub/D above (but with the parties being the sub-contractor and the sub-subcontractor).

The recitals

What are the recitals?

The recitals set out certain facts as a background to the parties entering into the sub-contract.

MPSub

The MPSub, has no recitals.

DBSub (and ICSub and ICSub/D)

The DBSub has eight recitals. Some of these recitals are virtually identical to recitals contained in the ICSub and the ICSub/D, as shown in Table 1 below.

Table 1 Recitals of DBSub/A, ICSub/A and ICSub/D/A.

DBSub/A	ICSub/A	ICSub/D/A
recital	recital	recital
First	First	First
Second	Second	Second
Third	N/A	Third
Fourth	N/A	Fourth
Fifth	N/A	Fifth
Sixth	N/A	Sixth
Seventh	Third	Seventh
Eighth	Fourth	Eighth

The first recital of DBSub/A, ICSub/A and ICSub/D/A

This recital refers to the main contract as described in item 1 of the schedule of information annexed to the relevant agreement (i.e. DBSub/A, ICSub/A, or ICSub/D/A, as applicable). The said schedule of information should be carefully completed by the contractor (the schedule of information is dealt with in detail later in this chapter).

The second recital of DBSub/A, ICSub/A and ICSub/D/A

This recital simply states that the contractor wishes the sub-contractor to execute the sub-contract works described in the numbered documents (item 16 of DBSub/A's sub-contract particulars; item 14 of both ICSub/A and ICSub/D/A. The sub-contract particulars are dealt with later in this chapter.)

The third recital of DBSub/A and ICSub/D/A

This recital is to be completed by the parties to identify the part of the sub-contract works (which may be all of the sub-contract works) that the sub-contractor is responsible for in respect of both design and construction. The entry made may be descriptive, or it may refer to the document(s) that more fully describes the relevant part of the sub-contract works.

The entry made in this recital is known as:

- the sub-contractor's designed works – under DBSub/A; and
- the sub-contractor's designed portion – under ICSub/D/A.

The fourth recital of DBSub/A and ICSub/D/A

This recital confirms that the contractor has provided documents showing and describing or otherwise stating the requirements of the contractor for the design and construction of the sub-contractor's designed works (or designed portion). These documents, which would probably include the Employer's Requirements and which may include the Contractor's Proposals, should be listed under item 16 of DBSub/A's sub-contract particulars (or item 14 of ICSub/D/A's sub-contract particulars) as numbered documents.

The said documents are known, collectively, as being the Contractor's Requirements.

To avoid future dispute, the documents comprising the Contractor's Requirements should be annexed to the sub-contract agreement and signed or initialled by or on behalf of each party.

The fifth recital of DBSub/A and ICSub/D/A

This recital confirms that in response to the Contractor's Requirement's (as identified in the fourth recital) the sub-contractor has supplied to the contractor:

- Documents showing and describing the proposals of the sub-contractor for the design and construction of the sub-contractor's designed portion. These documents are known collectively as the Sub-contractor's Proposals.

- Cost information in the form of a sub-contractor's designed works analysis under DBSub/A (or a sub-contractor's designed portion (SCDP) analysis when using ICSub/D/A) as applicable to the sub-contract sum (or sub-contract tender sum, if appropriate).

All documents comprising the Sub-contractor's Proposals and the DBSub/A's sub-contractor's designed works analysis (or an SCDP analysis when using ICSub/D/A) should be listed as numbered documents in the sub-contract particulars (see item 16 of DBSub/A; or item 14 when using ICSub/D/A). To avoid future dispute, these documents should also be annexed to the sub-contract agreement and signed or initialled by or on behalf of each party.

The sixth recital of DBSub/A and ICSub/D/A
This recital confirms that the contractor has examined the Sub-contractor's Proposals and, subject to the sub-contract conditions, is satisfied that they appear to meet the Contractor's Requirements.

The entire matter of design liability will be dealt with in more detail in Chapter 4 of this book. Note that the sixth recital does not require the contractor to check that the Sub-contractor's Proposals satisfy the Contractor's Requirements but merely that 'they appear to meet' them.

If, when built, the sub-contractor has not satisfied the Contractor's Requirements, then the sub-contractor could be required to alter his installation (at no cost to the contractor) to ensure that the Contractor's Requirements are met. Alternatively the contractor may accept the departure from the Contractor's Requirements (with a possible cost adjustment) but such subsequent acceptance should not be relied upon.

The seventh recital of DBSub/A and ICSub/D/A,
and the third recital of ICSub/A
This recital relates to the provision of a priced schedule of activities (the priced activity schedule).

This recital should be deleted if either:

- article 3B is to apply; or
- a priced activity schedule is not provided.

Article 3B is for use where the sub-contract works are to be completely re-measured and valued on a remeasurement basis.

In the activity schedule, each activity should be priced so that the sum of those prices equals the sub-contract sum excluding:

- Provisional Sums; and
- the value of work for which approximate quantities are included in any bills of quantities.

The eighth recital of DBSub/A and ICSub/D/A,
and the fourth recital of ICSub/A
This recital states that the sub-contractor has been given a copy of, or a reasonable opportunity to inspect, any other documents and information relating to the provisions of the main contract in so far as they relate to the sub-contract works and

which are listed in item 1.9 of the schedule of information (referred to generally in respect of the first recital of the agreement, and in more detail later within this chapter).

It is essential that both the contractor and the sub-contractor ensure that this requirement is properly fulfilled, and that item 1.9 of the schedule of information correctly lists *only* the other documents and information which relate to the sub-contract works and which the sub-contractor has been given a copy of or a reasonable opportunity to inspect.

Given its importance, a sub-contractor should resist the inclusion of such a list of documents and information in the sub-contract agreement until such time that copies are provided to him or until he has been given a reasonable opportunity to inspect the documents and information. Obviously, once this occurs the sub-contractor should then read and familiarise himself with same.

ICSub/NAM

The ICSub/NAM has only two recitals.

The first recital

This recital simply states that the contractor wishes the sub-contractor to execute the sub-contract works following the:

- sub-contractor's submission of his attached tender for the sub-contract works, as described in the attached invitation to tender and referenced tender documents therein; and
- submission of the priced documents pursuant to the invitation to tender.

Note the requirement to attach to ICSub/NAM/A the above mentioned documents.

The second recital

This recital states that for the purposes of the sub-contract agreement the sub-contract documents comprise:

- The Intermediate Named Sub-Contract Tender and Agreement (2005 edition).
- The conditions (ICSub/NAM/C) with the JCT Amendments as identified in the invitation to tender (ICSub/NAM/IT).
- The numbered documents. Each numbered document is to be listed in the sub-contract particulars, annexed to the sub-contract agreement and signed or initialled by or on behalf of each party.

MWSub/D

This sub-contract has three recitals, and the information to be inserted in each is self-explanatory.

The first recital

Four entries relating to the main contract are to be inserted as follows:

- date;
- identity of the employer;
- the main contract conditions (including any JCT amendments or other schedule of modifications);
- brief description of the main contract works.

The second recital
Three entries relating to the sub-contract are to be inserted as follows:

- Description of the sub-contract works. This should be brief but accurate, and must include that part of the sub-contract works that is to be designed by the sub-contractor in line with the documents provided by the contractor showing the sub-contractor's design portion.
- Pricing documents. Any documents recording the rates and prices for the sub-contract works are inserted here. If these do not apply (i.e. there is just a lump sum figure) this entry is deleted.
- Any further documents forming part of the sub-contract. It is essential that all documents (other than the pricing documents specifically listed elsewhere in recital 2) which are intended to be sub-contract documents are carefully listed here, such as for example, but not limited to, drawings, specifications, working limitations, etc. Those parts of the construction phase plan applicable to the sub-contract works should be listed under this section.

Whilst not stated, any relevant sub-contract document listed or referred to in the second recital should, to avoid any potential dispute, also be annexed to the sub-contract agreement and signed or initialled by or on behalf of each party.

The third recital
A copy of the main contract particulars should be annexed to this recital, and any alterations to those main contract particulars (for example a change in the date for completion of the main contract works) should be listed under this recital.

ShortSub

This sub-contract has only two recitals, and the information to be inserted in each is self-explanatory.

The first recital
Four entries relating to the main contract are to be inserted as follows:

- date;
- identity of the employer;
- the main contract conditions;
- brief description of the main contract works.

The second recital

Three entries relating to the sub-contract are to be inserted as follows:

- Description of the sub-contract works. This should be brief but accurate.
- Pricing documents. Any documents recording the rates and prices for the sub-contract works are inserted here. If these do not apply (i.e. there is just a lump sum figure) this entry is deleted.
- Any further documents forming part of the sub-contract. It is essential that all documents (other than the pricing documents specifically listed elsewhere in recital 2) which are intended to be sub-contract documents are carefully listed here, such as, for example, but not limited to, drawings, specifications, working limitations, etc.

Whilst not stated, any relevant sub-contract document listed or referred to in the second recital should, to avoid any potential dispute, also be annexed to the sub-contract agreement and signed or initialled by or on behalf of each party.

SubSub

This sub-subcontract has three recitals and the information to be inserted is listed below.

The first recital

Four entries relating to the sub-contract are to be inserted as follows:

- Date.
- Identity of the contractor.
- The sub-contract conditions. If the standard sub-contract contains amendments, or alternatively a non-standard sub-contract is being used, then a copy of the amendments, or the non-standard sub-contract conditions (excluding pricing information) should be annexed to the SubSub.
- Brief description of the sub-contract works.

The second recital

Four entries relating to the main contract are to be inserted as follows:

- date;
- identity of the employer;
- the main contract conditions;
- brief description of the main contract works.

The third recital

Three entries relating to the sub-contract are to be inserted as follows:

- Description of the sub-subcontract works. This should be brief but accurate.
- Pricing documents. Any documents recording the rates and prices for the sub-subcontract works are inserted here. If these do not apply (i.e. there is just a lump sum figure) this entry is deleted.
- Any further documents forming part of the sub-contract. It is essential that all documents (other than the pricing documents specifically listed elsewhere in the

third recital) which are intended to be sub-contract documents are carefully listed here, such as, for example, but not limited to, drawings, specifications, working limitations, etc.

Whilst not stated, any relevant sub-subcontract document(s) listed or referred to in the third recital should, to avoid any potential dispute, be annexed to the sub-contract agreement and signed or initialled by or on behalf of each party.

The articles

What are the articles?

The articles set out certain basic elements of the sub-contract (e.g. the sub-contract documents; the sub-contract conditions; the sub-contract sum (or the sub-contract tender sum); the final sub-contract sum; and the dispute resolution procedures).

What is included in the articles?

MPSub

Unlike the other sub-contracts considered in this book, MPSub has no articles of agreement.

DBSub, ICSub and ICSub/D

Article 1

This article lists the documents comprising the sub-contract documents.
   These documents being:

- The agreement (e.g. DBSub/A) and the sub-contract particulars.
- The schedule of information, plus all documents referenced within it and annexed to it.
- The sub-contract conditions (e.g. DBSub/C), incorporating thereto the JCT amendments stated in the sub-contract particulars (item 1), and any schedule of modifications referred to in the sub-contract particulars (item 1) and listed (and annexed) as a numbered document in the sub-contract particulars.
- The numbered documents annexed to the agreement.

Article 2

This article states the basic obligation of the sub-contractor, which is to carry out and complete the sub-contract works in accordance with the sub-contract.
   Where stated in the third recital (of the DBSub/A and ICSub/D/A only) the sub-contractor's obligations include the requirement to complete the design for the sub-contractor's designed works (or the sub-contractor's designed portion):

- in compliance with regulations 11, 12 and 18 of the CDM Regulations; and
- in accordance with such directions as the contractor may give for the integration of the design for the sub-contractor's designed works with the design for the main contract works as a whole.

Article 3A (Sub-contract sum and final sub-contract sum) and
Article 3B (Sub-contract tender sum and final sub-contract sum)
Either article 3A or article 3B is completed as appropriate. In both cases a sum of money in both words and figures needs to be inserted.

The article that is not completed is to be deleted.

Article 3A is used where a sub-contract sum is applicable (i.e. where the works are valued on an adjustment basis, which means that the sub-contract sum only changes if the sub-contract so provides – for example when variations are carried out).

Article 3B is used where a sub-contract tender sum is applicable (i.e. where the works are to be completely remeasured, usually referred to as the remeasurement basis).

Article 4

This article confirms that if any dispute or difference arises under the sub-contract either party may refer it to adjudication in accordance with clause 8.2 of the sub-contract conditions.

Article 5

This article relates to the position where arbitration is to be used to finally resolve any dispute or difference between the parties.

For arbitration to apply, the sub-contract particulars must be completed to record that article 5 and clauses 8.3 to 8.8 apply (i.e. by deleting the words 'do not apply'). If neither entry is deleted, *the default position is that legal proceedings will apply and not arbitration.*

Article 6

When article 6 applies, the English courts shall have jurisdiction over any dispute or difference between the parties which arises out of or in connection with the sub-contract.

As stated under article 5 above, unless the relevant entry is made within the sub-contract to select arbitration, the default position is that legal proceedings will apply (i.e. not arbitration).

The footnote for article 6 states that if the parties wish any dispute or difference to be determined by the courts of another jurisdiction (i.e. not the English courts) the appropriate amendment should be made to article 6.

IC/Sub/NAM

Article 1

Article 1 of ICSub/NAM/A virtually mirrors article 2 of DBSub/A. Where applicable (i.e. NAM designed works apply), the sub-contractor must complete the design in compliance with:

- the NAM design requirements;
- regulations 11, 12 and 18 of the CDM Regulations; and
- in accordance with such architect/contract administrator instructions given for the integration of that design with the design for the main contract works as a whole.

Article 2A (i.e. Sub-contract sum and final sub-contract sum) and
Article 2B (i.e. Sub-contract tender sum and final sub-contract sum)
Either article 2A or article 2B is completed as appropriate. In both cases a sum of money in both words and figures needs to be inserted.

The article that is not completed is to be deleted.

Article 2A is used where a sub-contract sum is applicable (i.e. where the works are valued on an adjustment basis, and the sub-contract sum only changes if the sub-contract so provides – for example when variations are carried out).

Article 2B is used where a sub-contract tender sum is applicable (i.e. where the works are to be completely remeasured, usually referred to as the remeasurement basis).

Article 3

The due date for the first interim payment is inserted here. Sensibly, the intention is to link its timing (and set future due dates for interim payments) to coincide with the timing of the due date for interim certificates for the main contract works. Note, however, that this first due date is to be no later than one month after the commencement on site of the sub-contract works.

Article 4

This article confirms that if any dispute or difference arises under the sub-contract either party may refer it to adjudication in accordance with clause 8.2.

Article 5

This article relates to the position where arbitration is to be used to finally resolve any dispute or difference between the parties.

A positive act is required for arbitration to apply as follows:

- the invitation to tender (item IT16) must state that arbitration applies; or
- if the invitation to tender does not state arbitration applies, but the contractor and sub-contractor are agreed that article 5 (i.e. arbitration) is to apply (which they are free to do), then:

  — they should state that article 5 applies at item 2 of the sub-contract particulars; and
  — the words 'if it is stated in the invitation to tender (item IT16) that article 5 applies, then' should be deleted.

Article 6

When article 6 applies, the English courts shall have jurisdiction over any dispute or difference between the parties which arises out of or in connection with the sub-contract.

As stated under article 5 above, unless the relevant entry is made within the sub-contract to select arbitration, the default position is that legal proceedings will apply (i.e. not arbitration).

The footnote to article 6 states that if the parties wish any dispute or difference to be determined by the courts of another jurisdiction (i.e. not the English courts) the appropriate amendment should be made to article 6.

MWSub/D, ShortSub and SubSub

Article 1

Article 1 requires the insertion of the sub-contract sum (or the sub-subcontract sum, as appropriate).

Article 2

Under this article, the *on-site* commencement date for the sub-contract works (or the sub-subcontract works, as appropriate) is to be inserted.

Article 3

The period for completion of the sub-contract works (or the sub-subcontract works, as appropriate) is to be inserted. That period must (where appropriate) take into account the planning and preparation period required as referred to in regulation 13(3) of the CDM Regulations.

Article 4

This article confirms that if any dispute or difference arises under the sub-contract either party may refer it to adjudication in accordance with clause 16.2.

Article 5

For arbitration to apply to finally resolve any dispute or difference between the parties, article 5 must be marked accordingly, i.e. by deleting the words 'does not apply'. If neither entry is deleted, legal proceedings will automatically apply, *not* arbitration.

Article 6

When article 6 applies, the English courts shall have jurisdiction over any dispute or difference between the parties which arises out of or in connection with the sub-contract.

As stated under article 5 above, unless the relevant entry is made within the sub-contract to select arbitration, the default position is that legal proceedings will apply (i.e. not arbitration).

The footnote for article 6 states that if the parties wish any dispute or difference to be determined by the courts of another jurisdiction (i.e. not the English courts) the appropriate amendment should be made to article 6.

The sub-contract particulars

The sub-contract particulars set out certain basic terms of the sub-contract.

Wherever the sub-contract particulars apply and wherever practicable, a default position has been provided in respect of the sub-contract particulars, and in such cases an entry only need be made where it is intended that the default position is not to apply.

Where an entry is made, and this entry requires a continuation sheet, then such continuation sheets should be identified as such, should be signed or initialled by or on behalf of each party, and should be annexed to the applicable agreement (e.g. DBSub/A or the ICSub/D/A, etc., as appropriate).

ICSub/NAM

ICSub/NAM takes a different approach, and does not have a detailed sub-contract particulars section, and only has two sub-contract particular items, being:

- Sub-contract particulars item 1: numbered documents.
- Sub-contract particulars item 2: details of changes and updates to particulars in the invitation to tender, tender and/or tender documents.

This different approach is because the invitation to tender (ICSub/NAM/IT) and tender (ICSub/NAM/T), both of which are comprised in the ICSub/NAM, are issued prior to the formation of the sub-contract agreement (ICSub/NAM/A), and already contain equivalent entries to the sub-contract particulars that are found in DBSub/A, ICSub/A and ICSub/D/A. Where relevant, comments on these entries are made against the sub-contract particular entries below.

MWSub/D, ShortSub and SubSub

The above do not have a sub-contract particulars section. However, comment, where relevant, is given below against particular items.

DBSub, ICSub and ICSub/D

Sub-contract particulars item number 1 – the conditions
(see also IT6 in ICSub/NAM/IT)

This relates to the conditions to apply, and confirms that those conditions are to be as DBSub/C, ICSub/C or ICSub/D/C, as appropriate.

The JCT amendments to the standard conditions are either those specifically noted, or if no JCT amendments are specifically noted then the default position is that those JCT amendments current at the sub-contract base date will apply.

In addition, there is the facility for a schedule of modifications to the standard conditions. If no such schedule applies the reference to a schedule of modifications should be deleted from article 1. If a schedule of modifications is provided then it should be listed (and annexed) as a numbered document in the sub-contract particulars.

The schedule of modifications can obviously considerably alter the entire basis of the standard conditions, and such schedules are usually drafted to reflect amendments to the main contract, or in an attempt to give an advantage to the contractor (and a disadvantage to the sub-contractor) in respect of particular clauses in the standard conditions printed text.

A sub-contractor must take careful note of the implications of the proposed schedule of modifications before agreeing to accept the said schedule as each proposed modification could have a serious implication upon the sub-contractor's contractual position.

Sub-contract particulars item number 2 – arbitration
(see also IT16 in ICSub/NAM/IT)

Under this item, the parties agree whether arbitration will or will not apply.

If the parties agree that disputes and differences are to be determined by arbitration then this entry must be completed to state that article 5 and clauses 8.3 to 8.8 will apply.

If it is decided that arbitration is not to apply then legal proceedings will apply.

If it is not decided by the parties that arbitration is, or is not to apply, then the default position is that legal proceedings will apply.

Sub-contract particulars item number 3 – base date
(see also IT17 in ICSub/NAM/IT)

The sub-contract base date is to be inserted here. The sub-contract base date will not necessarily be the same as the main contract base date. Usually it is a date that is close to the sub-contractor's tender or commencement date.

The sub-contract base date is not just some arbitrary or irrelevant date; it is relevant, amongst other things, to:

- matters relating to any divergences from statutory requirements;
- the applicable standard method of measurement to be used;
- the applicable definitions of the prime cost of daywork to be used.

Sub-contract particulars item number 4 – electronic communication

Any communications permitted to be made electronically (i.e. by email, etc.) and the format that they are to be made in are listed under sub-contract particulars item number 4, or alternatively in a document referenced at item 4 and annexed thereto.

If no electronic communications are identified here the default position is that all communications are to be in writing, unless subsequently agreed otherwise.

It should be noted that certain notices (e.g. default notices) must be given by actual, special or recorded delivery; therefore, in such a situation, it is unlikely that electronic communications would be appropriate.

Sub-contract particulars item number 5 – programme
Item 5.1 – preparation of sub-contractor's drawings
(not applicable to ICSub)

A time period in weeks is inserted here for the preparation of all necessary sub-contractor's drawings, etc. (coordination, installation, shop or builder's work or other drawings as appropriate), from receipt of the instruction to proceed with such preparation *and* from receipt of all other relevant drawings and specifications, etc., to the submission to the contractor of the drawings, etc., for comment.

Because the time period commences from:

- the receipt of the instruction to proceed with such preparation; and
- from receipt of all other relevant drawings and specifications, etc.

the contractor must ensure that both the required instruction and all relevant drawings and specifications are provided in sufficient time to meet the programme requirements.

(A near identical entry to the above is located within ICSub/NAM/T at T1.1, albeit the submissions are made to the architect/contract administrator (not the contractor).)

Item 5.2 – contractor's initial comments upon the sub-contractor's drawings (not applicable to ICSub)

Under sub-contract particulars item 5.2 a time period in weeks is to be inserted for the contractor's initial comments upon the drawings, etc. (provided by the sub-contractor in line with item 5.1 above), from when the drawings, etc., have been received by the contractor to when the drawings, etc., are returned to the sub-contractor.

Although not stated, the implication is that the time period would apply in respect of individual drawings, etc., rather than apply to the date of receipt by the contractor of a set of drawings.

Note that the periods entered under items 5.1 and 5.2 need to take into account the sub-contract clauses (clause 2.6.2 of DBSub/D/C, clause 2.6.3 of ICSub/D/C) which:

- Requires the sub-contractor to provide to the contractor design documents and other information to enable the contractor to observe and perform his obligations relating to the main contract/the contractor's design submission procedure.
- Prohibits the sub-contractor from commencing any work in respect of a design document or other information which is subject to the contractor's design submission procedure before that procedure has been complied with.

(A near identical entry to the above, save that the initial comments are made by the architect/contract administrator is found at IT10 of ICSub/NAM/IT.)

Item 5.3 – pre-site commencement – procurement of materials, fabrication, delivery, etc.
(item 5.1 of ICSub/A)

A time period in weeks is to be inserted here for the procurement of materials, fabrication (where appropriate) and delivery to site prior to commencing work on site or work on site in each Section.

The period inserted here would need to take into account any extended manufacturing/delivery times for any non-standard materials. It is conceivable that different periods may be required to be inserted for different Sections (although it should be noted that the entry makes no specific provision for this possibility).

(A near identical entry to the above is located within ICSub/NAM/T at T1.3.)

Consecutive total to sub-contract particular items 5.1 to 5.3
(of DBSub/A and ICSub/D/A)

It is important to note that the periods indicated under items 5.1, 5.2 and 5.3 above are consecutive periods. Each individual time period inserted against these items is then added together and inserted as a total period (for items 5.1 to 5.3) in the paragraph located between items 5.3 and 5.4 of DBSub/A and ICSub/D/A.

(An identical entry to the above is located within ICSub/NAM/T between T1.3 and T1.4.)

Item 5.4 – CDM planning period
(item 5.2 of ICSub/A)

Following amendment 1 (issued April 2007) the sub-contractor's CDM planning period (in weeks) is to be inserted here.

This represents the minimum amount of time for CDM planning and preparation, as required by regulation 13(3) of the CDM Regulations 2007.

Regulation 13(3) provides that there must be no requirement for the sub-contractor to commence construction work until he has received a sufficient opportunity to plan and prepare his work. To reflect this, the period of notice to commence work is stated to be inclusive of the sub-contractor's CDM planning period.

Therefore, if the sub-contractor requires, for example, a ten-week CDM planning period, the period of notice to commence (see entry at item 5.5 as discussed below) would have to be, as a minimum, ten weeks also. The potential long period for this notice means that contractors will need to be vigilant to ensure that the said notice is given at the correct time, as otherwise delays may occur before the sub-contractor has even commenced on site.

(An identical entry to the above is located within ICSub/NAM/T at T1.4.)

Item 5.5 – notice to commence
(item 5.3 of ICSub/A)
A time period in weeks is to be inserted for the period of notice required to commence works on site to enable a start to be made to the sub-contract works or to each Section (if applicable). This is to be inclusive of the CDM planning period.

This said time period (or periods) inserted under this item is not stated as being consecutive to the periods inserted for items 5.1 to 5.3 of DBSub/A and ICSub/D/A (or for the periods inserted for item 5.1 of ICSub/A) and, therefore, must be taken as being concurrent with those periods.

This position is clear at item 5.6 of DBSub/D/A and ICSub/D/A (item 5.4 of ICSub/A), which notes that the period (or periods) required for the carrying out of the sub-contract works on site commences after delivery of materials (i.e. the summated period of items 5.1 to 5.3 if using DBSub/D/A) *and* after the expiry of the period of notice to commence works (as item 5.5 of DBSub/D/A in the above example). The position is also confirmed under clause 2.3 of DBSub/C (clause 2.2 of ICSub/C and ICSub/D/C) that the on-site construction periods are 'subject to receipt by the Sub-contractor of notice to commence works'.

Therefore, using DBSub as an example:

- A five-week notice period is stated under item 5.5 of DBSub/A (includes CDM planning period at item 5.5 of five weeks).
- The total time period for items 5.1 to 5.3 of DBSub/A is eleven weeks.
- If, however, the notice under item 5.5 of DBSub/A was not given until week ten of the eleven weeks (total for items 5.1 to 5.3 of DBSub/A) then even if the materials were delivered to site after the eleven-week period, because of the *late* notice given under item 5.5 of DBSub/A, the five-week notice period runs from week ten. Accordingly, the sub-contractor could conceivably not commence works until the expiry of a fifteen-week period (i.e. week ten plus five weeks notice period required).

It should be noted of course that:

- There is nothing to prevent a sub-contractor from commencing prior to the stated notice period, and this may well frequently occur.
- However:

— the sub-contractor is not contractually obliged to commence prior to the stated notice period; and

— there should be no requirement for the sub-contractor to start works on site until he has had proper opportunity to plan and prepare for health and safety (as regulation 13(3) of the CDM Regulations, for which a period in item 5.4 is inserted).

The above highlights that the late issue of the notice required by item 5.5 of DBSub/A could have a major impact upon the programme for the main contract works. The contractor must ensure he provides this notice at a time to suit the required start date on site for the whole or a Section of the sub-contract works.

It should be noted that the notice required under sub-contract particulars item 5.5 of DBSub/A and ICSub/D/A relates only to the period of notice required to commence works on site to enable a start to be made to the sub-contract works or to each Section; it does not relate to any off-site works or design works.

(An identical entry to the above is located within ICSub/NAM/T at T1.5.)

(In respect of the MWSub/D and the ShortSub, clause 8.1 simply provides for the contractor to issue a written instruction to the sub-contractor instructing him to commence the sub-contract works, and on receipt of this instruction the sub-contractor must commence the sub-contract works on site within 14 days. In respect of the SubSub, the sub-contractor instructs the sub-subcontractor to commence works on site and the sub-subcontractor must commence his sub-subcontract works on-site within 10 days from receipt of the notice.)

Item 5.6 – period required for the carrying out of the sub-contract works on site
(item 5.4 of ICSub/A)

Under this item the period required for the carrying out of the sub-contract works on site is inserted and this will run from whichever is the latter of the following:

- after the delivery of materials (i.e. the summated period of items 5.1 to 5.3 of DBSub/A and ICSub/D/A, or the period of item 5.1 of ICSub/A); or
- after the expiry of the period of notice to commence works (as item 5.5 of DBSub/A and ICSub/D/A, or item 5.3 of ICSub/A).

The period or periods stated are the total 'on-site' working periods.

If Sectional completion does not apply, then only one period (in weeks) needs to be inserted.

If Sectional completion does apply, then for each Section a period (in weeks) needs to be inserted. If there is insufficient room on the entries provided in the sub-contract agreement for every required section, further sheets should be used and annexed to the sub-contract agreement as appropriate.

When agreeing to a period, the sub-contractor needs to take into consideration the information provided under item 3.2 and 3.3 of the schedule of information which forms part of the DBSub/A, ICSub/A and ICSub/D/A.

Item 3.2 of the schedule of information notifies the sub-contractor of the days that the site will be closed due to holidays, which may include, for example, that the site would not be open on Saturdays, Sundays and public holidays.

Item 3.3 of the schedule of information notifies the hours each day that the site will open for the sub-contractor to carry out the sub-contract works.

Items 3.2 and 3.3 noted above may have a major impact upon the sub-contract period required by the sub-contractor and obviously such matters need to be addressed and considered when reviewing any period to be inserted in the sub-contract particulars.

(An identical entry to the above is located within ICSub/NAM/T at T1.6.)

(In respect of the MWSub/D, the ShortSub and the SubSub, the footnote to article 3 (period for completion) states that the period inserted by the parties must duly take account of the planning and preparation period referenced in regulation 13(3) of the CDM Regulations. The period for completion is simply inserted under article 3.)

Item 5.7 – further details or arrangements that may qualify or clarify the period required for the carrying out of the sub-contract works on site (item 5.5 of ICSub/A)

By way of summary, sub-contract particulars item 5 (programme), as detailed above, has the following entries.

Table 2 Entries for sub-contract particulars item 5 for DBSub/A, ICSub/D/A and ICSub/A.

DBSub/A ICSub/D/A	ICSub/A	Time period inserted by the parties for:
5.1	–	Preparation of all necessary sub-contractor's drawings etc.
5.2	–	Contractor's initial comments upon the drawings etc.
5.3	5.1	Procurement of materials, fabrication (where appropriate) and delivery to site prior to commencing work on site.
5.4	5.2	CDM planning period.
5.5	5.3	Notice required to commence works on site.
5.6	5.4	The carrying out of the sub-contract works on site.

Obviously, the sub-contract works are unlikely to be undertaken in a vacuum. There will be other trades, obligations upon the contractor imposed by the main contract in respect of the main contract works (or sections), etc., access and coordination issues. Accordingly, the contractor will often seek to impose certain obligations on the sub-contractor (working to his programme, integration with other trades, sequence, etc.) to manage and meet his obligations under the main contract.

In terms of progress of the sub-contract works in relation to the main contract works the contractor has some protection by clause 2.3 of DBSub/C (clause 2.2 of ICSub/C and ICSub/D/C) which requires the sub-contractor to carry out and complete his work 'reasonably in accordance with the progress of the Main Contract works or each relevant Section of them'. However, it is always safer to rely on some specific requirement, rather than a more generalised obligation. This being the more so because the generalised obligation being relied upon in a similar clause in the DOM/1 form of sub-contract, was considered by Judge Gilliland as *not* requiring:

'The sub-contractor to plan his sub-contract work so as to fit in with either any scheme of work of the main contractor or to finish any part of the sub-contract works by a particular date so as to enable the main contractor to proceed with other parts of the work'[1].

Accordingly, under this final entry (item 5.7 of DBSub/A and ICSub/D/A, item 5.4 of ICSub/A) any further details or arrangements that may qualify or clarify the above or are otherwise relevant to the carrying out of the sub-contract works on site (or a Section thereof) should be inserted.

If there is insufficient room on the printed form itself, further sheets should be used and should be annexed to DBSub/A (or ICSub/A or ICSub/D/A, as appropriate).

All matters that may qualify or clarify the above should therefore be included here. Some limited examples may be:

- any particular programme requirement or constraints;
- long lead in time for procurement, e.g. hand made tiles from Italy;
- specific sequence of working required (i.e. first floor, third floor then basement);
- activities that must be completed by a certain date to accord with a section under the main contract;
- access restrictions;
- return visits, off-site periods;
- key milestones/specific dates to be met (i.e. completion of first fix; commissioning, etc.); and/or
- integration and interdependency of sub-contract works with other sub-contractors, etc.

It is important that all entries are carefully and accurately completed, not least because such entries may have a major impact upon some or all of the periods in the preceding entries, particularly the period for completion required by the sub-contractor.

(An identical entry to the above is located within ICSub/NAM/T at T1.7.)

Sub-contract particulars item number 6 – attendances
(See also IT11 in ICSub/NAM/IT. In addition, item T2 of ICSub/NAM/T permits the sub-contractor to list details of any other attendance items and/or other special requirements that the contractor is to provide free of charge.)
The entries for attendance are split into four sections as follows:

Table 3 Attendances under sub-contract particulars item 6.

6.1	Items of attendance which the contractor will provide to the sub-contractor free of charge
6.2	Joint Fire Code – additional items
6.3	Location of sub-contractor's temporary buildings
6.4	Clearance of rubbish resulting from the carrying out of the sub-contract works

---

[1] *Piggott Foundations Ltd* v. *Shepherd Construction Ltd* (1993) 67 BLR 53.

Item 6.1 Items of attendance which the contractor will provide to the sub-contractor free of charge

Items of attendance which the contractor is to provide to the sub-contractor free of charge are to be identified under this sub-contract particulars item.

A basic list of possible attendance items to be provided free of charge by the contractor is set out in the printed form, which is to be amended and/or added to (as appropriate). If there is insufficient room, further sheets should be used and annexed to the form.

It is important that this list is carefully considered and correctly amended because all attendances (other than those finally listed) are to be provided by the sub-contractor.

Item 6.2 Joint Fire Code – additional items

The main contract particulars included in the schedule of information should state whether or not the Joint Fire Code applies and, if it does, whether or not the requirements in respect of Large Projects apply.

If the Joint Fire Code does apply, it will be the Joint Fire Code that was current as at the main contract base date (not the sub-contract base date) that will apply.

If the Joint Fire Code does apply, the sub-contractor is expected to comply with that code, and the contractor is to provide the attendances listed under item 6.2.

If the requirements in respect of Large Projects are applicable, the contractor must also provide the service of an appropriate number of fire marshals.

Item 6.3 Location of sub-contractor's temporary buildings

Item 6.3 does not apply (and should be deleted) if:

- the sub-contractor is not to provide temporary buildings at all; or
- if the sub-contractor is to provide temporary buildings but these are to be located more than six metres from the building under construction.

If the Joint Fire Code applies to the project and the sub-contractor is to provide temporary buildings either within the building under construction and/or within six metres of the building under construction then:

- the sub-contractor is to ensure that the temporary buildings are constructed in compliance with the requirements of paragraphs 12.4 and 12.8 of the Joint Fire Code; and
- if the project is a Large Project under the Joint Fire Code, then, in line with item 6.3 of the sub-contract particulars, the contractor is to connect (if he is responsible for the connection) an installed fire detection system to a central alarm receiving station.

Item 6.4 Clearance of rubbish resulting from the carrying out of the sub-contract works

If all rubbish resulting from the carrying out of the sub-contract works is to be disposed of off-site (by the sub-contractor) then no entry needs to be made. Therefore, if no entry is made, then the default position is that all rubbish resulting from the carrying out of the sub-contract works is to be disposed of off-site by the sub-contractor.

If there are specific requirements for the manner in which rubbish is to be disposed of, then details should be inserted under item 6.4, for example, the sub-contractor is to dispose of his rubbish to a central skip provided by the contractor.

(In respect of the MWSub/D, clause 5.5 deals with attendances, and in respect of the ShortSub, clause 5.4 deals with attendances. In respect of both of those sub-contracts, except for any attendances specifically identified in the sub-contract documents (which the contractor provides free of charge) everything necessary to enable the sub-contractor to execute the sub-contract works is to be provided by the sub-contractor. Accordingly, if any such attendances are to be provided by the contractor these must be recorded under the second recital under the entry 'further documents forming part of this sub-contract'.

Clause 5.4 of SubSub is similar to clause 5.4 of ShortSub. All attendances are to be provided by the sub-subcontractor except for any specific attendances recorded in the sub-subcontract documents as being provided, free of charge, by 'others'. The third recital of SubSub is where any further sub-subcontract documents are recorded.)

Sub-contract particulars item number 7 (DBSub/A) – listed items
Sub-contract particulars item number 8 (ICSub/A and ICSub/D/A) – listed items
(NB: sub-contract particulars item 7 of ICSub/A and ICSub/D/A dealt with below)

If the contractor and the sub-contractor have agreed that certain goods and materials will be paid for prior to delivery to site, these should be listed under this sub-contract particulars item. Payment for such listed items is subject to the pre-conditions of clause 4.14 having been fulfilled (dealt with later within this book).

If a bond is required in respect of these items, this should be stated and the maximum liability of the surety should also be stated.

(An identical entry to the above is located within ICSub/NAM/IT at IT13.)

Sub-contract particulars item numbers 8 and 9 (DBSub/A) – retention percentage and retention bond
Sub-contract particulars item number 7 (ICSub/A and ICSub/D/A) – percentage of value

DBSub/A – sub-contract particulars item number 8 – retention percentage
An entry only needs to be made here if the retention percentage is to be different to the default percentage of 3%.

The retention percentage can be at any level that the parties agree, and it does not have to be the same percentage as applies to the main contract.

DBSub/A – sub-contract particulars item number 9 – sub-contractor's retention bond
An entry only needs to be made here if a retention bond is required in lieu of retention.

The form of the retention bond is contained under schedule 3 part 2 to DBSub/C.

The information to be provided under item 9 of the sub-contract particulars is:

• the maximum aggregate sum for the purposes of clause 2 of the retention bond; and
• the expiry date for the purposes of clause 6.3 of the retention bond.

A retention bond may be provided by a sub-contractor irrespective of whether or not a retention bond is provided by the main contractor under the terms of the main contract.

ICSub/A and ICSub/D/A – sub-contract particulars item number 7 –
percentage of value
An entry only needs to be made here if the retained percentage from the total value of interim payments is to be different to the default percentage of:

• 95% of total value up to practical completion (i.e. 5% withheld from interim payments);
• 97.5% of value between practical completion to the final payment (i.e. 2.5% withheld from interim payments).

There is no provision for a retention bond in ICSub/A or ICSub/D/A.

(An identical entry to the above entries for ICSub/A and ICSub/D/A is located within ICSub/NAM/IT at IT12.)

(Clause 12.2 of the MWSub/D and the ShortSub states that the amount of an interim payment due shall be 95% of the value of the works executed. The parties are, however, free to agree a different percentage in the sub-contract documents. There is no provision for a retention bond.

The SubSub does not provide for retention or for a retention bond.)

As a point of interest, the MPSub also makes no mention of either retention or a retention bond.

Sub-contract particulars item number 10 (DBSub/A) – fluctuations
Sub-contract particulars item number 9 (ICSub/A and ICSub/D/A) – fluctuations
ICSub/NAM/IT – IT14 and ICSub/NAM/T – T3 – fluctuations
Under DBSub/A, the fluctuations options available are options A, B or C
The only restriction regarding the choice made is that footnote 19 on page 10 of DBSub/A states that where option A or B applies in the main contract it is a requirement of that contract (see paragraphs A.3 and B.4 of schedule 7 of the Design and Build Contract) that the same option applies to any sub-contract.

Where option C applies to the main contract, the parties are free to agree that either options A, B or C apply to the sub-contract.

Where option A or B applies, a percentage addition to the fluctuation payments or allowances applicable is to be inserted. This percentage will relate to option A, paragraph A.12 or option B, paragraph B.13.

Where option B applies, a list of basic transport charges as referred to in paragraph B.1.5.1 of schedule 4 of DBSub/C is to be provided. This list may either be provided on the DBSub/A or it may be contained on a separate document that is simply referred to in the appropriate part of the DBSub/A. If a separate list is provided, it should be included as a numbered document.

The list provided under the sub-contract can be a different list to the equivalent list provided under paragraph B.1.5.1 of schedule 7 of the main contract.

Under ICSub/A and ICSub/D/A

There is only one fluctuations option available (i.e. for contribution, levy and tax fluctuations). Its provisions are recorded at schedule 2.

If sub-contract particulars item number 9 is deleted it will not apply.

Footnote 13 of ICSub/A (footnote 15 of ICSub/D/A) makes it clear that if the fluctuations option applies in the main contract it is a requirement of that contract (see paragraph 3 of schedule 7 of the Intermediate Building Contract) that the same option applies to any sub-contract.

Under ICSub/NAM/IT – IT14 and ICSub/NAM/T – T3

There are two alternative fluctuations options that may apply because of the potentially longer sub-contract period that may apply to named sub-contractors.

These options are option A (contribution, levy and tax fluctuations) and option C (formula adjustment).

In respect of option A, where applicable a percentage addition to the fluctuation payments or allowances applicable is to be inserted in the entry provided in IT14 (in respect of A.12 at schedule 2).

IT14 records that:

- If neither option A nor option C is deleted at IT14, then option A applies in default.
- If both options are deleted at IT14, then neither option will apply.
- Where option C applies, particulars are set out at T3 of the tender. Prior to the architect/contract administrator issuing T3 he should complete the required entries in T3 that are prescribed by the employer (see footnote 15 at IT14).

Sub-contract particulars item number 11 (DBSub/A) – dayworks
Sub-contract particulars item number 10 (ICSub/A and ICSub/D/A) – dayworks

This sub-contract particular item is used if the numbered documents do not include a schedule of daywork rates or prices (for labour, materials and plant) to be used to value Daywork items,

In such circumstances, Daywork is to be calculated in accordance with the definition or definitions identified under this sub-contract particulars item, together with the percentage entered for each section of the prime cost.

The three definitions referred to in this sub-contract particulars item (one or more of which may be chosen) are those agreed between the Royal Institution of Chartered Surveyors and:

- the Construction Confederation (i.e. RICS/CC)
- the Electrical Contractors Association (i.e. RICS/ECA)
- the Heating and Ventilating Contractors' Association (i.e. RICS/HVCA)

The three categories of prime cost are labour, materials and plant.

Labour

In respect of labour (which encompasses tradesmen, craftsmen and non-skilled labour) the options are either to:

- Use one or more of the three definitions noted above to obtain the labour daywork base rate at the time that the daywork was carried out, and then apply the appropriate percentage addition for labour as inserted against the appropriate definition (when using one of the three definitions to obtain the labour daywork base rate, then the wage fixing body or bodies whose labour rates are to be applied when charging for daywork are to be indicated under Sub-contract Particulars item 11).
  or
- Use the all-in rate inserted for labour in lieu of the cost of labour calculated in accordance with one of the definitions.

Materials

For materials, the only option is to apply the appropriate percentage addition as inserted onto the prime cost of materials.

Plant

In respect of plant the only option is to use one or more of the three definitions noted above to obtain the plant daywork base rate at the time that the daywork was carried out, and then apply the appropriate percentage addition onto the schedule of plant hire rates (if appropriate) as inserted against the appropriate definition.

(An identical entry to the above is located within ICSub/NAM/T at T4.)

Sub-contract particulars item number 12 (DBSub/A) – insurance – personal injury and property
Sub-contract particulars item number 10 (ICSub/A and ICSub/D/A) – dayworks
The insurance cover entered here is the minimum level required in respect of death or personal injury (other than to employees covered under the sub-contractors' statutory employers' liability policy) and in respect of damage to property (excluding Specified Perils damage to the main contract works and site materials) up to the terminal date.

The terminal date is defined under clause 6.1 of the sub-contract conditions.

The level of cover required should not normally exceed the level stated in the main contract particulars. However, the level could be less by agreement if the sub-contractor was unable to provide the level of cover required, and if the contractor's insurance covered for any shortfall on cover.

(An identical entry to the above is located within ICSub/NAM/IT at IT15. The insurance requirement in respect of the ICSub/NAM/E (i.e. the employer/sub-contractor direct agreement) is outside of the scope of this book.)

Sub-contract particulars item number 13 (DBSub/A) – incorporation of the sub-contract works into the main contract works
Sub-contract particulars item number 12 (ICSub/A and ICSub/D/A) – incorporation of the sub-contract works into the main contract works
The insertions here are to be for those elements of the sub-contract works that the contractor is prepared to regard as fully, finally and properly incorporated into the main contract works prior to practical completion of the sub-contract works or a section, as applicable, and the extent to which each of the listed elements needs to be carried out to achieve the status of being fully, finally and properly incorporated into the main contract works. If there is insufficient space to make the required entries, then a separate sheet may be used which should then be annexed to the sub-contract particulars.

It should be noted that once incorporated, clause 6.7.2 provides important protection to the sub-contractor in terms of loss or damage to same as follows:

> 'Where, during the progress of the Sub-contract works sub-contract materials or goods have been fully, finally and properly incorporated into the Main Contract works before the Terminal Date, then, in respect of loss or damage to any of the Sub-contract works so incorporated that is caused by the occurrence of a peril other than a Specified Peril, the sub-contractor shall only be responsible for the cost of restoration of such work lost or damaged and removal and disposal of any debris in accordance with clause 6.7.3 to the extent that such loss or damage is caused by the negligence, breach of statutory duty, omission or default of the sub-contractor or of any of the sub-contractors' Persons.'

The terminal date is defined under clause 6.1 of the sub-contract conditions and, for the purposes of this matter, ordinarily it will be the date of practical completion of the sub-contract works, or if applicable, a section thereof.

Obviously, the entries under this sub-contract particulars item 13 must be carefully and precisely set out in order that the contractor and sub-contractor can readily determine when certain elements of the sub-contract works have been fully, finally and properly incorporated into the main contract works prior to the terminal date.

(An identical entry to the above is located within ICSub/NAM/T at T5.)

Sub-contract particulars item number 14 (DBSub/A) – sub-contractor's designed works professional indemnity insurance
Sub-contract particulars item number 11 (ICSub/D/A) – sub-contractor's designed portion works professional indemnity insurance
ICSub/NAM/IT – IT15 – named sub-contractor's designed works professional indemnity insurance
Under DBSub/A
The entry here relates to the level of cover required in respect of professional indemnity insurance in respect of clause 6.10 of DBSub/C.

It should be particularly noted that if no amount of cover is inserted, clause 6.10 will not apply regardless of whether or not the other parts of this item are completed.
The amount of cover required either:

- relates to claims or a series of claims arising out of one event; or
- is the aggregate amount for any one period of insurance. DBSub/A sub-contract particulars item number 14 makes it clear that, unless stated otherwise, a period of insurance for these purposes is one year.

One or other of the above options is to be deleted.

If a level of cover that is different to the full indemnity cover noted above is required for pollution/contamination claims, this level of cover should be entered under item 14 of the sub-contract particulars. If no level of cover is stated, then the required level of cover will be the full amount of the indemnity cover stated above.

The period (in terms of years) that the professional indemnity insurance needs to be maintained from the date of practical completion of the sub-contract works, needs to be entered. The choices are six years, twelve years, or some other period in years. If no entry is made, the expiry date will be taken as being six years from the date of practical completion of the main contract works (not from the practical completion of the sub-contract works).

As a general guide, if the sub-contract is executed under hand a professional indemnity insurance period of six years after the date of practical completion of the sub-contract works would be adequate (because the limitation period for instituting proceedings is six years where a contract is executed under hand), but if the sub-contract is executed as a deed a professional indemnity insurance period of twelve years after the date of practical completion of the sub-contract works would be required (because the limitation period for instituting proceedings is extended to twelve years where a contract is executed as a deed).

Under ICSub/D/A
The entry relates to the level of cover required in respect of clause 6.14 of ICSub/D/C, but is otherwise as the comments in respect of DBSub/A above.

Under ICSub/NAM/IT
The entry relates to the level of cover required in respect of the professional indemnity insurance requirement in respect of the ICSub/NAM/E (i.e. the employer/sub-contractor direct agreement) and that said agreement is outside the scope of this book).

Sub-contract particulars item number 15 (DBSub/A) – settlement of disputes
Sub-contract particulars item number 13 (ICSub/A and ICSub/D/A) – settlement of disputes
ICSub/NAM/IT – IT16 – settlement of disputes
Relevant entries are made here in respect of adjudication, and if it is to apply, arbitration.

The entries are self-explanatory. However, in respect of adjudication, either an adjudicator may be named, or a nominator of an adjudicator may be selected.

If an adjudicator is named, that named adjudicator should be first approached when a party requires a reference to adjudication to be made.

Even if an adjudicator is named, it is recommended that a nominator of an adjudicator is selected in case the named adjudicator is unwilling or unable to act for some reason.

(An identical entry to the above is located within ICSub/NAM/IT at IT16.)

Sub-contract particulars item number 16 (DBSub/A) – numbered documents
Sub-contract particulars item number 14 (ICSub/A and ICSub/D/A) – numbered documents

Any numbered documents (and, as appropriate, any annexures to any numbered documents) should be listed under this sub-contract particulars item.

The said documents should be numbered sequentially, should be initialled or signed by each party, and should be annexed to the sub-contract particulars.

The numbered documents should comprise all the documents which are to be the sub-contract documents other than those already listed at article 1, namely the sub-contract agreement (e.g. DBSub/A) the sub-contract conditions (e.g. DBSub/C), the schedule of information and its listed annexures.

MPSub

Sub-contract particulars entry for clause 1 – base date

The sub-contract base date is to be inserted here.

Sub-contract particulars entry for clause 1 – period for completion

The sub-contract particulars clause 1 entry is where the period for completion is inserted.

MPSub makes clear that the period for completion commences from whichever of the following finishes latest:

- period for pre-site activities (sub-contract particulars, 20);
- period of notice to commence (sub-contract particulars, 20.1); or
- such other period which may be established under clause 22 (extension of time).

Accordingly, it appears that, notwithstanding, if access to the site is given earlier to the sub-contractor by the contractor, contractually the sub-contractor's obligation to complete within the period for completion does not commence until both the period of notice to commence and the period for pre-site activities have lapsed.

However, provided that the period of notice and duration for pre-site activities have both expired, once access to the site has been given to the sub-contractor the sub-contractor must complete the sub-contract works within the period for completion inserted under this section of the sub-contract particulars.

Sub-contract particulars entries for clause 1 – proposals, requirements, sub-contract sum and sub-contract works

Entries are to be made here for the sub-contractor's design proposals (if applicable), the Contractor's Requirements (if applicable), the sub-contract sum, and a brief (but accurate) description of the sub-contract works.

Sub-contract particulars entry for clause 5.1 – electronic communication

An entry is to be made here for what communications may be made electronically and the format in which those communications are to be made.

Sub-contract particulars entry for clause 13 – ground conditions

Clause 13 deals with ground conditions. However, it may be that the parties require alternative provisions for dealing with this matter.

If such alternative provisions are required:

- The alternative provisions are to be identified in the sub-contract particulars entry for clause 13.
- Clause 13 of the printed form is to be deleted and reference to the alternative provisions inserted against clause 13.

The sub-contract particulars state that clause 13 will apply unless alternative provisions are identified.

Sub-contract particulars entry for clause 14.1 – attendance

All items of attendance that will be provided free of charge by the contractor to the sub-contractor must be identified under this sub-contract particulars item. Any attendances not listed under this entry are the responsibility of the sub-contractor.

A basic list of common attendance items are set out under this entry, and this list is to be amended and/or added to (as appropriate). If there is insufficient room on the form, MPSub states that further sheets should be used to record such attendances, and that those sheets are signed or initialled by the parties and annexed to the MPSub.

Additionally, the common attendance items list utilises a notes column whereby (if applicable) for each item of attendance the parties have the facility to record any particular agreement or limitations in respect of any listed item of attendance or the duration that a particular item of attendance will be provided for.

Sub-contract particulars entry for clause 14.2 – clearance of rubbish arising from the carrying out of the sub-contract works

The two options for the sub-contractor's rubbish to be selected here are:

- option 1: rubbish to be taken to a central location for the subsequent disposal by the contractor; or
- option 2: removal and disposal of rubbish off site by the sub-contractor.

If this entry is not completed option 1 applies in default.

Sub-contract particulars entry for clause 15.3 – Joint Fire Code

A simple entry is made here to indicate whether or not the Joint Fire Code applies to the project. If no entry is made the default position is that it will apply.

Clause 15.3 of MPSub states that where the Joint Fire Code applies, both parties must comply with its requirements (failure to do so therefore being a breach of the sub-contract). Further, a defaulting party must, without delay, comply with any notified remedial measures considered necessary by an insurer arising from any non-compliance with the Joint Fire Code, and these remedial measures will not constitute a change.

Sub-contract particulars entry for clause 16.1 – provision of information
Responsibility for the provisions of design and/or production information is selected from the three options listed which relate to clauses 16.1.1 to 16.1.3 of MPSub. Where no option is selected clause 16.1.3 is deemed to apply.

Sub-contract particulars entry for clause 16.3 – dates to provide design and production information
Dates by which the contractor must provide design and/or production information are to be entered here.

Sub-contract particulars entry for clause 18.1 – design documents
Details of the format and quantity of any design documents the sub-contractor is to submit to the contractor are entered here.

Sub-contract particulars entry for clause 20 – period for pre-site activities
Following amendment 1 to MPSub (issued April 2007) the sub-contract particulars clause 20 entry now requires two entries to be made. These are:

- A period for the sub-contractor's pre-site activities.
- The date the above period will commence from.

Within this entry any necessary allowances for design and/or procurement, fabrication, delivery, etc., would need to be included. This period would also need to include for any time period required for any initial comments on the sub-contractor's drawings.

Amendment 1 to MPSub was made to reflect the changes introduced by the Construction (Design and Management) Regulations 2007. The sub-contract particulars clause 20 reflects regulation 13(3), by which every contractor (at every level in the contractual chain), must specify the minimum period available to his sub-contractors for planning and preparation. There must be no requirement for the sub-contractor to start site work until such time as he has received a sufficient opportunity to plan and prepare his work.

Sub-contract particulars entry for clause 20.1 – period of notice
This sub-contract particulars item deals with the period of notice that the sub-contractor is to be given of the date when access to the site will be given.

If no entry is made the default position is seven days.

It appears clear that the period of notice must be coordinated with the period stated for pre-site activities (sub-contract particulars, clause 20), especially in view of the requirements of CDM regulation 13(3) as noted above.

Sub-contract particulars entry for clause 20.2.2 – programme requirements
Clause 20.2 affirms (amongst other matters), that once access to the site is given (in accordance with clause 20.1) the sub-contractor must proceed with the

sub-contract works so as to achieve practical completion within the period for completion.

When inserting/reviewing the above period for completion both parties should be mindful of:

- MPSub clause 20.1. As well as other matters this records:

  'Access to the site shall be subject to any restrictions set out in the requirements. The sub-contractor shall not be entitled to exclusive possession of the site or any part of it and acknowledges that other works may be undertaken on the site at the same time as the sub-contract works, either by the contractor or by others.'

  Accordingly, when setting the period of completion for the sub-contract works both parties must be aware of:

  — Any restrictions affecting access in the requirements which will impact upon the sub-contract works.
  — Any works by others which will impact upon the sub-contract works.

- MPSub clause 20.2 (in particular clause 20.2.2) also records, amongst other things, that following access being given in accordance with clause 20.1, the sub-contractor:

  'Shall proceed with the Sub-Contract works:

  .2 in accordance with any specific dates or programme requirements identified in the sub-contract particulars; and . . .

  so as to achieve Practical Completion within the period for completion.'

  Accordingly, when setting the period of completion for the sub-contract works both parties must ensure that they are aware of the entries found at sub-contract particulars, clause 20.2.2. This is where any specific dates or programme require-ments (if any) are to be inserted, such as, for example:

  — Any key dates of the employer under the main contract, e.g. employer's fitting out contractor start date to a shop unit.
  — Commencement dates (anticipated) for certain areas, e.g. floor 1, floor 2.
  — Any constraints imposed by the main contract, such as certain sub-contract activities to be completed by a specific date to facilitate the contractor in achieving completion of a particular section under the main contract.
  — Any specific sequencing requirements of the sub-contract works.
  — Procurement periods, commissioning dates, etc.
  — Information in respect of other trades working concurrently, i.e.:

      — A programme coordinating various trade activities.
      — Interface requirements, i.e. any specific activities that must be completed by a certain date to permit following activities by following trades (and vice versa).

Obviously the above are but a few limited examples and are by no means a conclusive list. Nevertheless these items serve to underpin the fact that the sub-contract particulars clause 20.2.2 entry must be carefully and accurately completed, not least because such an entry may have a major impact upon the sub-contract period required by the sub-contractor, and obviously, such matters need to be addressed and considered when reviewing any period to be inserted in the sub-contract particulars.

Sub-contract particulars entry for clause 24 – daily rate for bonus
The sub-contract has the facility to provide a bonus payment to the sub-contractor for each day that practical completion occurs before the expiry of the period for completion.

The daily rate of bonus is inserted in the sub-contract particular's clause 24 entry.

If no rate is inserted it is deemed to be nil.

Sub-contract particulars entry for clauses 28.4 and 28.5 – cost savings and value improvements
One of the innovations of MPSub, is that the sub-contractor is encouraged to propose any changes that may result in either a cost saving and/or a value improvement and if this proposal is implemented the sub-contractor is entitled to a share of the financial benefit gained.

Inserted against the sub-contract particulars entries for clauses 28.4 and 28.5 is the percentage share that the sub-contractor will receive in respect of any such benefit obtained by the contractor (clause 28.4 entry) or the employer (clause 28.4 entry). If no percentage is inserted the sub-contractor's share, by default, is 50%.

Sub-contract particulars entry for clauses 31.2 – interim payment advice
Two entries are to be inserted in the sub-contract particulars as follows:

- Monthly date for issue of interim payment advices up to practical completion. If no date is inserted interim payment advices are to be issued on the 25th day of each month.
- Post practical completion – the minimum value that must be due to either party before an interim payment advice is issued. If no amount is inserted the sum is deemed to be £5,000.00.

Sub-contract particulars entry for clauses 36.9 – agreed non-compliances
If it is agreed that the sub-contractor is not obliged to comply with any insurance policies (other than the contract policies) detailed by the contractor within the requirements, this must be detailed here.

If no entry is made, clause 36.9 obliges the sub-contractor to comply with all insurance policies detailed within the requirements by the contractor.

Sub-contract particulars entry for clauses 37.1 and 37.2 – professional indemnity
Where the sub-contractor is undertaking design, MPSub contains an option to require the sub-contractor to take out and maintain professional indemnity insurance (clause 37 of MPSub refers).

This option will only apply if the entry against clause 37.1 in the sub-contract particulars is completed accordingly. If no entry is made the default position is that clause 37 will not apply.

Where this option applies, the minimum amount of indemnity is to be inserted against clause 37.2 of the sub-contract particulars and the type of cover required must be selected from the two alternatives listed.

Sub-contract particulars entry for clause 42 – qualifying period for suspension of notice

Clause 42 of MPSub deals with termination by either party.

The period of suspension before a notice under clause 42.1 can be issued is to be stated under this sub-contract particulars item. If no entry is made the period of suspension is deemed to be 13 weeks.

Sub-contract particulars entry for clause 46 – adjudication

Entries are to be made here in respect of adjudication.

The entries are self-explanatory, and either an adjudicator may be named, or a nominator of an adjudicator may be selected.

If an adjudicator is named, that named adjudicator should be first approached when a party requires a reference to adjudication to be made.

Even if an adjudicator is named, it is recommended that a nominator of an adjudicator is selected in case the named adjudicator is unwilling or unable to act for some reason.

It should be noted that MPSub does not provide for arbitration.

Sub-contract particulars entry for schedule 2 – pricing document

The parties must select from the four options within the pricing document (either rule A, rule B, rule C or rule D) that will apply for the payment of the sub-contract sum.

If no option is chosen rule A applies.

Rule A is in respect of interim valuations, rule B is in respect of stage payments, rule C is in respect of progress payments, and rule D is in respect of some other procedure (to be specified).

Summary of interim and final payment provisions

DBSub/A, ICSub/A and ICSub/D/A all provide a summary of the interim and final payment provisions for ease of reference.

Attestation forms

All of the sub-contracts covered in this book have an attestation form.

What is an attestation form?

Attestation simply means to certify the validity of the sub-contract agreement, and this validity is confirmed by the parties executing the sub-contract agreement in one of two ways:

- under hand; or
- as a deed.

What option is there for executing the sub-contract?

Table 4 Options for executing the sub-contracts.

Not all of the sub-contracts give both options, some only give one. This is set out below: Sub-contract	Options for execution
DBSub	Under hand or as a deed
MPSub	Deed only
	The guide to the MPSub states that as the sub-contract is intended for large scale projects, it is most unlikely that executing it under hand will be desirable, hence it is not provided for.
ICSub	Under hand or as a deed
ICSub/D	Under hand or as a deed
ICSub/NAM	Under hand or as a deed
MWSub/D	Under hand only
	A footnote on the attestation page notes that if the main contract is executed as a deed the parties may wish to consider amending MWSub/D to execute it as a deed also.
ShortSub	Under hand only
	Note that if the main contract is executed as a deed the parties may wish to consider amending ShortSub to execute it as a deed also.
SubSub	Under hand only
	Note: if the sub-contract is executed as a deed the parties may wish to consider amending SubSub to execute it as a deed also.

What difference does it make if the sub-contract is executed under hand or as a deed?

If a sub-contract is executed under hand the limitation period for commencing proceedings due to a breach of the contract is six years, whereas if it is executed as a deed, the limitation period is extended to twelve years.

The question of the limitation period often becomes an issue when latent defects become apparent. A *latent defect* is a defect that for some reason or another is concealed or lies dormant, as distinct from a *patent defect* which is open and obviously exists.

Normally a contractor would wish to have the same limitation period for any recourse against a sub-contractor as there exists against the contractor from the employer. Therefore, if the main contract was executed as a deed (as is usually the case) the contractor would normally require the sub-contract to be executed as a deed. However, if the main contract was executed under hand, the contractor would normally accept that the sub-contract should be executed under hand.

Why do we have limitation periods?

The principle behind limitation periods is to provide a cut-off date after which no claim can be brought (i.e. the claim becomes time barred). The reason behind this is that the law regards it as undesirable that liability could exist without time limit. Limitation periods are set down for both contract and tort in the Limitation Act 1980.

If a contractor's action is in contract against a sub-contractor in respect of defects, the position may depend upon whether the sub-contract contains an express clause whereby the sub-contractor indemnifies the contractor against breaches of contract. If it does (see for example DBSub which does so at clause 2.5.1 of DBSub/C) the contractor's action upon the indemnity clause may not be statute-barred until the expiry of the limitation period (i.e. six or twelve years, as appropriate) between the employer and the contractor under the main contract[2].

Does the Latent Damage Act 1986 affect the sub-contract's limitation period?

The Latent Damage Act 1986 does not affect the existing law on limitation in contract cases[3] where it has been found that section 14A of the Limitation Act could not be applied to actions in contract.

One particular area of note is that under section 32 of the Limitation Act 1980, the limitation period can be extended if the contractor (or a sub-contractor) has deliberately concealed the defect, and this may apply even where the employer (or the contractor, as appropriate) had the benefit of agents overseeing the works[4].

However, simply proceeding with the works does not necessarily give rise to a deliberate concealment; to establish deliberate concealment under section 32 of the Limitation Act 1980 the claimant must show that the defendant:

(1) has taken active steps to conceal a breach of duty after he has become aware of it; or
(2) is guilty of deliberate wrongdoing and has concealed or has failed to disclose such wrongdoing in circumstances where it is unlikely to be discovered for some time.

Execution under hand

Where this option is chosen all of the sub-contracts contain guidance notes on executing this sub-contract under hand.

---

[2] *County and District Properties* v. *Jenner* (1976) 3 BLR 38.
[3] *Iron Trades Mutual Insurance Co. Ltd* v. *JK Buckenham Ltd* [1990] 1 All ER 808.
[4] *Lewisham Borough Council* v. *Leslie & Co.* (1979) 12 BLR 22, CA.

Whilst the entries are relatively self-explanatory, they should nevertheless be meticulously followed and re-checked by the parties. All too often such important matters are not given the careful attention they deserve.

Execution as a deed

If executed as a deed, the parties may execute the agreement either as a company (or other body corporate) or as an individual, as appropriate. In all cases the contractor's name and the sub-contractor's name should be inserted in the appropriate place on the page headed Execution as a Deed.

Again, where this option is chosen all of the above sub-contracts contain detailed notes on executing the sub-contract as a deed. Whilst the entries are relatively self-explanatory, as stated above, they must be meticulously followed and re-checked by the parties.

Foreign companies

Where foreign companies are involved, they can execute deeds (under the Companies Act 1989, as applied by the Foreign Companies (Execution of Documents) Regulations 1994 and the 1995 amendments to those regulations). However, because of the complications that may arise when foreign companies are involved, it is often best to obtain professional advice regarding the proposed execution method.

Alternative attestation forms

In cases where the forms of attestation set out in the sub-contracts are not appropriate (for example, in the case of certain housing associations and partnerships), then the appropriate forms may be inserted by the parties in lieu of the forms provided. Again, professional advice should be sought if this is required.

Schedule of information

DBSub/A, ICSub/A and ICSub/D/A

Item 1 of the schedule of information provides the following information (in respect of the main contract):

- The name of the:
  - employer;
  - employer's agent (DBSub/A); or architect/contract administrator (if using ICSub/A or ICSub/D/A);
  - CDM Coordinator (if named in the main contract).
- Confirmation that the Principal Contractor is the contractor, or if this is not the case, the name of the Principal Contractor is inserted here.
- Identification of the main contract, plus (where relevant):

— any applicable JCT amendments to it (i.e. published by the JCT);
— details of any schedule of modifications, i.e. any employer (non-standard) amendments to the main contract's published text. Where applicable, the schedule of modifications should be annexed to the sub-contract agreement (i.e. DBSub/A, or, ICSub/A, or ICSub/D/A as applicable).

- A copy of the main contract particulars must be annexed to the sub-contract agreement (i.e. DBSub/A, or, ICSub/A, or ICSub/D/A as applicable). This is directly copied from the main contract between the employer and the contractor.
- Any changes to the main contract particulars. For example, any change to the completion date for the main contract works or of any Section(s).
- Confirmation as to whether or not the main contract works is divided into Sections, and, if applicable, details of those Sections (if the information is not already provided in the main contract particulars).
- Finally, under item 1.9 of the schedule of information, any other documents and information relating to the provisions of the main contract of which the sub-contractor has had a copy or a reasonable opportunity to inspect.

Item 2 of the schedule of information notes that a copy of the Construction Phase Plan, or more accurately, those parts of the Construction Phase Plan for the main contract that is applicable to the sub-contract works (together with any developments to the Construction Phase Plan by the Principal Contractor notified to the sub-contractor before or during the progress of the sub-contract works) should be annexed to the schedule of information. As noted in the footnote to item 2, item 2 is to be deleted if the project is not notifiable under the CDM Regulations 2007.

Item 3 of the schedule of information sets out certain programme information.

Any documents annexed to the schedule of information should be signed or initialled by or on behalf of each party.

MPSub

MPSub's philosophy is, as far as possible, to provide a comprehensive set of conditions regulating the contractor's and sub-contractor's relationship. It seeks to avoid excessive reliance upon the main contract provisions being incorporated by reference.

Consequently, MPSub does not contain a schedule of information. However, at schedule 1, MPSub requires the contractor to enter:

- Item 1: the relevant particulars of the main contract, i.e. the date and identification of the parties (the contractor and employer).
- Item 2: any amendments to the standard main contract.
- Item 3: the contract particulars to the main contract (excluding financial details).
- Item 4: any other relevant information concerning the main contract is inserted at item 4.

MPSub does not give the sub-contractor an express right to inspect the actual main contract itself. This should be unnecessary, because the relevant main contract particulars are to provided by the contractor at schedule 1.

ICSub/NAM

ICSub/NAM does not contain a schedule of information. This is because such information has already been conveyed to the sub-contractor within the invitation to tender (ICSub/NAM/IT) and the tender documents, both of which form part of the sub-contract documents appended to the sub-contract agreement (see ICSub/NAM/A first and second recitals.)

ICSub/NAM/IT includes, for example:

- General information section – details of the main contract works, main contract sections (if applicable), description of the sub-contract works, identity of certain parties (e.g. the employer, architect/contract administrator, contractor, CDM Coordinator, the principal contractor, etc.), are to be inserted here.
- Main contract information section – see entries IT1 to IT5.
- Sub-contract information section – see entries IT6 to IT17.
- Tender documents – footnote 1 (page 2 of ICSub/NAM/IT) provides guidance on what information should be provided (if available).

MWSub/D

The MWSub/D does not contain a schedule of information. However:

- Provision for brief entries identifying the main contract conditions are included in the first recital.
- No specific entries for CDM 2007 (e.g. identifying the CDM Coordinator, etc.) are included. However, the guidance notes state that the contractor is to agree with the sub-contractor how health and safety issues will be dealt with.
- Articles 2 and 3 deal with the date for commencement and the period for completion, respectively. The guidance notes state that any information beyond this, for example site working times, or any working restrictions, should, if necessary, be set out in the pricing document or further documents inserted in the second recital.
- Details of the main contract particulars and the amendments to those particulars should be provided under the third recital.

ShortSub

All as MWSub/D, except that the ShortSub does not have a third recital.

SubSub

As MWSub/D above, but additionally note:

- Given that the contractual relationship is between a sub-contractor and a sub-subcontractor, brief entries also identifying the sub-contract conditions are included in the second recital.
- Rather than details of the main contract particulars and the amendments to those particulars being provided in the third recital, any further applicable documents should be inserted instead.

Is it important that the information outlined above is accurate?

DBSub/A, ICSub/A and ICSub/D/A

It is important that the schedule of information is accurately provided by the contractor and noted by the sub-contractor particularly because clause 2.5.1 of DBSub/C (clause 2.4.1 of ICSub/C and ICSub/D/C) states:

> 'Insofar as the Contractor's obligations under the Main Contract, as identified in or by the Schedule of information, relate and apply to the Sub-Contract works (or any part of them), the Sub-contractor shall observe, perform and comply with those obligations, including (without limitation) those under clauses 2.10 (Levels and setting out), 2.21 (Fees or charges legally demandable), 2.22 and 2.23 (Royalties and patent rights) and 3.22 and 3.23 (antiquities) of the Main Contract Conditions, and shall indemnify and hold harmless the Contractor against and from:
>
> .1   any breach, non-observance or non-performance by the Sub-contractor or his employees or agents of any of the provisions of the Main Contract; and
>
> .2   any act or omission of the Sub-contractor or his employees or agents which involves the Contractor in any liability to the Employer under the provisions of the Main Contract.'

ICSub/NAM

All as above, except that in clause 2.4.1 of ICSub/NAM/C reference to the 'schedule of information' is substituted by the words 'Sub-Contract document', for the reasons explained above.

MPSub

Care must be taken when completing schedule 1 to accurately identify the relevant parts of the main contract that apply to the sub-contract, particularly as MPSub (within its definitions) deems the relevant main contract provisions to be those recorded in schedule 1 of MPSub. The effect of this deeming provision is, in the event of any discrepancy, to give the information recorded in schedule 1 precedence over the actual main contract itself.

MWSub/D

Despite its brevity, it is in the sub-contractor's interest to ensure that it is provided with clear, accurate and sufficient information in respect of the main contract, and any matters relevant or affecting the sub-contract works, especially given the obligations placed on the sub-contractor at clause 7.

Whilst clause 7.1 facilitates the sub-contractor requesting a copy of the main contract (omitting details of the contractor's pricing), sensibly, any relevant information affecting the sub-subcontract needs to be in the sub-contractor's possession prior to the parties entering into a contract.

ShortSub

All as MWSub/D above.

SubSub

All as MWSub/D above (except that the MWSub/D deals with a contractor/ sub-contractor relationship and the SubSub deals with a sub-contractor/sub-subcontractor relationship).

Supplementary particulars (containing information needed for formula adjustment)

DBSub/A (only)

These particulars only need to be given when fluctuation option C (formula adjustment) applies.

When fluctuation option C does apply, the relevant insertions should be made by the contractor or the sub-contractor as appropriate.

Chapter 3
Definitions and Interpretations

Definitions

In the JCT sub-contracts being considered within this book, the definition of certain terms is provided.

In this regard, clause 1.1 of DBSub/C, ICSub/C, ICSub/D/C and ICSub/NAM/C states that:

'Unless the context otherwise requires or this Sub-contract specifically provides otherwise, the following words and phrases, where they appear in capitalised form, in the Sub-contract Agreement or these Conditions, shall have the meanings stated or referred to below.'

Clause 1 of MPSub states that:

'Unless the context otherwise requires or this Sub-contract specifically provides otherwise, the following words and phrases, where they appear in capitalised form in this Sub-contract (excluding the Requirements and the Proposals), shall have the meanings stated or referred to below.'

Clause 1 of the MWSub/D, the ShortSub and the SubSub states that:

'Unless the context otherwise requires or this Sub-contract specifically provides otherwise, the following words and phrases, where they appear in capitalised form, in the Agreement or these Conditions, shall have the meanings stated or referred to below.'

The above clauses are very important because, when dealing with the sub-contracts in question, any defined terms must be reported in the capitalised form rather than in a lower case form, so that there is no doubt about what is actually being referred to.

Therefore, when corresponding about, for example, the sub-contract works the phrase the 'Sub-contract Works' should be used rather than the more generalised phrase the 'sub-contract works', to make it clear that the sub-contract works *as defined in the sub-contract* is being specifically referred to.

Despite the importance of the above definition clauses, this book does not generally follow this convention.

Therefore, generally, a word that is defined in the sub-contract is included in the text of this book in lower case – for example 'Variation' is simply included as variation.

The only exception to the above convention is where it is considered that this may cause unnecessary ambiguity to the reader.

With the above point in mind, the reader is strongly advised to refer to the various defini-
tion clauses as noted above when reading the various chapters of this book.

Interpretation

Reference to clauses, etc.

Clause 1.2 of DBSub/C, ICSub/C, ICSub/D/C and ICSub/NAM/C simply clarifies, that, unless otherwise stated, a reference in the sub-contract agreement or in
the sub-contract conditions to a clause or schedule is to that clause in, or that
schedule to, the appropriate sub-contract agreement or conditions.

Also, unless the context otherwise requires, a reference in a schedule to a paragraph is to that paragraph of that schedule.

The MPSub, the MWSub/D, the ShortSub and the SubSub do not make any
mention of this issue at all.

Sub-contract to be read as a whole

Clause 1.3 of DBSub/C, ICSub/C, ICSub/D/C and ICSub/NAM/C states that the
sub-contract is to be read as a whole, but then sets out the priority of documents
in the event that there are any inconsistencies.

This is a very important clause and must be used to resolve disputes about which docu-
ment should be relied upon in the event that there is any inconsistency between the
documents.

The MPSub, the MWSub/D, the ShortSub and the SubSub do not make any
mention of this issue at all.

Headings, references to persons, legislation, etc.

Clause 1.4 of DBSub/C, ICSub/C, ICSub/D/C and ICSub/NAM/C, and clause 2
of MPSub clarify that, unless the context requires otherwise, in the sub-contract
agreement, conditions and schedules (as appropriate):

- The headings in the sub-contract are included for convenience only and shall not
  affect the interpretation of the sub-contract.
- The singular includes the plural and vice versa.
- A gender includes any other gender.
- A reference to a 'person' includes any individual, firm, partnership, company
  and any other body corporate.
- A reference to a statute, statutory instrument or other subordinate legislation
  ('legislation') is to such legislation as amended and in force from time to time,
  including any legislation which re-enacts or consolidates it, with or without
  modification.

The MWSub/D, the ShortSub and the SubSub do not make any mention of these
issues at all.

Reckoning periods of days

Clause 1.5 of DBSub/C, ICSub/C, ICSub/D/C and ICSub/NAM/C, and clause 2 of MWSub/D, ShortSub and SubSub makes clear that where an act is required to be done within a specified period of days after or from a specified date, the period shall begin immediately after that date. Where the period would include Christmas Day, Good Friday or a day which under the Banking and Financial Dealings Act 1971 is a bank holiday in England and Wales, that day shall be excluded.

Clause 3 of MPSub is similarly worded but makes the qualification that the stated rule does not apply where the period in question is stated as being in weeks, months or years (rather than in days).

Contracts (Rights of Third Parties) Act 1999

The Contracts (Rights of Third Parties) Act 1999, which came into force on 11 November 1999, gives a third party the right to enforce a term of a contract, *if* there is an express provision that he should, and *if* he is named in a contract which purports to confer a benefit on him.

Clause 1.6 of DBSub/C, ICSub/C, ICSub/D/C and ICSub/NAM/C, clause 4 of MPSub, and clause 3 of MWSub/D, ShortSub and SubSub makes it clear that notwithstanding any other provision of the sub-contract, nothing in the sub-contract confers or is intended to confer any right to enforce any of its terms on any person who is not a party to it.

Therefore, none of the sub-contracts featured in this book allow for any third party rights.

Giving of service of notices and other documents, and electronic communications

Clause 1.7 of DBSub/C, ICSub/C, ICSub/D/C and ICSub/NAM/C sets out the default position in respect of the serving of notices and other documents if there are no other specific provisions set out in the conditions. This default position being:

- Any notice or other document may be given or served by any effective means (which presumably means delivery by hand, by courier, etc., but not by email, etc., unless specifically stated) and shall be deemed to be duly given or served if addressed and given by actual delivery or sent by pre-paid post to the party to be served at the address stated in the sub-contract agreement or such other address as may from time to time be agreed.
- If there is no agreed address for the notice to be served, the notice or other document shall be effectively served if given by actual delivery or sent by pre-paid post to the party's last known principal business address, or if a body corporate, its registered or principal office.

It is vitally important that certain notices are properly served (e.g. withholding notice from the final payment, and default notices), and great care needs to be taken in such situations that the requirements of clause 1.7 are met.

Clause 1.8 of DBSub/C, ICSub/C, ICSub/D/C and ICSub/NAM/C deals with electronic communications and appears, on face value, to relate to communications other than notices, etc. However, that clause then states that such communications may be made in accordance with any procedures stated or identified in the sub-contract particulars or otherwise agreed in writing by the parties.

The parties can state in the sub-contract particulars the communications that may be made electronically, and the format that those communications are to take. It is therefore submitted that the parties could agree (and certainly the phrase 'or otherwise agreed in writing by the Parties' would seem wide enough to permit) that notices could be served by electronic communication.

Clause 5 of MPSub notes that notices required to be given under clauses 40 to 43 (termination) are to be given by actual delivery, registered post or recorded delivery and shall take effect upon delivery.

Except for these notices all communications required to be made by one party to the other under the MPSub shall be:

- in writing;
- by the procedure specified in the sub-contract particulars for electronic communications (if applicable); or
- by such other means as shall have been agreed in writing by the parties.

All communications are to be sent to either the address notified from time to time by a party for the purposes of communications or, if no address has been notified, the address given in the sub-contract conditions.

Clause 4 of MWSub/D, ShortSub and SubSub states that a notice or other document may be served by any effective means, and adds that it shall be effectively served if it is addressed, pre-paid and delivered by post:

- to the addressee's last known principal business address, or, if he is or has been carrying on trade, profession or business, his last known principal business address; or
- where the addressee is a body corporate, to the body's registered or principal office.

The MWSub/D, the ShortSub and the SubSub do not address the possibility of electronic communications at all.

Effect of the final payment notice or final payment advice

DBSub, ICSub, ICSub/D and ICSub/NAM

The effect of the final payment notice is dealt with by way of clause 1.9 of DBSub/ C, ICSub/C, ICSub/D/C and ICSub/NAM/C.

Clause 1.9.1 of these sub-contracts states that (other than in the case of fraud) the notice of the final payment (the final payment notice) shall have effect in any proceedings (whether by adjudication, arbitration or legal proceedings) under or arising out of the sub-contract, as:

- Conclusive evidence that where and to the extent that any of the particular qualities of any materials or goods or any particular standard of an item of workmanship was described expressly to be for the approval of the architect/contract administrator (except for in respect of DBSub/C), in any of the sub-contract documents, or in any instruction issued by the architect/contract administrator (the employer under the DBSub/C) under the main contract that affects the sub-contract works, the particular quality or standard was to the reasonable satisfaction of the architect/contract administrator (or the employer under the DBSub/C).

    Despite the above, the final payment notice is not conclusive evidence that the materials or goods or workmanship noted above (or indeed that any other materials or goods or workmanship) comply with any other requirement or term of the sub-contract.

- Conclusive evidence that the necessary effect has been given to all of the terms of the sub-contract which require that an amount is to be taken into account in the calculation of the final sub-contract sum. The sub-contracts incorporate what is, in effect, a 'slip rule' in that the final sub-contract sum can be subsequently adjusted where there has been an accidental inclusion or exclusion of any work, materials, goods or figure in any computation or any arithmetical error in any computation.

    Unfortunately no guidance is given as to how far this slip rule can extend, but it is submitted that in the context of the significance of clause 1.9 the slip rule would be construed narrowly and that any accidental inclusion or exclusion would need to be self-evident to an independent third party.

- Conclusive evidence that all and only such extensions of time, if any, as are due have been given.

- Conclusive evidence that the reimbursement of direct loss and/or expense, if any, due to the sub-contractor is in final settlement of all and any claims which the sub-contractor has or may have arising out of the occurrence of any of the relevant sub-contract matters, whether such claim be for breach of contract, duty of care, statutory duty or otherwise.

- Conclusive evidence that the reimbursement of direct loss and/or expense, if any, due to the contractor as a result of the progress of the main contract works being materially affected by any act, omission or default of the sub-contractor is in final settlement of all and any claims which the contractor has or may have so arising, whether such claim be for breach of contract, duty of care, statutory duty or otherwise.

In view of the above, it can be seen that the effect of the final payment notice is far reaching for both parties. It means that if it remains unchallenged, the final payment notice cannot be 'opened-up' in respect of the quality of any materials or goods or of the standard of an item of workmanship which had expressly been for the approval of the employer or the architect/contract administrator, the valuation of the sub-contract works, the sub-contractor's entitlement to extensions of time, the sub-contractor's entitlement to direct loss and expense, or the contractor's reimbursement for direct loss and expense.

The final payment notice must be in writing and must be sent to the sub-contractor by special or recorded delivery, and the effect of the final payment notice should not be underestimated by sub-contractors.

If the sub-contractor disagrees with the final payment notice he should commence adjudication, arbitration or other legal proceedings within ten days of receipt of the notice if he wishes to avoid the notice being deemed conclusive evidence as detailed above. It should be noted that if the proceedings only relate to one issue (for example, the sub-contractor's entitlement to loss and expense) then the final payment notice will still be taken as being conclusive evidence of all other matters.

In the situation where either party has already commenced proceedings before the final payment notice is issued, unless the notice is to be taken as being conclusive evidence of all matters outlined above, either or both parties will have to take a further step in such proceedings within 12 months from or after the date the final payment notice was given.

At the conclusion of such proceedings, the final payment notice will be taken as being conclusive evidence of the matters outlined above, subject only to the terms of any decision, award or judgment in, or settlement of those proceedings.

If an adjudicator gives his decision on a dispute or difference after the date of the final payment notice and either party then wishes to have the dispute determined by arbitration or legal proceedings, such proceedings must be commenced within 28 days of the date on which the adjudicator gives his decision if the decision is not to be final in respect of the matters outlined above.

MPSub

Under the MPSub, clause 31.7 notes that the final payment advice is final and binding upon the parties in relation to amounts due from the contractor to the sub-contractor under or in connection with the sub-contract, including any sums due to the sub-contractor as a consequence of claims for breach of contract, breach of statutory duty, negligence or otherwise, unless within 28 days of the final payment advice being issued the sub-contractor disputes any aspect of it by reference to adjudication or litigation.

MWSub/D, ShortSub and SubSub

The MWSub/D, the ShortSub and the SubSub do not state that the final payment notice has any effect in respect of being conclusive evidence of any matter.

Applicable law

Clause 1.10 of DBSub/C, ICSub/C, ICSub/D/C, ICSub/NAM/C and clause 6 of MPSub state that the default position is that the sub-contract shall be governed by and construed in accordance with the law of England. If the parties do not wish the law of England to apply, then the parties should make the appropriate amendment to the sub-contract.

Although the MWSub/D, the ShortSub and the SubSub do not specifically refer to this matter, they do relate directly back to the main contract and the sub-contract (where applicable), and therefore, normally, the MWSub/D, the ShortSub and the SubSub will be governed and construed in accordance with the law of England.

Chapter 4
Sub-contractors' General Obligations

Most sub-contracts set out the general obligations of the sub-contract and the JCT sub-contracts are no different.

What are the general obligations in the JCT sub-contracts?

DBSub

Clause 2.1 of DBSub/C sets out the general obligations of the sub-contractor. These obligations are:

- to carry out and complete the sub-contract works;
- to carry out and complete the sub-contract works in a proper and workmanlike manner;
- to carry out and complete the sub-contract works in compliance with the sub-contract documents;
- to carry out and complete the sub-contract works in compliance with the Construction Phase Plan (previously termed the Health and Safety Plan); and
- to also carry out and complete the sub-contract works in compliance with other statutory requirements, and to give all notices required by the statutory requirements in relation to the sub-contract works.

MPSub

Clause 7.1 of MPSub sets out the general obligations of the sub-contractor. These obligations are:

- to execute and complete the sub-contract works; and
- to execute and complete the sub-contract works in accordance with the sub-contract, including (if required by the sub-contract), the completion of the design, the specification or selection of materials and the execution of the construction works.

ICSub

Clause 2.1 of ICSub/C is identical to clause 2.1 of DBSub (as above).

ICSub/D

In the first part of clause 2.1 of ICSub/D the wording of the general obligations is identical to clause 2.1 of DBSub. Therefore, the stated general obligations of the sub-contractor are not repeated here.

However, note that clause 2.1 of ICSub/D contains further additional obligations in respect of the sub-contractor's design portion as follows:

(1) In compliance with the numbered documents (to the extent that these are relevant to the sub-contract) the sub-contractor shall:

    (a) complete the design for the sub-contractor designed portion (SCDP) works;

    (b) select any specifications for the kinds and standards of the materials, goods and workmanship to be used in the SCDP works, so far as not stated in the Contractor's Requirements or the Sub-contractor's Proposals.

(2) Comply with the directions of the contractor for the integration of the design of the SCDP with the design of the main contract works as a whole, subject to the provisions of clause 3.5.2 of ICSub/D/C.

(3) Comply with regulations 11, 12 and 13 of the CDM Regulations.

The obligations in respect of the sub-contractor's design portion are considered and dealt with later in this chapter.

ICSub/NAM

Clause 2.1 of ICSubNAM/C is identical to clause 2.1 of ICSub/D (as above).

MWSub/D

MWSub/D does not have a specific general obligations clause as per the JCT sub-contracts considered above.

Indeed, presumably because of its nature and intended use[1] (i.e. low-risk straightforward sub-contract packages, namely a small sub-contract package for a single trade or ones of straightforward design content), clause 5 is simply headed 'sub-contractor's obligations'. However, note that the guidance notes found at the rear of the sub-contract qualifies clause 5 as being the 'main obligations', i.e. by implication, not all.

The obligations expressed at clause 5 are set down below:

• Clause 5.1 contains several obligations which are broken down as follows:
    Clause 5.1 commences by stating that the sub-contractor is to carry out and complete the sub-contract works:

    — in a proper and workmanlike manner; and
    — in compliance with the:

        — sub-contract documents;
        — Construction Phase Plan (where applicable); and
        — the statutory requirements (including the CDM Regulations).

In addition to the above, clause 5.1 obliges the sub-contractor to give all notices required by the statutory requirements in relation to the sub-contract works and to pay all applicable fees and charges related to the sub-contract works provided that the sub-contract documents require the sub-contractor to do so.

---

[1] As noted elsewhere, MWSub/D is also only intended for use where the main contract is the JCT Minor Building Contract with contractor's design.

Clause 5.1 concludes by setting out the sub-contractor's obligations in relation to the sub-contractor's designed portion stating that:

— The sub-contractor shall, using reasonable skill, care and diligence, complete the design for the sub-contractor's designed portion (clause 5.1.1). Note that this obligation on the sub-contractor includes, where not described or stated in the Contractor's Requirements, the selection of any specifications for the kinds and standards of the materials, goods and workmanship to be used in the SCDP works.
— The sub-contractor shall not be responsible for the Contractor's Requirements or for verifying the adequacy of any design contained in the Contractor's Requirements. However, if an inadequacy is discovered in the Contractor's Requirements the Contractor's Requirements must be altered or modified as necessary (clause 5.1.2).
— Subject to clause 9.1 of MWSub/D, the sub-contractor shall comply with any direction issued by the contractor relating to the integration of the design of the Sub-contractor's designed portion with the design of the main contract works (clause 5.1.3).
— The sub-contractor shall comply with regulations 11, 12 and 18 of the CDM Regulations (clause 5.1.4).
— As and when necessary, the sub-contractor must, without charge, provide the contractor with two copies of such drawings or details and specifications of materials, goods and workmanship, and (if requested) related calculations and information, as are reasonably necessary to explain the sub-contractor's designed portion and so that the contractor complies with his obligations under the main contract to supply the architect/contract administrator with such information (clause 5.1.5); and
— must not before the expiration of ten days from the date of delivery commence any work to which the documents referred to in clause 5.15 relate (clause 5.1.6).

- Clause 5.2 of MWSub/D deals with the position regarding the sub-contractor's discovery (clause 5.2.1) or non-discovery (clause 5.2.2) of divergences between the statutory requirements and either the sub-contract documents or instructed variations issued under clause 10.1 of MWSub/D. This is discussed in more detail later in this chapter.
- Clause 5.3 of MWSub/D states that the sub-contractor is to provide goods and materials in accordance with the standard stated in the sub-contract documents. Where no standard is stated therein, any goods and materials provided by the sub-contractor must be of a satisfactory quality.
- Clause 5.4 of MWSub/D deals with the sub-contractor's obligations relating to the construction skills certification scheme (CSCS). This is dealt with later in this chapter.
- Clause 5.5 of MWSub/D states that the sub-contractor is required to provide 'everything' necessary to facilitate the carrying out and completion of the sub-contract works. However, this excludes any items of attendances specifically stated in the sub-contract documents as being supplied by the contractor to the sub-contractor free of charge.
- Clause 5.6 of MWSub/D states that unless the sub-contractor has received the prior written consent of the contractor, he must not assign any benefit arising from the sub-contract, nor sub-let any part of the sub-contract works.

ShortSub

Like MWSub/D, ShortSub does not have a specific general obligations clause (unlike DBSub/C, MPSub, etc., considered earlier).

Indeed, presumably because of its nature and intended use (i.e. a small sub-contract package and/or a low-risk straightforward sub-contract package) clause 5 is simply headed 'sub-contractor's obligations'. However, note that the guidance notes found at the rear of the sub-contract qualifies clause 5 as being the 'main obligations' i.e. by implication, not all.

The obligations expressed at clause 5 are set out below. The sub-contractor:

- Is to carry out and complete the sub-contract works (clause 5.1).
- Is to carry out and complete the sub-contract works in compliance with the sub-contract documents (clause 5.1).
- Is to carry out and complete the sub-contract works with due diligence and in a good workmanlike manner (clause 5.1).
- Is to provide goods and materials in accordance with the standard stated in the sub-contract documents. Where no standard is stated therein, any goods and materials provided by the sub-contractor must be of a satisfactory quality (clause 5.2).
- Has obligations relating to the construction skills certification scheme (CSCS)(clause 5.3). This is dealt with later in this chapter.
- Is required to provide 'everything' necessary to facilitate the carrying out and completion of the sub-contract works. However, this excludes any items of attendances specifically stated in the sub-contract documents as being supplied by the contractor to the sub-contractor free of charge (clause 5.4).
- Must not assign any benefit arising from the sub-contract, nor sub-let any part of the sub-contract works unless he has received the prior written consent of the contractor (clause 5.4).
- Is to comply with any applicable statutory requirements (including the CDM Regulations) relevant to the sub-contract works (expressed as being statute, statutory instrument(s), rule, order, regulation, by-law, etc.), and to give all notices required by the statutory requirements in relation to the sub-contract works (clause 5.6). This clause further obliges the sub-contractor to pay all applicable fees and charges related to the sub-contract works provided that the sub-contract documents require the sub-contractor to do so.

SubSub

As ShortSub.

What is the sub-contractor's obligation to carry out and complete the sub-contract works?

In any building sub-contract, the sub-contractor's primary obligation is to carry out and complete the sub-contract works in accordance with the sub-contract documents.

This usually takes the form of an express term in the sub-contract to that effect (as is the case under the above JCT sub-contracts). Were this obligation not an express term it would, in any event, be a term implied into the sub-contract.

It is generally accepted, following the judgment in *London Borough of Merton* v. *Stanley Hugh Leach*[2], that there is an implied term that a contractor will not hinder or prevent the sub-contractor from carrying out his obligations in accordance with the terms of the sub-contract.

In that case, Judge Vinelott said:

'Where in a written contract it appears that both parties have agreed that something should be done which cannot effectively be done unless both concur in doing it, the construction of the contract is that each agrees to do all that is necessary to be done on his part for the carrying out of that thing though there may be no express words to that effect.'

What is meant by the sub-contractor's obligation to carry out and complete the sub-contract works in a proper and workmanlike manner?

To the extent that this is not an express clause it would be implied as a term into the sub-contract.

This means that the sub-contractor must do the work with all proper skill and care[3]. Normally when deciding what level of skill and care is required the court will consider all the circumstances of the contract, including the degree of skill professed (expressly or impliedly) by the sub-contractor[4].

Breach of this duty includes the use of materials containing patent defects, even where the sub-contractor had not chosen the source of those materials. It may also include relying uncritically on an incorrect plan supplied by the contractor where an ordinarily competent sub-contractor should have had serious doubts about the accuracy of the plan[5].

This also links into a sub-contractor's duty to warn.

What is the sub-contractor's duty to warn?

Two cases in 1984[6] placed a duty on sub-contractors to warn their employer of design problems that they knew about, and, in certain circumstances, irrespective of whether or not the sub-contractor had any design liability.

In respect of this matter, it has been found by the Court of Appeal that in a case[7] where there was a major roof collapse the sub-contractor had not done enough to discharge his duty of care even though he had worked to a design instructed by the client, had discussed the matter with the contractor's engineer and had suggested an alternative solution which was unacceptable to the client. In that case there was a risk of personal injury to the sub-contractor's employees, and this may

---

[2] *London Borough of Merton* v. *Stanley Hugh Leach* (1985) 32 BLR 51.
[3] *Young & Marten Ltd* v. *McManus Childs Ltd* (1969) 9 BLR 77.
[4] *Young & Marten Ltd* v. *McManus Childs Ltd* (1969) 9 BLR 77.
[5] *Lindenberg* v. *Canning* (1992) 62 BLR 147. In that case the plan incorrectly showed obviously load bearing walls as non-load bearing walls.
[6] *Equitable Debenture Assets Corporation Ltd* v. *William Moss* (1984) 2 Con LR 1 and *Victoria University of Manchester* v. *Wilson* (1984–1985) 1 Const LJ 162.
[7] *Plant Construction plc* v. *Clive Adams Associates and Others* [2000] BLR 137, CA.

be why the Court of Appeal considered that the sub-contractor should have pro-
tested more vigorously and pressed his objections on the grounds of safety, perhaps
even to the degree that he should have refused to continue to work until the safety
of his workmen was addressed.

Of course, it is open to question whether the Court of Appeal would have
reached the same decision if the sub-contractor's employees were not at risk of
personal injury.

With this in mind and in a similar case[8], where it was alleged that the sub-
contractor was under a duty to warn in respect of works to be carried out by others
after he had satisfactorily completed his work, the court found that the sub-contractor
did *not* have a duty to warn.

The entire question of duty to warn is far from clear, and, in particular, in the
situations where there is a design defect that does not amount to something dangerous, and where a sub-contractor should have known of the problem of design,
but did not, the law is clearly not yet fully developed.

What are the sub-contract documents that the sub-contractor has to comply with when carrying out and completing the sub-contract works?

DBSub, ICSub, ICSub/D and ICSub/NAM

The DBSub/C, ICSub/C, ICSub/D/C and ICSUB/NAM/C all define the sub-
contract documents at clause 1.1 as being 'the documents referred to in Article 1'
of the agreement that is to be used in connection with these particular provisions
(e.g. DBSub/A when using DBSub/C; ICSub/A when using ICSub/C, etc.).

By reference to the applicable sub-contract agreement (DBSub/A, ICSub/A,
ICSub/D/A and ICSub/NAM/A) the documents referred to in article 1 therein
are:

- the agreement (as stipulated for use with the conditions being used, e.g. ICSub/
  D/A when using ICSub/D/C, etc.), the sub-contract particulars and the schedule of information;
- the documents referred to in and annexed to the schedule of information;
- the relevant sub-contract conditions stated in the agreement (these will be
  expressed as being the DBSub/C, ICSub/C, ICSub/D/C and ICSub/NAM/C
  as applicable for use with that particular agreement) incorporating the JCT
  amendments stated in the sub-contract particulars, and subject to any schedule
  of modifications included in the numbered documents; and
- the numbered documents annexed to the agreement.

MPSub

MPSub's terminology is slightly different to the above sub-contracts. Clause 7.1 of
MPSub states that the sub-contractor is to execute and complete the sub-contract
works in accordance with the 'sub-contract'.

The sub-contract is defined as being:

---

[8] *Aurum Investments Ltd v. Avonforce Ltd and Others* (2001) 17 Const LJ 145.

- the sub-contract conditions;
- the sub-contract particulars;
- schedule 1 (relevant particulars of the main contract);
- schedule 2 (pricing document);
- any annexures;
- the requirements and the proposals.

MWSub/D

Clause 5.1 states that the sub-contractor is to carry out and complete the sub-contract works in compliance with the sub-contract documents. Clause 1 defines the sub-contract documents as 'this form of sub-contract together with any other documents identified in the Second Recital'.

ShortSub

All as MWSub/D above.

SubSub

Clause 5.1 states that the sub-contractor is to carry out and complete the sub-subcontract works in compliance with the sub-subcontract documents. Clause 1 defines the sub-subcontract documents as 'this form of sub-subcontract together with any other documents identified in the Third Recital'.

What are the sub-contract obligations to carry out and complete the sub-contract works in compliance with the Construction Phase Plan (previously called the Health and Safety Plan)?

DBSub, ICSub, ICSub/D and ICSub/NAM

As noted above, clause 2.1 requires the sub-contractor to carry out and complete the sub-contract works in compliance with the Construction Phase Plan.

The Construction Phase Plan is defined under clause 1.1 as being:

'Where the project is notifiable under the CDM Regulations, those parts of construction phase plan for the Main Contract applicable to the Sub-contract Works and annexed to the Schedule of Information, together with any developments of it by the Principal Contractor notified to the Sub-contractor before or during the progress of the Sub-contract Works.'

MPSub

Under the MPSub, a sub-contractor would be obliged to comply with the Construction Phase Plan if it was a sub-contract document (which it will almost invariably be). Note further that a 'breach of the CDM Regulations' is listed as one of the material breaches as defined at clause 1.

MWSub/D

At clause 5.1 the sub-contractor is obliged to carry out and complete the sub-contract works in compliance with the Construction Phase Plan (which is defined at clause 1).

ShortSub

The ShortSub does not contain an express term requiring the sub-contractor to carry out and complete the sub-contract works in compliance with the Construction Phase Plan. However, the guidance notes state that the contractor will need to agree with the sub-contractor how health and safety issues will be dealt with. No doubt the size and respective resources of the contractor and the sub-contractor will be an influencing factor.

Clause 5.6 also deals with the sub-contractor's obligation to comply with the CDM Regulations as applicable to the sub-contract works.

SubSub

As ShortSub.

What is the Construction Phase Plan (previously called the Health and Safety Plan)?

The Construction Phase Plan is developed in two parts:

- The pre-construction Construction Phase Plan, which is usually put together by the CDM Coordinator and which brings together the health and safety information from all parties involved at the pre-construction stage. The pre-construction Construction Phase Plan must include information from the employer about inherent risks which reasonable enquiry would reveal; and this plan forms the basis of the development of a construction stage Construction Phase (construction phase plan) by the Principal Contractor.
- The construction stage Construction Phase, which is developed by the Principal Contractor. This plan is developed to include:
  — risk and other assessments prepared by the Principal Contractor and other contractors/sub-contractors;
  — the health and safety policy of the Principal Contractor;
  — safe method of work statements, etc.

The construction stage Construction Phase (construction phase plan) forms the basis for the health and safety management of the project and continues to evolve through construction. There is usually a requirement that the construction stage Construction Phase (previously termed the Health and Safety Plan) is updated at regular intervals.

Where are statutory requirements defined in the sub-contracts?

DBSub, ICSub, ICSub/D, ICSub/NAM and MWSub/D

The statutory requirements are defined under clause 1.1 (clause 1 MWSub/D) as being:

> 'Any statute, statutory instrument, regulation, rule or order made under any statute or directive having the force of law which affects the Sub-contract Works or performance of any obligations under this Sub-contract and any regulation or by-law of any local authority or statutory undertaker which has any jurisdiction with regard to the Sub-contract Works or with whose systems they are, or are to be, connected.'

MPSub

Under MPSub the statutory requirements are defined under clause 1.1 of MPSub as being:

- any act of parliament and any instrument, rule or order made under any act of parliament;
- any regulation or by-law of any local authority or of any statutory undertaker which has jurisdiction with regard to the project or with whose systems those of the project are or will be connected; and
- any directive of the European Community having the force of law.

ShortSub and SubSub

The statutory requirements are defined in clause 5.6 as being any statute, statutory instrument, rule or order or any regulation or by-law applicable to the sub-contract works (or sub-subcontract works in the SubSub).

What are the statutory requirements?

These statutory requirements would include:

- Defective Premises Act 1972;
- Health and Safety at Work Act 1974;
- Sale of Goods Act 1979;
- Supply of Goods and Services Act 1982;
- Building Act 1984;
- Latent Damage Act 1986;
- Insolvency Act 1986;
- Consumer Protection Act 1987;
- Town and Country Planning Act 1990;

- Building Regulations Act 1991;
- Housing Grants, Construction and Regeneration Act 1996;
- Party Wall Act 1996;
- Scheme for Construction Contracts (England and Wales) Regulations 1998 (SI 1998/649);
- Human Rights Act 1998;
- Late Payment of Commercial Debts (Interest) Act 1998;
- Freedom of Information Act 2000;
- Enterprise Act 2002.

It should be noted that nearly all building work has to comply with the Building Regulations which are updated, amended or changed from time to time, and which stipulate the standards that must be met when carrying out building works.

The Public Health Acts 1875 and 1936 enabled local authorities to make by-laws regulating the construction of buildings. The Public Health Act 1961 provided for the replacement of local building by-laws by the Building Regulations which, when they came into force in 1966, applied throughout England and Wales with the exception of Inner London for which the London Building Acts remained in force. The current consolidating statute is the Building Act 1984 and the principal Building Regulations are the Building Regulations 1991 (which came into force on 1 June 1992) and the Building Regulations (Amendment) Regulations 1992 (which came into force on 26 June 1992). The Public Health Act 1936 remains in force in relation to drains and sewers.

Certain buildings are or may be exempt from the Building Regulations, and a requirement of the Building Regulations may be relaxed or dispensed with upon application to the Secretary of State.

The Building Regulations are currently split into 13 different approved documents covering different areas of construction, as follows:

(1) Structure, loadings, ground movement and disproportionate collapse, etc.
(2) Fire safety, etc.
(3) Dealing with site preparation and resistance to moisture, etc.
(4) Toxic substances, cavity insulation, etc.
(5) Resistance to passage of sound, etc.
(6) Ventilation, etc.
(7) Hygiene, sanitary conveniences and washing facilities, etc.
(8) Drainage and waste disposal, etc.
(9) Heat producing appliances and combustion appliances, etc.
(10) Protection from falling, collision and impact, etc.
(11) Conservation of fuel and power, etc.
(12) Access and facilities for disabled people, etc.
(13) Glazing, etc.

It is an express requirement that the sub-contractor complies with the requirements of the Building Regulations, and therefore a sub-contractor who builds in contravention of the Building Regulations may be in breach of the sub-contract. It is therefore sensible for the sub-contractor to request written confirmation at a pre-contract meeting that all of the required sub-contract works have been approved by building control (building control either being the relevant department of the appropriate local authority, or an equivalent department within an approved external agency (e.g. the NHBC)).

Where a sub-contractor builds in accordance with the contractual design but in contravention of the Building Regulations, it is considered that his liability may turn on whether or not he was aware of the contravention[9].

What are the sub-contractor's obligations where he sub-lets the works and/or design to another?

The fact that the sub-contractor has sub-let the whole or any part of the sub-contract works (including the design of same) does not affect his obligations under the sub-contract. Legally, the sub-contractor remains wholly responsible for carrying out and completing the sub-contract works in all respects in accordance with the sub-contract.

Sub-contractor's design

Design liability can be assumed by either an express term in a sub-contract, or in certain circumstances (despite no express obligation) a sub-contractor may nevertheless assume a design obligation by his actions. The latter possibility is a very involved and complicated area of law which is outside the scope of this book.

Some of the JCT sub-contracts considered in this book can be used in circumstances where a sub-contractor is required to complete the design of all or part of the sub-contract works (when used in conjunction with the appropriate main contract form); others cannot.

Where the JCT sub-contracts do facilitate the option of design by the sub-contractor, the relevant section of the recitals, contract particulars, etc., need completion accordingly.

Does the sub-contract include provision for sub-contractor design?

DBSub

This sub-contract does contain provision for sub-contractor's design. Indeed, whether or not the sub-contractor is to complete the design of all or part of the sub-contract works this sub-contract can still be used.

Where the sub-contractor is to complete the design of all or part of the sub-contract works such works are referred in DBSub as the sub-contractor's designed works.

The sub-contractor's basic obligations in respect of design are covered under clause 2.2 of DBSub, and may be summarised as follows.

The sub-contractor is to:

---

[9] *Equitable Debenture Assets Corporation Ltd* v. *William Moss* (1984) 2 Con LR 1.

- Complete the design for the sub-contractor's designed works (in accordance with the numbered documents to the extent that they are relevant).
- Select any specifications for the kinds and standards of the materials, goods and workmanship to be used in the sub-contractor's designed works, so far as not stated in the Contractor's Requirements or the Sub-contractor's Proposals.
- Comply with the directions of the contractor for the integration of the design of the sub-contractor's designed works with the design of the main contract works as a whole, subject to the provisions of clause 3.5.1.3 of DBSub/C which relates to the sub-contractor's objections to an instruction.
- Comply with regulations 11, 12 and 18 of the CDM Regulations.

What are the sub-contractor's designed works under DBSub?
The third to sixth recitals require careful completion where the sub-contractor undertakes sub-contractor's designed works, and particular attention must be given to the footnotes in DBSub/A in respect of the completion of the third to sixth recitals and items 5.1, 5.2 and 14 of the sub-contract particulars.

Further, as the definitions clause (clause 1.1) defines the sub-contractor's designed works by reference to the third recital (where design responsibility is intended to be transferred to the sub-contractor) the importance of carefully completing the third recital in DBSub/A cannot be overstated.

In respect of the third recital footnote 3 states the following:

- The third recital is to be deleted if the sub-contractor is not required to design any part of the sub-contract works.
- If the sub-contractor is required to complete the design of all the sub-contract works, the words 'all the sub-contract works' is to be inserted here.
- Should the sub-contractor be required to complete only part of the sub-contract works, the relevant part should be clearly identified in the third recital.

The process normally followed in respect of the design documents referred to in the third to sixth recitals is:

- The contractor provides documents (included as numbered documents under sub-contract particulars item 16 of DBSub/A) showing and describing or otherwise stating the requirements of the contractor for the design and construction of the sub-contractor's designed works. These documents together comprise the Contractor's Requirements.

  Obviously, the Contractor's Requirements must be carefully and accurately prepared by the contractor because this is the information upon which the sub-contractor will prepare his Sub-contractor's Proposals (as part of the compilation of his tender).

  There is no set specified form for the Contractor's Proposals, which vary on a project basis according to the contractor's specific needs. The Contractor's Requirements can, therefore, range from a simple brief summary of what is required up to a very detailed set of advanced drawings and detailed specifications.

- In response to the above, the sub-contractor provides documents (which should also be included as numbered documents under sub-contract particulars item 16 of DBSub/A) showing and describing the proposals of the sub-contractor for the design and construction of the sub-contractor's designed works. These said sub-contractor documents together comprise the Sub-contractor's Proposals.

MPSub

This sub-contract may contain provision for the sub-contractor's design. The MPSub is suitable for sub-contract works (where the main contract is the Major Project construction contract) and can be used as a sub-contract irrespective of whether or not the sub-contractor undertakes design.

Clause 7 of MPSub states the general obligations of the sub-contractor. In terms of the sub-contractor's design responsibility (if applicable under the sub-contract), this records, amongst other things, that:

- The sub-contractor's obligation to execute and complete the sub-contract works in accordance with the sub-contract includes (if required by the sub-contract) the completion of the design, the specification or selection of materials as well as the execution of the construction works (clause 7.1).
- The sub-contract warrants, amongst other things, that he has both the competence and will allocate resources necessary to fulfil the role, if required under the sub-contract, of designer in the manner referred in the CDM Regulations.

MPSub seeks to be flexible regarding the allocation of design responsibility. It provides the parties with various alternative options for the completion of the design, ranging from the sub-contractor having no design responsibility to an obligation upon the sub-contractor to complete the design for all of the sub-contract works.

These alternative options are listed at clauses 16.1.1 to 16.1.3 of MPSub. The decision of which alternative applies is activated by the parties completing the relevant entry in the sub-contract particulars. The requirements set out the extent of the design input (if applicable) on the sub-contractor.

The options are as follows:

- Clause 16.1.1: under this option the sub-contractor becomes responsible for completing the design for all of the sub-contract works, and for the preparation of all corresponding design documents.
- Clause 16.1.2: under this option the contractor is deemed responsible for providing all design and all production information for the sub-contract works to the sub-contractor, unless the requirements specifically state otherwise, as follows:

  — Clause 16.2.1: the requirements call for the Sub-contractor's Proposals in respect of certain specifically identified elements of the sub-contract works.
  Where this scenario applies the sub-contractor is responsible for completing the design for the identified elements of the sub-contract works (contained in the requirements), and, likewise, for the preparation of all corresponding design documents relating to same. This design obligation in respect of specific elements of the sub-contract works is similar to the sub-contractor's design portion (SCDP) found in other JCT sub-contracts (e.g. SBCSub/D and ICSub/D).
  — Clause 16.2.2: the requirements set out that the sub-contractor is to prepare particular (limited) design documents. Unlike clause 16.2.1, no proposals are

called for; the sub-contractor just provides the information called for in the requirements.

Obviously, the requirements may seek to employ any combination of the above options set out at clauses 16.2.1 and 16.2.2 where the sub-contract works contain numerous discreet elements or activities.

- Clause 16.1.3: under this option the contractor is responsible for providing all of the design and/or production information necessary for the execution of the sub-contract works. The sub-contractor has no design obligation.

What are the design documents?

As noted above, clause 16 refers to the sub-contractor's obligation to produce design documents. The sub-contract at clause 1 defines this term as meaning:

> 'The design and/or production information that is required to be prepared by the sub-contractor in accordance with the sub-contract, including drawings, specifications, details, schedules of levels, setting out dimensions and the like.'

How is the sub-contractor's design integrated under MPSub?

Under MPSub, where the sub-contractor is required to prepare design documents, clause 16.5.1 provides that the contractor remains responsible for the design and integration of any interfaces between the sub-contract works and other elements of the project, unless the requirements identify the sub-contractor as having responsibility for the design and integration of any particular interface. To facilitate such integration, clause 16.5.2 places a duty on the sub-contractor to cooperate with the contractor and/or any other designers and/or other sub-contractors engaged by the contractor to enable the integration of the design of the sub-contract works with the design of other elements of the project.

The sub-contractor's compliance with any reasonable instructions issued by the contractor in relation to such cooperation does not constitute a change (clause 16.5.2). Perhaps anticipating that this may cause disputes in practice, the guide to the MPSub gives the following guidance in relation to the extent of cooperation envisaged, which is worth summarising below.

Cooperation under clause 16.5.2:

- Includes, for example, the sub-contractor attending meetings, exchanging information.
- Excludes the sub-contractor having to provide something different from that required by the requirements and/or proposals (if applicable). In such circumstances this would be greater than cooperation, and would give rise to a change.

ICSub

This sub-contract is not suitable where the sub-contractor is to design any part of the sub-contract works.

ICSub/D

This sub-contract is appropriate where the main contract is the Intermediate Building Contract with contractor's design, and the sub-contractor is required to design all or part of the sub-contract works (referred to as the sub-contractor's designed portion).

The sub-contractor's obligations in relation to the sub-contractor's design portion (SCDP) are covered under clause 2.1 of ICSub/D/C, and may be summarised as follows.

The sub-contractor is to:

- Complete the design for the SCDP works (in accordance with the numbered documents to the extent that they are relevant).
- Select any specifications for the kinds and standards of the materials, goods and workmanship to be used in the SCDP works, so far as not stated in the Contractor's Requirements or the Sub-contractor's Proposals.
- Comply with the directions of the contractor for the integration of the design of the SCDP with the design of the main contract works as a whole, subject to the provisions of clause 3.5.2 of ICSub/D/C which relates to the sub-contractor's objections to an instruction.
- Comply with regulations 11, 12 and 18 of the CDM Regulations.

What is the sub-contractor's designed portion under ICSub/D?

The SCDP works are described in the third recital of ICSubD/A. This is done either by stating the nature of work in the SCDP works or by referring to the document(s) (i.e. the design document(s)) that more fully describe it.

The process normally followed (in respect of the said design documents) is:

- The contractor provides documents (included as numbered documents under sub-contract particulars item 14 of ICSub/D/A) showing and describing or otherwise stating the requirements of the contractor for the design and construction of the SCDP works. These documents together comprise the Contractor's Requirements.
- In response to the above, the sub-contractor provides documents (which should also be included as numbered documents under sub-contract particulars item 14 of ICSub/D/A) showing and describing the proposals of the sub-contractor for the design and construction of the SCDP works. These documents together comprise the Sub-contractor's Proposals.

ICSub/NAM

This sub-contract can be used whether or not the named sub-contractor is required to design all or part of the sub-contract works (referred to as the NAM designed works), provided the main contract is either the Intermediate Building Contract, or Intermediate Building Contract with contractor's design.

The contract is not suitable, however, if:

- the sub-contractor is not named in the main contract; and/or
- the proposed sub-contract work is part of the contractor's design portion.

Article 1 of the agreement ICSub/NAM/A leaves no room for doubt that where NAM designed works apply, the sub-contractor is to complete the design for them in compliance with:

- the NAM design requirements;
- regulations 11, 12 and 18 of the CDM Regulations; and
- in accordance with such architect/contract administrator instructions given for the integration of that design with the design for the main contract works as a whole.

The sub-contractor's obligations in relation to the sub-contractor's NAM designed works are also stated at clause 2.1 of ICSub/NAM/C, and may be summarised as follows.

The sub-contractor is to:

- Complete the design for the NAM designed works in accordance with the NAM design requirements (included in the numbered documents).
- Select any specifications for the kinds and standards of the materials, goods and workmanship to be used in the NAM designed works, so far as not stated in the NAM design requirements.
- Comply with the instructions of the architect/contract administrator for the integration of the design of the NAM designed works with the design of the main contract works as a whole, subject to the provisions of clause 3.5.2 of ICSub/NAM/C which relates to the sub-contractor's objections to an instruction.
- Comply with regulations 11, 12 and 18 of the CDM Regulations.

What are NAM designed works?

The sub-contractor's NAM designed works are defined at clause 1.1 as follows:

> 'Such of the Sub-contract Works as under NAM Design Requirements the sub-contractor is required to design or for which he is required to complete the design.'

Also, the NAM design requirements are defined at clause 1.1 as follows: 'The requirements relating to any NAM Designed Works included in the numbered documents'.

Where such design input is required, the invitation to tender and accompanying tender documents sent to the proposed named sub-contractor must include:

- Within the general information section, the extent of the design and the sub-contract works to which it relates and this must be clearly identified under the description of the sub-contract works or in the tender documents.
- Within the tendered documents, *inter alia*, any requirements for design work to be undertaken by the named sub-contractor prepared by or on behalf of the employer. Any such requirements will form part of the documentation to be subsequently included within the numbered documents for the purposes of the sub-contract agreement.

When the agreement is entered into between the parties, the sub-contractor is then obliged to complete the design for the NAM designed works in accordance with the NAM design requirements (as included within the numbered documents).

MWSub/D

This sub-contract is appropriate where the main contract is the minor works building contract with contractors' design 2005 edition, revision 1, 2007, and the sub-contractor is required to design all or part of the sub-contract works (referred to as the sub-contractor's designed portion).

The SCDP works are described in the second recital of MWSub/D.

The sub-contractor's obligations in relation to the sub-contractor's designed portion (SCDP) are covered under clause 5.1. as discussed in detail earlier in this chapter.

ShortSub

This sub-contract is for use where small simple sub-contract works are applicable, and the sub-contractor has no design responsibility.

SubSub

Whilst this sub-contract is both simple, short and intended for use where small simple (not complex) sub-subcontract works are required, the guidance notes appended to it do nevertheless contemplate (see 'other matters' at item E) that the sub-subcontractor may be required to undertake design and therefore record the following:

'It is necessary that a number of other matters applicable to the sub-subcontract shall be in the pricing document or further documents referred to in the third recital. These should include:

E Design – if the sub-subcontractor is required to complete the design of all or any of the sub-subcontract works, the sub-contractor should provide his requirements in respect of the sub-subcontract works for which the sub-subcontractor is required to complete the design.'

Design obligations are a complex area of law.

If the sub-subcontractor is required to design all or part of the sub-subcontract works the sub-subcontractor obviously needs to be qualified and competent to do so.

Moreover, before entering into the sub-subcontract the sub-contractor and the sub-subcontractor should obtain professional legal advice and give careful consideration to the suitability of any such design obligations being imposed and any ad hoc design provisions that are required to be incorporated into the sub-subcontract to accommodate same.

What information is to be provided by the contractor in respect of his design obligation?

Compliance with CDM Regulations 11, 12 and 18; DBSub, MPSub, ICSub/D, ICSub/NAM and MWSub/D

In the context of the DBSub/C, MPSub (and other applicable JCT sub-contracts), where a sub-contractor is responsible for design works, the sub-contractor is a designer.

All of these sub-contracts (as discussed above) oblige the sub-contractor to comply with CDM Regulations 11, 12 and 18 (which deal with duties on designers). Also of note is that regulation 4 of CDM 2007 states that a designer must not accept an appointment unless they are competent, and regulation 5 deals with cooperation with others.

The designer's duties under CDM Regulations 11, 12 and 18 are discussed further in Chapter 7.

DBSub

Sub-contractor's designed works information

Clause 2.6 of DBSub/C details the information that the sub-contractor is required to provide to the contractor in respect of the sub-contractor's designed works and also deals with the timing of its supply.

Over and above this, clause 2.6.1 notes that the sub-contractor is also required to comply with regulations 11, 12 and 18 of the CDM Regulations.

In addition to CDM Regulations 11, 12 and 18 what is stated to be provided by the sub-contractor, and when, under clause 2.6?

Clause 2.6.1 of DBSub/C obliges the sub-contractor to provide the contractor (free of charge) with three copies of:

- such sub-contractor's design documents and (if requested) calculations as are reasonably necessary to explain or amplify the Sub-contractor's Proposals; and
- all levels and setting out dimensions which the sub-contractor prepares or uses for the purposes of carrying out and completing the sub-contractor's designed works.

Clause 2.6.2 of DBSub/D/C deals with the timing of the issue of the sub-contractor's design documents.

This sub-clause states that the sub-contractor is to provide the documents 'as and when necessary from time to time' to enable the contractor to observe and perform his obligations:

- under clause 2.8 of the main contract conditions and schedule 5 of DBSub/C (i.e. the contractor's design submission procedure); or
- as otherwise stated in the sub-contract documents (if any other such procedure is stated in the sub-contract documents, these would need to be listed and attached as a numbered document under sub-contract particulars item 16 of DBSub/A).

Clause 2.6.2 further prohibits the sub-contractor from commencing any relevant sub-contract designed works until the requirements of the relevant design submission procedure have been satisfied.

The contractor's design submission procedure is discussed later in this chapter.

MPSub

Do regulations 11, 12 and 18 of the CDM Regulations apply under MPSub?

As noted above, where the sub-contractor has design responsibility, clause 7.2 states, amongst other things, that the sub-contract warrants that he has both the competence and will allocate resources necessary to fulfil the role of designer in the manner referred to in the CDM Regulations.

What design document information is to be provided by the sub-contractor and when?
MPSub deals with this at clause 18.1 (as part of the design submission procedure). Any sub-contractor prepared design documents must be submitted to the contractor (in the number and format identified in the sub-contract particulars) for the contractor's review either:

- on the dates, or on or before the expiry of the period shown on any design programme contained either in the requirements or the proposals, or should this information not exist;
- in sufficient time to facilitate the incorporation of all comments identified by the contractor (pursuant to clause 18) within the relevant sub-contractor's design document prior to the procurement/execution of the sub-contract works it applies to.

ICSub

Not applicable.

ICSub/D

The provisions in ICSub/D mirror those discussed in DBSub above except for the following minor differences:

- ICSub/D refers to SCDP works (whereas DBSub uses the term 'Sub-contractor's Designed Works').
- Clause 2.5.3 of ICSub/D details the information that the sub-contractor is required to provide in relation to the SCDP works and clause 2.6.3 of ICSub/D addresses the timing of its supply. Note, the obligations are virtually identical to the obligations in clause 2.6.1 of DBSub as follows.

  Clause 2.5.3 of ICSub/D provides that the sub-contractor, in addition to complying with regulations 11, 12 and 18 of the CDM Regulations, is to provide the contractor (free of charge) with three copies of:

  — such sub-contractor's design documents and (if requested) calculations as are reasonably necessary to explain or amplify the Sub-contractor's Proposals (ICSub/D clause 2.5.3.1); and
  — all levels and setting out dimensions which the sub-contractor prepares or uses for the purposes of carrying out and completing the SCDP (ICSub/D clause 2.5.3.2).

  Clause 2.6.3 of ICSub/D (timing of the issue of the sub-contractor's design documents referred to in clause 2.5.3) states that the sub-contractor is to provide the documents 'as and when necessary from time to time' to enable the contractor to observe and perform his obligations:

- In respect of the contractor's design submission procedure as stated in the sub-contract documents. Accordingly, these would need to be listed and attached as a numbered document under sub-contract particulars item 14 of ICSub/D/A.

ICSub/NAM

Where there are NAM designed works, clause 2.5.3 and clause 2.6.3 of ICSub/NAM/C refer. These are identically worded to clause 2.5.3 and clause 2.6.3 of ICSub/D except:

- ICSub/NAM uses the term NAM design documents (referred to as sub-contractor design documents in ICSub/D/C).
- ICSub/NAM uses the term NAM designed works (whereas ICSub/D/C uses the term SCDP works).
- Clause 2.5.3.1 of ICSub/NAM/C uses the term NAM design requirements works (whereas clause 2.5.3.1 of ICSub/D/C uses the term Sub-contractor's Proposals).
- Clause 2.6.3 of ICSub/NAM/C (timing of the issue of the sub-contractor's NAM design documents referred to in clause 2.5.3) requires the NAM design documents to be issued to both the architect/contract administrator and the contractor (clause 2.6.3 of ICSub/D/C refers to this information being provided to the contractor only).

MWSub/D

In relation to the sub-contractor's designed portion, the obligations of the sub-contractor are dealt with at clauses 5.1.1 to 5.1.6.

Clause 5.1.4 obliges the sub-contractor to comply with regulations 11, 12 and 18 of the CDM Regulations.

Clause 5.1.5 further obliges the sub-contractor to provide the contractor (free of charge), as and when necessary, with two copies of such drawings or details and specifications of materials, goods and workmanship, and (if requested) related calculations and information, as are reasonably necessary to explain the sub-contractor's designed portion and so that the contractor complies with his obligations under the main contract to supply the architect/contract administrator with such information.

Clause 5.1.6 also obliges the sub-contractor not to commence any work to which the documents referred to in clause 5.15 relate before the expiration of ten days from the date of delivery.

SubSub

As noted earlier, SubSub does not contain an express term dealing with this matter and any obligations on the sub-subcontractor relating to design would need to be carefully considered and recorded in the sub-subcontract documents at the third recital.

What is the contractor's design submission procedure?

In circumstances where a sub-contractor is obliged to provide design documents and other information the sub-contracts will, invariably, include reference to a design submission procedure.

DBSub

Clause 2.6.2 of DBSub/C obliges the sub-contractor to issue his design documents to the contractor in sufficient time to permit the contractor (in respect of the sub-contractor's design works) to comply with his obligations under clause 2.8 of the main contract and schedule 5 (the contractor's design submission procedure), or as alternatively stated in the sub-contract documents.

The contractor's design submission procedure as detailed under schedule 5 of the DBSub/C is as follows:

• The contractor prepares and submits two copies (in such format as is stated in the Employer's Requirements or the Contractor's Proposals) of the relevant design documents (including, where appropriate, the sub-contractor's design documents) to the architect/contract administrator in sufficient time to allow any comments of the architect/contract administrator to be incorporated prior to the relevant design document being used for procurement and/or for the carrying out of the works.

No actual time limit is set for the submission of the design documents, but the contractor (and the sub-contractor) needs to be aware of the submission procedure to be followed and needs to ensure that the various design documents are issued at an early enough date to allow the submission procedure to be completed without causing a delay to the main contract works and/or the sub-contract works.

This is particularly relevant because clause 2.6.2 of DBSub/C makes it clear that the sub-contractor shall not commence any work until the relevant design document has satisfactorily completed the submission procedure; and clause 2.14 of DBSub/C states that no extension of time will be given to the sub-contractor where he has failed to provide in time any necessary design documents in line with clause 2.6.2 of DBSub/C.

• Within 14 days from the date of receipt of the design documents referred to above, (or, if later, 14 days from either the date of or the expiry of the period for submission of the design documents as stated in the contract documents), the architect/contract administrator shall return one copy of the contractor's design documents to the contractor marked either 'A', 'B' or 'C'.

What is the status of a contractor's design document marked 'A', 'B' or 'C'?
• Documents marked with an 'A'.
  If documents are marked with an 'A' that means that the sub-contractor can carry out the work strictly in accordance with those documents.

• Documents marked with a 'B'.
  If documents are marked with a 'B' that means that the sub-contractor can carry out the work in accordance with the submitted documents, provided that the architect/contract administrator's comments are incorporated and provided that an amended copy of the document in question is promptly submitted to the architect/contract administrator.

• Documents marked with a 'C'.
  If documents are marked with a 'C' that means that the sub-contractor cannot carry out the work in accordance with the submitted documents without following a further procedure, as outlined below.

Must the architect/contract administrator justify why a contractor's design document is marked as status 'B' or 'C'?
Where design documents are marked with a 'B' or a 'C', the architect/contract administrator is to identify why he considers that the document is not in accordance with the main contract (paragraph 4 of the contractor's design submission procedure refers).

Following his receipt what happens if the architect/contract administrator does not respond to a design documents in the required time period?
In such circumstances, following the expiry of the time period, the documents will be regarded as though they had been marked with an 'A'.

Clearly, because of this positive default position, it is vitally important that sub-contractors maintain a strict control over the management and register of documents issued and received.

What happens subsequently where a drawing is returned marked 'C'?

- The sub-contractor is not to carry out any work in accordance with a design document marked with a 'C'.
- The employer (and the contractor, by inference) shall not be liable to make payment for any such work that is executed by the sub-contractor.
- If the sub-contractor agrees with the architect/contract administrator's comments, he simply amends the design document and re-submits it to the contractor to allow it to go through the contractor's design submission procedure (as outlined above) again.
- Alternatively, if the sub-contractor does not agree with the architect/contract administrator's comments, then:
  - The sub-contractor is to notify the architect/contract administrator with reasons (through the contractor), within seven days of receipt of the architect/contract administrator's comments, why he considers that the compliance with the said comments would give rise to a variation.
  - Upon receipt of such a notification, the architect/contract administrator shall, within seven days, either confirm or withdraw the comment.
  - The submission procedure is silent on what happens if a comment is withdrawn. However, by virtue of paragraph 3 it is likely that where the comment is withdrawn by the architect/contract administrator, the design document (previously marked with a 'C') will assume the status of a design document marked with an 'A'.
  - Alternatively, where the comment is confirmed by the architect/contract administrator, the sub-contractor is to amend the design document and is then to re-submit it to the contractor to allow it to go through the contractor's design submission procedure (as outlined above) again.

Of course, if the sub-contractor still does not agree with the architect/contract administrator's comment, the option of adjudication to resolve the dispute is available to the sub-contractor.

Does the architect/contract administrator or the contractor 'approve' the sub-contractor's design documents?
Nowhere in the above submission procedure does the architect/contract administrator or the contractor 'approve' the sub-contractor's design documents.

The sub-contractor continues to have liability for the sub-contractor's design irrespective of the application of the design submission procedure to it. Any comments by the architect/contract administrator (including, where applicable, the subsequent confirmation or withdrawal by the architect/contract administrator of such comments) under this procedure does not:

- signify acceptance by the employer or the architect/contract administrator that the design document (or the amended design document, as appropriate) is in accordance with the main contract (or the sub-contract) requirements; or
- that it gives rise to a variation (refer to DBSub/C, schedule 5, paragraph 8.1).

The sub-contractor's strict liability for design is reinforced by clause 8.3 of schedule 5 (of DBSub/C) which states:

'Neither compliance with the design submission procedure in this Schedule nor with the Architect/Contract Administrator's comments shall diminish the Contractor's obligations to ensure that the Contractor's Design Documents and CDP Works are in accordance with the Main Contract.'

Because of the way that the Contractor's Design Submission Procedure has been incorporated into the DBSub/C, it is submitted that in the above extract the word Contractor is interchangeable with the word Sub-contractor, the CDP Works (contractor's design portion works) must include the sub-contractor's designed works, and the Main Contract must incorporate the sub-contract.

On a similar vein, clause 2.13.4 of DBSub/C makes it clear that, although the contractor is to give notice to the sub-contractor specifying anything which appears to him to be an inadequacy in the sub-contractor's design documents, no such notice (nor any failure to give such a notice) shall relieve the sub-contractor of his obligations in connection with the design.

Finally, and in any event, it is reasonably well established in law that, even if some form of 'approval' were given, this would be unlikely to detract from the sub-contractor's general liability as the designer of the sub-contractor's design works.

What if the comments made by the architect/contract administrator upon the sub-contractor's design documents (marked 'C') actually constitute a variation?
In such circumstances the sub-contractor needs to be aware that, if the sub-contractor (via the contractor) does not state that he considers that the architect/contract administrator's comments constitute a variation (on a design document marked with a 'C') within seven days of receipt of the architect/contract administrators' comments, then the comments in question shall *not* be treated as giving rise to a variation (refer to DBSub/C, schedule 5, paragraph 8.2). It should be noted that this requirement is a requirement under the main contract, the time period for the sub-contractor may, in effect, need to be less.

MPSub

MPSub includes a design submission procedure at clause 18 of MPSub, being applicable where the sub-contractor is required to prepare design documents.

Whilst not dissimilar to the DBSub procedure (dealt with above), no reference to the main contract procedure or the involvement of the architect/contract administrator is included. Comments on the sub-contractor's design document are provided by the contractor.

The design submission procedure detailed at clause 18 of MPSub is as follows:

- Clause 18.1 deals with the content and timing of the sub-contractor's design documents submitted to the contractor. The sub-contractor is required to submit design documents to the contractor:

  — in the quantities and format stated in the sub-contract particulars; and
  — at such time to comply with any date or period on a design programme contained in the requirements or the proposals, or if this information does not exist, in sufficient time to allow any comments of the contractor (under clause 18) to be incorporated prior to the relevant design document being used for procurement and/or for the carrying out of the works.

- Within 21 days from the date of receipt of the design documents referred to above, or 21 days from the date of or the expiry of the period for submission of the design documents stated in the design programme contained in the requirements or the proposals (whichever is the later date), the contractor shall return one copy of the contractor's design documents to the contractor marked either 'A', 'B' or 'C'.
- Clause 18.5 sets out the status of 'A', 'B' or 'C' and is not dissimilar from DBSub in this respect (see comments previously made above).
- Both parties need to be aware that:

  — if the contractor receives a design document, but fails to respond within the timeframe set by the procedure, the submitted design document automatically attains an 'A' status in default (clause 18.3 refers);
  — the contractor's liability to pay the sub-contractor is expressly limited to work executed in accordance with design documents marked 'A' or 'B' but not 'C' (clause 18.6 refers).

- Where design documents are marked with a 'B' or 'C', the contractor must identify why he considers the document is not in accordance with the sub-contract (clause 18.4 refers). Following receipt of the contractor's comments the sub-contractor can:

  — Accept the contractor's comment, forthwith amending the design document as appropriate and resubmitting it; or
  — notify the contractor within four days:

    — that he disagrees with the contractor's comment that his design document is not in accordance with the sub-contract. The sub-contractor's notice must state why the contractor's comment would constitute a change, together with a supporting statement enclosing the sub-contractor's reasons; and/or
    — that (if applicable) compliance with the contractor's comment affects the sub-contractor's obligations under clause 17.2 (compliance with statutory requirements; satisfaction of performance specifications, selection of materials) and/or clause 17.3 (reasonable skill and care obligations).

      Sub-contractors need to be aware of the requirements of clause 18.7 upon them, and its four-day timetable. MPF clause 18.9 makes plain that in the absence of a sub-contractor's notice being issued (under clause 18.7) the contractor's comments are deemed not to give rise to a change; hence any procedural failings by the sub-contractor may have financial consequences and should be avoided.

What must the contractor do if he receives a clause 18.7 notice?
Clause 18.8 sets out the procedure as follows:

- Upon receipt of the sub-contractor's notice the contractor has ten days to confirm or withdraw his comment.
- If the contractor confirms his comment the sub-contractor must amend his design document as necessary and resubmit it forthwith.
- It is important to be aware that neither confirmation nor withdrawal of a contractor's comment in respect of a design document signifies the contractor's acceptance that:

  — the design document (or amended design document) is in accordance with the sub-contract; or
  — the sub-contractor's compliance with the contractor's comments would constitute a change. Reference to the sub-contract itself will be needed to determine this.

Does the design submission procedure relieve the sub-contractor of his obligations with regard to the design documents he has prepared?
Clause 18.10 makes it clear that the sub-contractor's compliance with the design submission procedure and/or the contractor's comments issued under the procedure do not relieve the sub-contractor of his responsibility to ensure:

- all design documents are prepared in accordance with the sub-contract; and
- the completed sub-contract works are in accordance with the sub-contract.

The only exception to the above (as stated at the very beginning of clause 18.10) is where the sub-contractor notifies the contractor (see clause 18.7) that compliance with a contractor's comment affects the sub-contractor's obligations under:

- clause 17.2 (compliance with statutory requirements; satisfaction of performance specifications, selection of materials); and/or
- clause 17.3 (reasonable skill and care obligations).

ICSub

Not applicable.

ICSub/D

ICSub/D differs from DBSub with clause 2.6.3 of ICSub/D/C making no reference to the main contract design submission procedure, but instead a design submission procedure contained in the sub-contract documents. There is no standard design submission procedure, and the sub-contractor therefore needs to familiarise himself with any ad hoc design submission procedure that is incorporated into the sub-contract.

ICSub/NAM

Like ICSub/D, clause 2.6.3 of ICSub/NAM/C makes no reference to the main contract design submission procedure. Instead, it requires the sub-contractor to comply with the NAM design requirements (the requirements relating to any NAM design works included in the numbered documents) when submitting NAM design documents. There is no standard design submission procedure, and the sub-contractor therefore needs to familiarise himself with any ad hoc design submission procedure that is incorporated into the sub-contract.

MWSub/D

MWSub makes no reference to a design submission procedure. Accordingly, if a design submission procedure is required this would need to be recorded in the sub-contract documents at the second recital. As there is no standard design submission procedure the sub-contractor would need to familiarise himself with any ad hoc design submission procedure that is incorporated into the sub-contract.

ShortSub

Not applicable.

SubSub

As MWSub/D above except that the design submission procedure would need to be recorded in the sub-subcontract documents at the third recital.

How is the provision of further drawings, details and directions dealt with in relation to sub-contract works and/or the sub-contractor's design?

Although, in theory, the numbered documents are intended to show and describe the sub-contract works in full, in practice there is often the need for further drawings and details to be issued to the sub-contractor, or to be issued by the sub-contractor.

DBSub

Clauses 2.7.1 and 2.7.2 of DBSub/C deal with this situation. These clauses require:

- The contractor to provide without charge to the sub-contractor:
  - — two copies of such further drawings or details as are reasonably necessary to explain and amplify the numbered documents; and/or
  - — such directions (including those for or in regard to the expenditure of Provisional Sums) as are necessary to enable the sub-contractor to carry out and complete the sub-contract works in accordance with the sub-contract.
- The sub-contractor to provide to the contractor:

— working/setting out drawings and other information necessary for the contractor to make appropriate preparations to enable the sub-contractor to carry out and complete the sub-contract works in accordance with the sub-contract.

Although clause 2.7.2 of DBSub/C does not state that the information to be provided by the sub-contractor is to be free of charge to the contractor, this would probably be the case by implication. Also, as no number of copies is expressly stated, it could be argued that only one copy needs to be provided, although it is more likely that two copies would be implied in keeping with the more general contractor's design submission procedure.

Clauses 2.7.3 and 2.7.4 of DBSub/C require that:

- such further drawings, details, information and directions referred to in clauses 2.7.1 and 2.7.2 shall be provided or given at the time it is reasonably necessary for the recipient party to receive them, having regard to the progress of the sub-contract works and the main contract works; and

- where the recipient party has reason to believe that the other party is not aware of the time by which the recipient needs to receive such further drawings, details, information or directions, he shall, so far as is reasonably practicable, advise the other party sufficiently in advance to enable him to comply with the requirements of clause 2.7 of DBSub/C.

If the sub-contractor has any doubts that the contractor is aware of when the sub-contractor needs to receive such information, the sub-contractor should identify to the contractor what he needs, and by when, at his earliest opportunity. The same also applies to the contractor, i.e. where the date of receipt of the sub-contractor's information noted at clause 2.7.2 affects the execution of work by others.

MPSub

Clause 16.2 obliges the contractor to provide the sub-contractor with any information necessary for the sub-contractor's preparation of design documents by:

- The date set out in any design programme contained within the requirements or the proposals.
- If there is no design programme or any relevant dates stipulated, the contractor must provide the information at a date that permits the sub-contractor to comply with his obligations under clause 18 (design submission procedure) and clause 20 (commencement and completion).

Clause 16.3 also requires the sub-contractor to provide design and/or production information by the dates stated in the sub-contract particulars at a date that the sub-contractor reasonably requires in order to comply with his obligations under clause 20 (commencement and completion).

Clause 16.4 contains a similar provision to clause 2.7.4 of DBSub/C. Should the sub-contractor have reasonable grounds to believe the contractor is unaware of his obligations to provide such information (in accordance with clauses 16.2 and 16.3), he must notify the contractor accordingly of what he needs and in time to permit the contractor to perform his obligations under these clauses.

ICSub

Issue of construction information and directions

(1) By the contractor
Except for being subject to clause 2.10 and 2.11 of the intermediate main contract conditions, clause 2.5.1 of ICSub/C mirrors the wording of clause 2.7.1 of DBSub/C (as above).

(2) By the sub-contractor
Clause 2.5.2 of ICSub/C is identical to clause 2.7.2 of DBSub/C (as above).

Time of supply
Clause 2.6.1 and clause 2.6.2 of ICSub/C are identical to clauses 2.7.3 and 2.7.4 of DBSub/C (as above).

ICSub/D

Issue of construction information and directions

(1) By the contractor
Except for being subject to clause 2.10 and 2.11 of the intermediate main contract conditions, clause 2.5.1 of ICSub/D/C mirrors the wording of clause 2.7.1 of DBSub/C (as above).

(2) By the sub-contractor
Clause 2.5.2 of ICSub/D/C is identical to clause 2.7.2 of DBSub/C (as above). Clause 2.5.3 of ICSub/D/C relates to the issue of the contractor to the sub-contractor's design documents (clause 2.5.3.1) and levels and setting out dimensions (clause 2.5.3.2).

Time of supply
Clause 2.6.1 and clause 2.6.2 of ICSub/D/C are virtually identical to clauses 2.7.3 and 2.7.4 of DBSub/C (as above).

Clause 2.6.3 of ICSub/D/C deals with the timing of the issue of the sub-contractor's design documents referred to in clause 2.5.3, and is discussed earlier in this chapter.

ICSub/NAM

Issue of construction information and directions

(1) By the contractor
Clause 2.5.1 of ICSub/NAM/C virtually mirrors the wording of clause 2.7.1 of DBSub/C (as above), except for the following differences:

(a) the further drawings or details forwarded by the contractor are those supplied by the architect/contract administrator; and
(b) are relevant to the sub-contract works for the purposes of clause 2.1.2. This clause relates to architect/contract administrator's instructions regarding the integration of the NAM design works within the design of the main contract works as a whole.

(2) By the sub-contractor

Clause 2.5.2 of ICSub/NAM/C is identical to clause 2.7.2 of DBSub/C (as above).

Clause 2.5.3 of ICSub/NAM/C is dealt with earlier in this chapter and relates to the issue to the contractor of the sub-contractor's NAM design documents (clause 2.5.3.1) and levels and setting out dimensions (clause 2.5.3.2).

Time of supply

Clause 2.6.1 and clause 2.6.2 of ICSub/NAM/C are identical to clauses 2.7.3 and 2.7.4 of DBSub/C (as above).

Clause 2.6.3 of ICSub/NAM/C deals with the timing of the issue of the sub-contractor's NAM design documents (where applicable) referred to in clause 2.5.3, and is discussed earlier in this chapter.

MWSub/D and ShortSub

Neither of these sub-contracts include any express provisions dealing with the issue of further construction information, documentation, setting out details, etc., between the contractor and the sub-contractor. Such additional procedures, if required, would need to be identified as a sub-contract document in the second recital.

MWSub/D (not ShortSub) at clause 5.1.5 does deal with the issue and timing of SCDP information from the sub-contractor to the contractor.

SubSub

SubSub does not contain any express provisions dealing with the issue of further construction information or design documentation, and therefore if such additional procedures are required by the parties these would need to be carefully considered and identified as a sub-subcontract document in the third recital.

Is the sub-contractor responsible for the adequacy of the design contained in the Contractor's Requirements?

Until relatively recently, it was commonly thought that a contractor (or sub-contractor) undertaking design did not assume any responsibility for the design (by others) contained within the employer's (or contractor's) requirements. The contractor's design responsibility was limited to the discreet element of the design he specifically carried out, i.e. the work he undertook to finalise the design in the employer's (or contractor's) requirements.

However, in the case of *Co-op* v. *Henry Boot*[10], it was found that in a standard design and build contract the contractor was not only responsible for any design specifically carried out by him, but more onerously was also liable for the completed scheme as a whole, even where the design work for a specific element was carried out earlier by others.

---

[10] *Co-operative Insurance Society Ltd* v. *Henry Boot (Scotland) Ltd* [2002] EWHC 1270 (TCC).

Accordingly, unless there is a specific clause included in the contract to the contrary, the current position of a sub-contractor undertaking design appears to be that held in *Co-op* v. *Henry Boot*[11].

However, where a sub-contractor undertakes design, the relevant JCT sub-contracts (i.e. the ones considered in this book) which expressly facilitate some design input from the sub-contractor now include a specific clause which appears largely to protect the sub-contractor from the *Co-op* v. *Henry Boot* position noted above. The exception to this is SubSub, and accordingly we deal with this first below.

SubSub

As noted earlier, SubSub does not include express provision for design by the sub-subcontractor but records in the guidance note this possibility (see other matters, item E) occurring by any design obligations and ad hoc provisions being added as a sub-subcontract document by their reference at the third recital.

If the sub-subcontractor is required to design all or part of the sub-subcontract works, the sub-subcontractor obviously needs to be qualified and competent to do so.

Moreover, the sub-contractor and sub-subcontractor need to be aware that design obligations are a complex area of law. Accordingly, and before entering into the sub-subcontract, both parties should obtain professional legal advice and must give careful consideration to any ad hoc provisions required to facilitate design by the sub-subcontractor, i.e. that they are acceptable, suitable and adequate in relation to the design obligations and risks that the sub-subcontract will impose upon the sub-subcontractor.

As part of this process it is suggested that the sub-subcontractor will need to check, amongst other matters, that any ad hoc provisions:

(1) contain a specific clause protecting the sub-contractor from the *Co-op* v. *Henry Boot* position noted above;
(2) limit the sub-subcontractor's design liability to that of 'reasonable skill and care' and not a 'fitness for purpose' obligation (as discussed in more detail later in this chapter);
(3) can be covered by professional indemnity insurance.

The above is certainly not a conclusive list and as stated above, expert legal advice is necessary.

DBSub

Clause 2.8.2 of DBSub/C makes it clear that, subject only to clause 2.12 'divergences from statutory requirements' (dealt with below) the sub-contractor is not responsible for the Contractor's Requirements or for verifying the adequacy of the design contained within them.

MPSub

Clause 17.1 of MPSub also makes plain that the sub-contractor is not responsible for either the content or the adequacy of the design contained within either the

---

[11] *Co-operative Insurance Society Ltd* v. *Henry Boot (Scotland) Ltd* [2002] EWHC 1270 (TCC).

requirements, or any information provided by the contractor (which includes design and all production information).

Clause 17.2 (stated to be subject to clause 17.1) contains the warranties by the sub-contractor that his design of the sub-contract works will comply with: the statutory requirements (clause 17.2.1); satisfy any performance specification contained within the requirements (clause 17.2.2); and that the sub-contractor will select materials in accordance with a publication entitled 'good practice in the selection of construction materials'.

ICSub

Not applicable. This sub-contract is not suitable where a sub-contractor is required to design any part of the sub-contract works.

ICSub/D

Clause 2.21.5 is identically worded to clause 2.8.2 of DBSub/C (as noted above), and the sub-contractor's liability is the same, subject only to clause 2.10 'divergences from statutory requirements'.

ICSub/NAM

This matter is not specifically dealt with in the ICSub/NAM but it is considered that where design is required to be carried out by a named sub-contractor similar provisions as ICSub/D would apply.

MWSub/D

Clause 5.1.2 makes clear that the sub-contractor shall not be responsible for the Contractor's Requirements or for verifying the adequacy of any design contained in the Contractor's Requirements.

ShortSub

Not applicable. This sub-contract is not suitable where a sub-contractor is required to design any part of the sub-contract works.

What is the standard of design expected by the sub-contractor under the sub-contract?

If a sub-contractor is required to design some or all of the sub-contract works under the sub-contract, the sub-contractor must identify precisely, pre-contract, what his design liability will be under the sub-contract. A crucial point is whether the sub-contractor's design liability is either a 'reasonable skill and care' obligation or a 'fitness for purpose' obligation. The sub-contractor must be aware of this, because legally the differences between these two obligations are fundamental to the sub-contractor's liability.

A 'fitness for purpose' obligation is much stricter than a 'reasonable skill and care' obligation and sets a much higher level of design liability.

In addition to the contract, it needs to be remembered that a designer (including the sub-contractor, where the sub-contractor carries out design work) may also have tortious obligations, but these obligations are outside the scope of this book.

Reasonable skill and care

Most standard form contracts produced by the professional bodies of construction consultants contain an express term which stipulates that the professional will carry out his duties using 'reasonable skill and care'.

One of the best definitions for skill and care has been provided by the Court of Appeal in the *Eckersley* v. *Binnie & Partners*[12] case, where Lord Justice Bingham stated:

> 'The law requires of a professional man that he live up in practice to the standards of the ordinary skilled man exercising and professing to have his specialist professional skills. He need not possess the highest expert skill; it is enough if he exercises the ordinary skill of an ordinary competent man exercising his particular art. In deciding whether a professional man has fallen short of the standards observed by ordinary skilled competent members of his profession, it is the standard prevailing at the time of the acts or omissions which prove the relevant yardstick. He is not to be judged by the wisdom of hindsight.'

With regard to the professional man the standard of reasonable skill and care is also implied by section 13 of the Supply of Goods and Services Act 1982.

Fitness for purpose

Despite the above comments regarding a designer's normal obligation, where there is not an express obligation to the contrary in the sub-contract, the sub-contractor's liability would normally be far more onerous. This liability would be to meet the standard of 'fitness for purpose', which means that the sub-contractor would be liable to ensure that the finished product was reasonably fit for its intended purpose. Obviously, this is a far higher and more onerous obligation than that normally imposed on professional designers of using 'reasonable skill and care'.

Indeed, where a 'fitness for purpose' design liability applies, it is no defence for the sub-contractor to state he has used 'reasonable skill and care' in carrying out his design work.

Because of the 'usual' position of fitness for purpose in the absence of an express term, it is not unusual for sub-contracts (and contracts) to include an express term that the liability of the sub-contractor (or main contractor) for design liability is one of 'reasonable skill and care'. Sub-contractors should, however, not assume that such an express term is in their contract; rather they must check to see that such an express term is included early in the pre-contract process.

DBSub

Clause 2.13.1 of DBSub/C makes it clear that when carrying out design work under DBSub/C, the level of care expected of the sub-contractor is one of 'reasonable skill and care' rather than one of 'fitness for purpose'.

---

[12] *Eckersley* v. *Binnie & Partners* (1988) 18 Con LR 1, CA.

MPSub

MPSub at clause 17.3 includes an express term stating that the standard of care expected of the sub-contractor is the same as a designer, i.e. 'reasonable skill and care'.

However, the published guidance notes to MPSub (see footnote 4 to clause 17.3) also sets out an alternative option, which if agreed to and adopted by the parties is used to amend clause 17.3, by which the sub-contractor warrants that the completed sub-contract works shall be 'suitable for the purpose stated in the requirements', making the level of care expected of the sub-contractor one of 'fitness for purpose'.

The MPSub guidance notes add that the JCT does not consider that such a fitness for purpose provision should be a standard clause, albeit the parties may, in some circumstances, be prepared to adopt it. At item 38 of the guidance notes, a four bullet point guidance where a fit for purpose obligation 'may be considered' is provided as follows:

- The purpose can be clearly defined and is for a well-established type of use.
- The construction process utilises tried and tested methods of construction and the sub-contract works are of a non-complex nature, limited to technical knowledge known or capable of being known by a competent design at the time of design.
- The sub-contractor has addressed the availability of professional indemnity insurance with both his broker and the contractor. It would also be appropriate for the parties to discuss the possibility of project based insurance at the same time.
- The sub-contractor has been given control of the design process.

ICSub

Not applicable. This sub-contract is not suitable where a sub-contractor is required to design any part of the sub-contract works.

ICSub/D

All as DBSub above. Clause 2.21.1 of ICSub/D/C refers.

ICSub/NAM

This matter is not specifically dealt with in the ICSub/NAM but it is considered that where design is required to be carried out by a named sub-contractor similar provisions as ICSub/D would apply.

MWSub/D

Clause 5.1.1 states that the obligation is that of reasonable skill, care and diligence when the sub-contractor undertakes the design for the sub-contractor's designed portion (clause 5.1.1).

ShortSub

Not applicable. This sub-contract is not suitable where a sub-contractor is required to design any part of the sub-contract works.

SubSub

See earlier comments above regarding the incorporation of design responsibility under ShortSub.

How does the Defective Premises Act 1972 affect the sub-contractor's design liability?

As well as contract and tort the law imposes a statutory obligation on house builders (and designers thereto). This statute is the Defective Premises Act 1972.

The Defective Premises Act sets down obligations in respect of new homes for occupation. The important aspect of the Act is that anyone designing a new dwelling owes a duty to anyone who later purchases the dwelling to ensure that his work is done in a professional manner. Accordingly, under the DB/Sub, for example, a sub-contractor's liability to such third parties via this statute may be higher than the reasonable skill and care liability he owes to the contractor.

However, it should be noted that in order to be in breach of the duty the failure must cause the dwelling to be unfit for habitation when completed.

Is there a financial limitation of liability in the sub-contract?

DBSub

Clause 2.13.3 of DBSub/C deals with the financial limitation of liability and states that if a financial limit of liability has been included in the main contract (under clause 2.17.3) then that same limit will also apply to the sub-contractor. In the situation where the sub-contractor is not solely liable in respect of this matter, then the sub-contractor would only be expected to make a proportional contribution towards any financial limit set under the main contract.

MPSub

This matter is not specifically dealt with by the MPSub but is something that the parties (particularly the sub-contractor) should consider including.

ICSub

Not applicable.

ICSub/D

As DBSub above, ICSub/D at clause 2.21.3 includes a virtually identical financial limitation of liability clause, which is subject to one having been included in the main contract (under clause 2.34.3).

ICSub/NAM

This matter is not specifically dealt with by the ICSub/NAM but is something that the parties (particularly the sub-contractor) should consider including.

MWSub/D

As MPSub above.

ShortSub

Not applicable.

SubSub

This is not expressly dealt with. See earlier comments above regarding the incorporation of design responsibility under SubSub.

Design development

One of the major areas of dispute in respect of changes to design works relates to what is called 'design development'.

Generally, a sub-contractor may be called upon to design an entire component, such as a cladding system where he will be required to consider all relevant matters (e.g. location, wind forces, anticipated snow loadings, etc.), or he may be called upon to design to a performance specification (where another designer has already made the basic design considerations and calculations), for example a central heating system that complies with a performance specification provided.

In the latter case, in particular, if the sub-contractor does not meet the required performance specification, he cannot rely on the defence that the product supplied is 'fit for purpose'. In other words, the central heating system designed by the sub-contractor may achieve an acceptable level for the building in question (i.e. it is 'fit for purpose'), but if it does not reach the specification level required by the contractor, then it will be unacceptable.

However, what frequently happens is that a design put forward by a sub-contractor is rejected by a contractor (normally because it has been rejected by the employer) because it is not to the contractor's liking or does not meet the contractor's perceived requirements.

In such a situation, the contractor will often ask for a change to the sub-contractor's design on the basis that this is simply design development (i.e. that the sub-contractor will not get paid any extra for the development to his design to meet the requirements of the contractor). The sub-contractor will naturally resist on the basis that the changes required by the contractor are a variation to the scope of the sub-contract works.

In such a situation, it will be for the sub-contractor to demonstrate that his submitted design satisfied the requirements of the design brief, or for the contractor to demonstrate that the sub-contractor's submitted design would not satisfy the design brief. Of course, the contractor's objection to the sub-contractor's design would be unsuccessful if it was based upon considerations of price (i.e. savings effected by the sub-contractor that were not being passed onto the contractor) rather than on any technical failings of the sub-contractor's design.

There is not a great deal of case law on this particular subject, but the *Skanska* v. *Egger*[13] Court of Appeal case considered the matter of design development. In that particular case the parties entered into a contract whereby Skanska agreed, for a so-called guaranteed maximum price, to develop the design of, manage, procure and construct a factory building.

Disputes arose as to various claims by Skanska (in particular in respect of additional steelwork). Skanska argued that the claims arose from changes in the Employer's Requirements, whilst Egger contended that they were merely instances of design development comprehended within the Employer's Requirements.

Skanska was responsible for the completion of the design of the project as outlined in the contract documents and the contract drawings.

The Court of Appeal found that although Skanska was obliged to install more steel than was indicated at tender stage, the tender drawings *did show a requirement for steel* and there was sufficient evidence for the first instance judge to have decided that the contract provided for more detailed information to be provided post-contract. No further payment was therefore due for the additional work.

The Skanska case illustrates the dangers to a contractor of entering into a design and build contract on a guaranteed maximum price basis where design issues have not been fully formulated at the time of contracting.

Of course, much turns upon the individual contract terms and the facts of a given case, but the principle that a contractor (or a sub-contractor) who takes on design responsibilities may find himself responsible for a good deal of development costs is of universal application.

Finally, sometimes, a sub-contractor is required to carry out a design that is 'to the entire satisfaction' of the contractor (or the employer). This would appear to be an entirely open-ended obligation on the sub-contractor. However, it is submitted that in respect of any such requirement the contractor's (or the employer's) scope for satisfaction is limited by the design specification provided to the sub-contractor, and that, unless there are any express provisions in the sub-contract to the contrary, the sub-contractor will be entitled to an additional payment, and, if appropriate, additional time, if the design level that satisfies the contractor (or the employer) exceeds the design specification level that was initially provided to the sub-contractor. This principle was established in the case of *Dodd* v. *Churton*[14].

Errors and failures by the sub-contractor – other consequences

DBSub

Clause 2.14 of DBSub/C states that no extension of time shall be given, and no loss and expense will be allowed, and no (contractual) termination for non-payment will be permitted, where the cause of the progress of the sub-contract works having been delayed, affected or suspended is:

- any error, divergence, omission or discrepancy in the Sub-contractor's Proposals;

---

[13] *Skanska Construction UK Ltd* v. *Egger (Barony) Ltd* [2002] BLR 236.
[14] *Dodd* v. *Churton* [1897] 1 QB 562.

- any error, divergence, omission or discrepancy in any of the sub-contractor's designed works information (clause 2.6.1);
- failure of the sub-contractor in completing the sub-contractor's design documents to comply with regulations 11, 12 and 18 of the CDM Regulations;
- failure of the sub-contractor to provide in due time any sub-contractor's design documents or related calculations or information either:

  — as required by clause 2.6.2 of DBSub/C (see commentary above); or
  — in response to a written application from the contractor specifying the date (having due regard to the progress of the sub-contract works) when certain relevant documents or other information is required.

MPSub

This matter is not specifically dealt with in MPSub.

However, clause 22.1 (extensions of time) is clear that no adjustment to the period for completion is permitted where MPSub specifically states that a particular matter does not constitute a change. Each clause will, therefore, have to be read on its own merit to ascertain this fact.

For example, clause 11.4 (discussed later) deals with the discovery of a discrepancy between the requirements, proposals and/or the statutory requirements and notes that the instruction from the contractor as to which is to be adopted will not, ordinarily, constitute a change, i.e. the risk here rests with the sub-contractor.

ICSub

Not applicable.

ICSub/D

ICSub/D at clause 2.9 includes a similar provision to clause 2.14 of DBSub noted above. Clause 2.9 deals with matters that will not constitute an extension of time or an adjustment in the calculation of the final sub-contract sum, namely:

- clause 2.9.1 directions issued for:

  — the correction of any error, divergence, omission or discrepancy within or between the SCDP documents (excluding the Contractor's Proposals); or
  — a variation of work not forming part of the SCDP works but necessitated due to any error, divergence, omission or discrepancy within or between the SCDP documents (excluding the Contractor's Proposals) or its correction.

- any delay or suspension caused by a failure of the sub-contractor to comply with regulations 11, 12 and 18 of the CDM Regulations (clause 2.9.2); or
- failure of the sub-contractor to provide in due time any sub-contractor's design documents or related calculations or information either (clause 2.9.2):

  — clause 2.9.2.1 as required by clause 2.6.3 of ICSub/C (see commentary above); or
  — clause 2.9.2.2 in response to a written application from the contractor specifying the date (having due regard to the progress of the sub-contract works) when certain relevant documents or other information are required.

ICSub/NAM

Clause 2.9 of ICSub/Nam/C includes a similar provision to clause 2.9 of ICSub/D/C noted above. However, a clause similar to clause 2.9.2.2 of ICSub/D/C does not appear in ICSub/NAM/C.

MWSub/D

MWSub/D does not include a similar provision to clause 2.14 of DBSub/C.

ShortSub

Not applicable.

SubSub

This is not dealt with in SubSub.

Materials, goods and workmanship

What is the sub-contractor's liability for materials, goods and workmanship?

Before looking at the specific sub-contracts, there are some common general principles discussed below.

What is workmanship?

The normal understanding of the word 'workmanship' in the construction industry relates to the skill and/or care exercised by a sub-contractor (or a contractor) in the physical execution of the works.

Ordinarily, to the extent that the choice of materials is left to the sub-contractor, 'workmanship' may also mean design (or at least suitability of the materials for the purpose for which they have been used). In the absence of express terms to the contrary, the law has, for many years, implied a term that building work will be carried out in a proper and workmanlike manner[15].

What is the position in respect of materials and goods?

The position generally is that a sub-contractor is to ensure compliance with the specification where materials or goods are specified in the contract. If the specification states a brand name or a particular supplier of the material, the sub-contractor would still be under a warranty that those materials or goods are of good quality when used[16].

A warranty of fitness for purpose (i.e. a warranty that the materials or goods will be fit for the purpose for which they are intended to be used) may also apply

---

[15] *Test Valley Borough Council* v. *Greater London Council* (1979) 13 BLR 63.
[16] Refer generally to the Sale of Goods Act 1979, and the Supply of Goods and Services Act 1982.

if the circumstances indicate that there was a reliance on the sub-contractor's skills regarding the suitability of the materials or goods[17].

However, where the circumstances indicate that there is no reliance whatsoever on the skill and care on the part of the sub-contractor on the issue of quality (for example, the terms of the sub-contract compel the sub-contractor to accept particular materials or goods from a particular supplier that, for example, had exclusion clauses in respect of liability); the sub-contractor will not be liable for defects in them (i.e. will not be liable if those materials or goods are not of good quality)[18].

This latter position is, however, very much the exception, and the general position remains that the sub-contractor retains liability for materials and goods that are not of good quality when used, and, where there was a reliance on the sub-contractor's skills regarding the suitability of the materials or goods, the sub-contractor retains liability that the materials or goods are reasonably fit for the purpose for which they will be used.

The reason for this is because the law considers that there is a need to maintain a chain of liability from the employer down to the manufacturer. Without such a chain of liability, a sub-contractor would only be able to recover nominal damages from the next party down the chain (the supplier or manufacturer). Generally, the law adopts the view that society is not well served by allowing those causing loss or damage to escape liability whilst those who suffer the loss are denied any remedy. Wherever possible, therefore, the courts interpret contracts in such a way that the chain of liability is maintained.

Against this background we can now consider the contractual position under the specific contracts considered in this book.

What is the applicable standard of materials and goods?

DBSub

Clause 2.4.1 of DBSub deals with materials and goods and states that:

- Except for the sub-contractor's design works, all of the materials and goods for the sub-contract works shall, *so far as procurable*, be in accordance with the kinds and standards described within the sub-contract documents.
- In respect of the sub-contractor's designed works, all of the materials and goods are, *so far as procurable*:

  — to be of the kinds and standards described in the contractor's requirement; or
  — where not specifically described in the contractor's requirement, they are to be as described in the Sub-contractor's Proposals or documents referred to in clause 2.6.1.

- Clause 2.4.1 includes a facility for the sub-contractor to substitute any materials or goods; however, the sub-contractor must receive the written consent of the contractor before doing so. The contractor's consent is not to be unreasonably delayed or withheld. The contractor's consent does not relieve the sub-contractor of his other obligations.

---

[17] *Young & Marten Ltd* v. *McManus Childs Ltd* (1969) 9 BLR 77.
[18] *Gloucestershire CC* v. *Richardson* (1968) 1 AC 480.

Clause 2.4.4 states that if the quality of materials or standards of workmanship are not described in the manner stated in clause 2.4.1 or 2.4.2 they must be to a standard appropriate to the sub-contract works.

Note also that clause 2.4.5 of DBSub requires the sub-contractor (upon the request of the contractor) to provide to the contractor reasonable proof that the materials and goods used by him comply with clause 2.4. as appropriate.

The requirement to work to the sub-contract documents implies that, in respect of materials and goods, the fact that the contractor consents to substituted materials or goods does not give the sub-contractor a defence to his liability that those materials or goods will be of good quality when used.

From the general principles already explained:

- The sub-contractor is not under a warranty of fitness for purpose in respect of materials or goods specified in the sub-contract documents (but is normally liable that the said materials or goods are of good merchantable quality).

  However, some specifications contain lists of different types of materials or goods for the same purpose, with the choice left to the sub-contractor as to which type to use. In such a case, a sub-contractor would not take on the fitness for purpose liability simply by making a choice of one of the listed materials or goods, since, in such a case, it could not be considered that any reliance had been placed on the sub-contractor's skills regarding the suitability of the materials or goods in question[19].

- The position is somewhat different to the extent that materials and goods form part of the sub-contractor's designed works as the sub-contractor would almost certainly be liable for both the good quality and the fitness for purpose of those materials and goods.

MPSub

Clause 12.1 deals with the standards of materials and states that:

- The sub-contractor must use materials and goods which comply with the kinds and standards described in the sub-contract.
- Where the sub-contract does not describe the kinds or standards of materials and goods, the sub-contractor must use materials and goods that are *reasonably fit for their intended purpose.*
- All of the materials and goods used by the sub-contractor must be of satisfactory quality.
- Where materials or goods described in this sub-contract *are not procurable* the sub-contractor must propose to the contractor for his acceptance an alternative that is, wherever possible, of an equivalent or better kind or standard. The contractor's acceptance of the proposed alternative is not to be unreasonably delayed or withheld.
- The use of any alternative is stated not to constitute a change unless the alternative accepted by the contractor is of a lesser kind or standard to that described in this sub-contract, whereupon the provisions of clause 29 (changes) shall apply as though the contractor had instructed a change.

---

[19] *Rotherham Metropolitan Borough Council* v. *Frank Haslam Milan & Co. Ltd and Others* (1996) 78 BLR 1.

ICSub

Unlike clause 2.4.1 of DBSub, the ICSub does not contain a specific clause that materials and goods (and workmanship) are to be in accordance with that described in the contract documents.

However, clause 2.1 of ICSub obliges the sub-contractor to carry out and complete the sub-contract works in compliance with the sub-contract documents.

ICSub/D

As ICSub above.

ICSub/NAM

As ICSub above.

MWSub/D

Clause 5.3 of MWSub/D obliges the sub-contractor to provide goods and materials that comply with the standards stated in the sub-contract documents. If no standard is recorded, goods and materials provided by the sub-contractor must be of a satisfactory quality.

ShortSub

Clause 5.2 of ShortSub is as per MWSub/D above.

SubSub

As ShortSub above.

What if the goods are not procurable?

As noted above, clause 2.4.1 of DBSub and 12.1 of MPSub qualify that the sub-contractor's obligation to provide materials or goods as described applies in so far as those materials and goods are procurable.

Because the sub-contract does not specify any geographical limits of procurability, an argument could be raised that materials and goods that are only procurable abroad are still 'procurable' in line with clause 2.4.1 of DBSub (or clause 12.1 of MPSub). Whether or not the materials or goods are procurable (at all) will depend on the facts and what is reasonable.

What is the standard of workmanship?

DBSub

Clause 2.4.2 of DBSub states that:

- Excluding the sub-contractor's designed works, workmanship for the sub-contract works shall be of the standards described in the sub-contract documents.

- Workmanship for the sub-contract's designed works shall be of the standards described in the Contractor's Requirements or, if not there specifically described, as described in the Sub-contractor's Proposals.

The sub-contractor's obligation on standards of workmanship is not qualified by the phrase 'so far as procurable', which, in theory at least, means that it is no defence to the sub-contractor's obligation that the human skills or equipment to achieve that standard of workmanship are not procurable.

MPSub

Clause 12.2 of MPSub states that all workmanship is to be of the standards described in the sub-contract. However, if the sub-contract does not specify a standard, the sub-contractor's workmanship must be undertaken in a good and workmanlike manner.

ICSub

Unlike clause 2.4.2 of DBSub, ICSub does not contain such a specific clause, albeit a similar state of affairs is likely to be implied as a term to the contract. Further, clause 2.1 of ICSub obliges the sub-contractor to carry out and complete the sub-contract works in a 'proper and workmanlike manner'.

ICSub/D

As ICSub above.

ICSub/NAM

All as ICSub above.

MWSub/D

Clause 5.1 of MWSub/D obliges the sub-contractor, *inter alia*, to carry out and complete the sub-contract works in a 'proper and workmanlike manner'.

ShortSub

Clause 5.1 of ShortSub obliges the sub-contractor to execute sub-contract works in accordance with the sub-contract documents and, *inter alia*, in a 'good and workmanlike manner'.

SubSub

As ShortSub above.

What if the approval of the quality of materials or the standards of workmanship is a matter for the opinion of the architect/contract administrator?

DBSub

Not applicable. However, clause 2.4.3 obliges the sub-contractor to provide the contractor with samples in respect of the standard of workmanship or quality of the goods or materials he intends to provide as referred to in either the Contractor's Requirements or in the Sub-contractor's Proposals. Such samples are to be provided before the sub-contractor carries out the relevant work and/or orders the relevant goods or materials.

MPSub

Where the sub-contract does not describe the kinds or standards of materials and goods, the sub-contractor must use materials and goods that are 'reasonably fit for their intended purpose'.

ICSub

Clause 2.3.1 of ICSub/C specifically deals with this issue, and where it applies states that such quality and standards shall be to the architect's/contract administrator's reasonable satisfaction.

Obviously, in such a situation, the sub-contractor should obtain from the contractor confirmation of the architect's/contract administrator's reasonable satisfaction where it applies to any materials, goods or workmanship in the sub-contract works, as the sub-contract does not provide for any direct communication between the sub-contractor and the architect/contract administrator. It is often appropriate for a sub-contractor to suggest that a sample of its work is approved by the architect/contract administrator so that a 'yardstick' for the standard of future works is provided.

In respect of the architect's/contract administrator's reasonable satisfaction, the architect/contract administrator must use an objective standard and it is therefore submitted that he has no power to demand quality or workmanship of the highest standard, unless that is reasonable in the circumstances of the case.

Interestingly, a decision of the Court of Appeal[20] suggests that the architect/contract administrator may be entitled (or may be required) under certain circumstances to take into account the competitiveness of the contractor's prices for the relevant materials or goods in deciding whether they are to his reasonable satisfaction. Unfortunately, given the competitive nature of the construction industry, this rather tenuous argument is unlikely to have any impact on a sub-contractor that has given a very competitive price in the circumstances where the contractor's price is not so competitive.

Another point to note is that the sub-contract refers to materials or goods being to the reasonable satisfaction of the architect/contract administrator and not to the reasonable satisfaction of the contractor.

---

[20] *Cotton v. Wallis* [1955] 1 WLR 1168, CA.

In addition to the above, clause 2.3.1 of ICSub/C also states that, to the extent that the quality or materials or the standards of workmanship are neither described in the sub-contract documents, nor stated to be a matter for the opinion or satisfaction of the architect/contract administrator, they shall be of a standard appropriate to the sub-contract works.

ICSub/D

As ICSub above. Additionally, clause 2.3.1 of ICSub/D also deals with the sub-contract being equally silent in respect of the quality of materials or the standards of workmanship relating to the SCDP, stating that in such circumstances these shall be of a standard appropriate to the sub-contractor's designed portion works (SCDP works).

ICSub/NAM

All as ICSub/D above. Note, however, that clause 2.3.1 of ICSub/NAM refers to NAM designed works under the sub-contract, not SCDP works as ICSub/D does.

MWSub/D

This position is not dealt with. However, as noted earlier, clause 5.3 states that where the sub-contract documents are silent as to the standard of goods and materials they shall be of satisfactory quality.

ShortSub

Clause 5.2 of ShortSub is as per MWSub/D above.

SubSub

As ShortSub above.

What is the Construction Skills Certification Scheme (CSCS)?

All of the above sub-contracts require the sub-contractor (or sub-subcontractor in the case of SubSub) to take all reasonable steps to encourage the sub-contractor's persons to be registered cardholders under the Construction Skills Certification Scheme (CSCS) or to be qualified under an equivalent recognised qualification scheme.

The Construction Skills Certification Scheme (CSCS) is managed by CSCS Ltd. The purpose of the CSCS is to ensure that those involved in the construction industry are competent in their occupation and have health and safety awareness. The CSCS operates by issuing cards (CSCS cards) to those that are qualified to receive the cards, in different colours to suit different occupations.

Compliance with main contract and indemnity

Is the sub-contractor expressly obliged to comply with any specific obligations the contractor has under the main contract?

DBSub

Clause 2.5.1 of DBSub/C requires the sub-contractor to observe, perform and comply with the contractor's obligations under the main contract, as:

- identified in or by the schedule of information (included within DBSub/A); and
- in so far as those obligations relate and apply to the sub-contract works (or any part of them).

Those obligations are specifically stated to include (without limitation):

- those under clause 2.18 of the main contract conditions (relating to fees or charges legally demandable);
- those under clause 2.19 and 2.20 of the main contract conditions (relating to royalties and patent rights);
- those under clause 3.15 and 3.16 of the main contract conditions (relating to antiquities).

In addition, the sub-contractor is to indemnify and hold harmless the contractor in respect of his obligations under the above main contract liabilities, against and from:

- any breach, non-observance or non-performance by the sub-contractor or his employees or agents of any of the provisions of the main contract (refer to clause 2.5.1.1 of DBSub/C); and
- any act or omission of the sub-contractor or his employees or agents which involves the contractor in any liability to the employer under the provisions of the main contract (refer to clause 2.5.1.2 of DBSub/C).

Also, under clause 2.5.2 of DBSub/C (subject only to the exceptions contained in insurance clauses 6.4 and 6.7.1 of DBSub/C as dealt with under Chapter 11 of this book) the sub-contractor is to indemnify and hold harmless the contractor against and from any claim, damage, loss or expense due to or resulting from any negligence or breach of duty on the part of the sub-contractor, his employees or agents (including any misuse by him or them of scaffolding or other property belonging to or provided by the contractor).

As the sub-contract deems the relevant information concerning the main contract by reference to schedule 1 it is important that the contractor takes care when completing this information. Likewise, the specific links in this clause to particular clauses in the main contract make it essential that the correct sub-contract published for use with a particular main contract is selected and used.

MPSub

Clause 8 of MPSub deals with the relationship of the sub-contract to the main contract.

Clause 8.1 provides that the sub-contractor must, as if he were the contractor, observe, comply and conform with all relevant provisions of the main contract, provided that:

- they relate and apply to the sub-contract works; and
- they are not inconsistent with the provisions of the sub-contract.

Clause 8.2 further provides (subject only to the provision of clause 35.1 of MPSub in respect of insurances) that the sub-contractor is to indemnify the contractor against the consequences of:

- any breach by the sub-contractor of clause 8.1 of MPSub;
- any act of omission by the sub-contractor which results in the contractor incurring liability to the employer under the main contract;
- any loss or expense, damage, claim resulting from any negligence or breach of duty by the sub-contractor.

ICSub

Clause 2.4 of ICSub/C is virtually identical to clause 2.5 of DBSub/C, but note that the clauses referred to in the main contract are different, as follows:

- those under clause 2.3 of the main contract conditions (relating to fees or charges legally demandable);
- those under clause 2.9 of the main contract conditions (relating to levels and setting out).

ICSub/D

Clause 2.4 of ICSub/D/C. All as ICSub above.

ICSub/NAM

Clause 2.4 of ICSub/D/C. All as ICSub above.

MWSub/D

Except for the contractor's pricing information, clause 7.1 of MWSub/D deems the sub-contractor to be aware of the main contract provisions in so far as they apply to the sub-contract works.

Note that the contractor, if requested by the sub-contractor, must provide the sub-contractor with a copy of the main contract (omitting details of the contractor's pricing). Given the obligations placed on the sub-contractor by clause 7 of MWSub/D the sub-contractor is well advised to request this information (if possible, at tender stage so that he can identify any onerous conditions and/or risks that may be passed down to him under this provision before considering entering into a contract).

Clause 7.2 provides that the sub-contractor must carry out and complete the sub-contract works so that no act or omission by the sub-contractor places the contractor in breach of the main contract.

In so far as the contractor's obligations and liabilities under the main contract relate to the sub-contract works, clause 7.3 further provides that the sub-contractor must perform the obligations and assume the liabilities of the contractor.

Note that no indemnity provision is stated.

ShortSub

Clause 7 of ShortSub refers and is all as MWSub/D above.

SubSub

Clause 7 of SubSub. All as ShortSub above, except the wording of this clause is amended to reflect that the contractual relationship under SubSub is between a sub-contractor and its sub-subcontractor (whereas ShortSub is a contractor and his sub-contractor contractual relationship).

What is an indemnity clause?

An indemnity clause is a clause where one party agrees to make good a loss suffered by the other party in respect of damage or claims arising out of various matters.

It should be noted that where the above sub-contract contains a general indemnity to the contractor (e.g. clause 2.5.1 and 2.5.2 of DBSub/C) the sub-contractor's liability could include the contractor's costs, liquidated damages due to the employer, and payments to third parties to whom the contractor is liable. Further, being an indemnity, this may increase the liability of the sub-contractor to the contractor for costs, losses and expenses beyond that which might otherwise be recoverable for breach of the sub-contract and is, potentially, unlimited.

Can the sub-contractor's liability be limited?

The contractor and the sub-contractor may expressly agree that the sub-contractor's liability is limited, and this may be done by way of an overall cap on liability of the contractor's costs and damages.

However, there is no provision for such a cap in the sub-contracts considered in this book. If any overall cap on liability is agreed this would need to be carefully recorded; using DBSub as an example it would need to be recorded on a numbered document listed and attached to sub-contract particulars item 16 of the DBSub/A.

In addition, the indemnity as set out above has the effect of potentially extending the sub-contractor's period of liability because the cause of action (that sets the liability period running) under an indemnity does not generally arise until the loss in question is actually suffered.

Errors, discrepancies and divergences

Given the volume and complexity of the documentation generally forming part of a building contract, most include express provisions for dealing with any errors or discrepancies between them. The way that the DBSub, the MPSub, the ICSub, the ICSub/D and the ICSub/NAM deal with these issues is covered in the following sections. The way that the MWSub/D, the ShortSub and the SubSub deal (or do not deal) with this matter is first noted below.

MWSub/D

MWSub/D only contains limited provisions dealing with the position where:

- an inadequacy is discovered in the Contractor's Requirements, stating that the Contractor's Requirements must be altered or modified as necessary (clause 5.1.2); or
- there is a sub-contractor's discovery (clause 5.2.1) or non-discovery (clause 5.2.2) of divergences between the statutory requirements and either the sub-contract documents or instructed variations issued under clause 10.1 of MWSub/D.

ShortSub and SubSub

Despite the above (and unlike the other sub-contracts considered in this book), ShortSub and SubSub do not include any express provision for resolving any errors, discrepancies or divergences and are not therefore considered below. Whether any terms similar to those found in other building contracts dealing with these matters would be implied into ShortSub or SubSub remain unclear.

Bills of quantities provided by the contractor

DBSub

Clause 2.8.1 of DBSub/C indicates that, unless specifically stated otherwise, bills of quantities (including any bills of quantities prepared for the purpose of obtaining a schedule 2 quotation) must have been prepared in accordance with the Standard Method of Measurement.

Clause 1.1 of DBSub/C defines this as: 'The Standard Method of Measurement of Building Works, 7th Edition . . . [e]current, unless otherwise stated in the Sub-contract Documents, at the Sub-contract Base Date.'

MPSub

The MPSub does not have an express clause dealing with this issue.

ICSub

Clause 2.7.1 of ICSub/C. All as DBSub above, except reference to 'schedule 2 quotation' is omitted because it does not apply to ICSub.

ICSub/D

Clause 2.7.1 of ICSub/D/C. All as ICSub above.

ICSub/NAM

Clause 2.7.1 of ICSub/NAM/C applies. All as ICSub above.

What if the bills of quantities are not in accordance with SMM7 or contain errors?

DBSub

Clause 2.9.1 of DBSub/C refers. If, in respect of any bills of quantities (as referred to in clause 2.8.1), any of the following list applies, the departure, error or omission shall not vitiate (i.e. shall not make invalid or ineffectual) the sub-contract but shall be corrected:

- SMM7 has not been used in the preparation of the bills of quantities, but this departure has not been stated;
- there is any error in description or quantity;
- there is any omission of items that should have been measured; or
- there is any error in, or omission of information in any item which is the subject of a Provisional Sum for defined work.

Also, where the description of a Provisional Sum for defined work does not provide the information required by SMM7, the description shall be corrected so that it does provide that information.

Clause 2.9.3 of DBSub/C provides that, subject to clause 2.12 of DBSub/C (notices to be issued by the sub-contractor and directions to be given by the contractor in respect of divergences from statutory requirements) any correction, alteration or modification under clause 2.9.1 of DBSub/C shall be treated as a variation.

ICSub

Clause 2.7.2 of ICSub/C repeats the provisions of DBSub above, with clause 2.8.2 also stating that the departure, error or omission shall not vitiate the sub-contract.

ICSub/C at clause 2.9 has a 'sweeper' clause dealing with variations arising from clause 2.8 directions (errors, omissions and inconsistencies).

ICSub/D

All as ICSub above.

ICSub/NAM

All as ICSub above.

What about inadequacies/discrepancies in the Contractor's Requirements?

DBSub

Inadequacy

Clause 2.9.2 of DBSub/C refers. If an inadequacy is found in any design in the Contractor's Requirements then if, or to the extent that, the inadequacy is not dealt with in the Sub-contractor's Proposals, the Contractor's Requirements shall be altered or modified accordingly.

Clause 2.9.3 of DBSub/C provides that, subject to clause 2.12 of DBSub/C, any correction, alteration or modification under clause 2.9.2 of DBSub/C shall be treated as a variation.

Discrepancy or divergence within the Contractor's Requirements

These are dealt with under clause 2.11.2 of DBSub/C as follows:

- The Sub-contractor's Proposals deal with that discrepancy or divergence:

  — in such circumstances the Sub-contractor's Proposals prevail (on the assumption that they comply with statutory requirements); and
  — no adjustment will be taken into account in the calculation of the final sub-contract sum, i.e. the sub-contract sum (or the sub-contract tender sum) is deemed to include for the resolution of the discrepancy or divergence within the Contractor's Requirements.

- The Sub-contractor's Proposals do not deal with the discrepancy or divergence within the Contractor's Requirements:

  — The sub-contractor must inform the contractor in writing of his proposed amendment to deal with discrepancy or divergence within the Contractor's Requirements.
  — The contractor will either agree to the proposed amendment or decide how he wishes the discrepancy or divergence to be dealt with.
  — In either event, the agreement or the decision of the contractor will be treated as a variation.

Of course, the above comments need to be considered against the background of the sixth recital of DBSub/A which confirms that the contractor has examined the Sub-contractor's Proposals and, subject to the sub-contract conditions, is satisfied that they appear to meet the Contractor's Requirements. It is important to note that the sixth recital does not require the contractor to check that the Sub-contractor's Proposals satisfy the Contractor's Requirements, but merely that 'they appear to meet' them.

MPSub

Discrepancy identified within the requirements

MPSub clause 11.2 refers.

If a discrepancy is identified within the requirements (e.g. oak skirting and MDF skirting are both recorded as being installed in a particular room) the sub-contractor chooses between the alternatives given and notifies the contractor accordingly (e.g. MDF skirting).

If the Contractor then requires something different to be installed (e.g. oak skirting) then he must instruct the sub-contractor accordingly and this will constitute a change.

Discrepancy between the requirements, the proposals and/or the statutory requirements

Clause 11.4 refers.

If a discrepancy is identified between any pair, or all three, of the above documents the contractor has authority to instruct which of the discrepant provisions the sub-contractor must adopt. This instruction will not constitute a change, unless the change to the requirements (or the proposals or both) is necessitated by a discrepancy arising after the base date due to an alteration to the statutory requirements (see clause 10.2 of MPSub).

Other than the limited exception relating to clause 10.2, the responsibility for any inconsistency between the requirements, the proposals and the statutory requirements falls on the sub-contractor.

ICSub

Not applicable. This sub-contract is not suitable where the sub-contractor undertakes design.

ICSub/D

Clause 2.8.3 deals with notices by the sub-contractor upon finding any departure, error, omission or in an inconsistency referred to in clause 2.7 (bills of quantities – see above) or 2.8.1 (see later) of ICSub/D.

In respect of the principles discussed here the following apply.

Clause 2.8.3.2 of ICSub/D provides that, should the clause 2.8.3.1 notice relate to an inconsistency within the Contractor's Requirements which the Sub-contractor's Proposals fail to deal with in a manner consistent with the statutory requirements, the sub-contractor must, as soon as possible following its discovery, also issue written proposals for any necessary amendments.

Clause 2.8.4 further provides that:

● the Sub-contractor's Proposals will take precedence in circumstances where any discrepancy in the Contractor's Requirements is dealt with in the Sub-contractor's Proposals consistent with the statutory requirements.

ICSub/NAM

Not applicable. There are generally no Contractor's Requirements under this sub-contract. However, if the named sub-contractor was liable for design it is considered that the provisions of ICSub/D, above, would apply.

What is the position if there are errors or inadequacies in the sub-contractor's design documents?

DBSub

Discrepancies in the sub-contractor's designed works documents are further dealt with at clause 2.11.1 of DBSub/C.

Where there are discrepancies or divergences within or between the sub-contractor's designed works documents (excluding the Contractor's Requirements), the procedure is:

- The sub-contractor is to give a notice to the contractor (under clause 2.10 as noted above).
- The sub-contractor is to send a statement setting out his proposed amendments to remove the said discrepancy or divergence.
- The contractor is then to issue his directions accordingly.
- The sub-contractor shall comply with those directions.
- To the extent that those directions relate to the removal of a discrepancy or divergence, they will not be taken into account in the calculation of the final sub-contract sum.

Clause 2.9.4 of DBSub/C is clear that the final sub-contract sum must not be adjusted to take into account any variations that are issued arising from errors in description or in quantity in the Sub-contractor's Proposals or in the sub-contractor's design works analysis. The same position applies equally for errors relating to the omission of items from these documents.

MPSub

Clause 11.3 of MPSub provides that any discrepancy identified within the proposals is resolved by the contractor instructing the sub-contractor which of the alternatives in the proposals is to be adopted. The instruction will not constitute a change.

ICSub/C

Not applicable. This sub-contract is not suitable where the sub-contractor undertakes design.

ICSub/D

Clause 2.8.3.2 of ICSub/D provides that where a notice issued under clause 2.8.3.1 relates to an inconsistency within or between the SCDP documents (excluding the Contractor's Requirements) the sub-contractor must, as soon as practical following its discovery, also issue written proposals for any necessary amendments.

Clause 2.9 .1 is clear that no extension of time, nor addition to the calculation of the final sub-contract sum will be made in respect of directions concerning errors, omissions or inconsistencies within or between design documents prepared by the sub-contractor, or variations arising from this.

ICSub/NAM

All as ICSub/D above, but clauses 2.8.3 and 2.9.2 of ICSub/NAM refer.

Notification of discrepancies

DBSub

Clause 2.10 of DBSub/C deals with 'notification of discrepancies, etc.'

The sub-contractor must immediately give a written notice to the contractor if the sub-contractor finds any departure, error or inadequacy as referred to in clause 2.9 (i.e. Bills of Quantities, etc., as above) or any error or discrepancy in or between any of the following documents:

- clause 2.10.1: the sub-contract documents (i.e. the documents referred to in article 1 of DBSub/A as appropriate);
- clause 2.10.2: the main contract;
- clause 2.10.3: any direction issued by the contractor under the DBSub/C (except for a direction requiring a variation); and
- clause 2.10.4: the sub-contractor's designed works documents as detailed under clause 2.6.1 of DBSub/C.

The contractor must then issue a direction as to how the error or discrepancy is to be dealt with.

Whether these constitute a variation or not has been considered in the relevant sections above.

MPSub

Clause 11.1 deals with this matter. The party identifying the discrepancy within or between the requirements, the proposals and/or the statutory requirements must give immediate notice of it to the other.

Whether these constitute a change or not is dealt with at clause 11.2 to 11.4 of MPSub and has been commented upon in the relevant sections above.

ICSub

Clause 2.8 of ICSub/C 'directions on errors, omissions and inconsistencies' deals with this.

Clause 2.8.1 is drafted on a similar basis to clause 2.10 of DBSub/C but there are some subtle differences.

The contractor's duty to issue necessary directions under clause 2.7 (bills of quantities) or in respect of the following is not first dependent upon a notice from the sub-contractor:

- the sub-contract documents (clause 2.8.1);
- the main contract (clause 2.8.2);
- any direction issued by the contractor under the ICSub/C, except for a direction requiring a variation (clause 2.8.3).

However, is the sub-contractor required to give notice?
Yes, clause 2.8.3 refers. The sub-contractor must immediately give written notice (with appropriate details) if he finds any departure, error, omission or inconsistency as referred to in clause 2.7 or clause 2.8.1 of ICSub/C (see above).

Also, is the subsequent direction a variation?
Clause 2.9.1 states that if a direction under clause 2.8.1 of ICSub/C varies the 'quality or quantity of the work included within the sub-contract sum and constitutes a variation' it is to be valued under the provisions of section 5 of ICSub/C.

Clause 5.1 of ICSub/C defines what shall constitute a variation and, accordingly, this must be referred to determine if a variation has occurred.

ICSub/D

All as ICSub above except (because this sub-contract deals with design by the sub-contractor) additional clause 2.8.4 is added in ICSub/D/C to the list of relevant categories of documents as follows: clause 2.8.4: the documents detailed under clause 2.5.3 of ICSub/D/C relating to SCDP Works.

Is the direction a variation?
Clause 2.9.1 of ICSub/D/C. All as ICSub/C above, except a further qualification is added which states that no direction under clause 2.8.1 of ICSub/C for the following will be grounds for an extension of time or the adjustment in the calculation of the final sub-contract sum:

- the correction of any error, divergence, omission or discrepancy within or between the SCDP documents (excluding the Contractor's Proposals); or
- a variation of work not forming part of the SCDP works but necessitated due to any error, divergence, omission or discrepancy within or between the SCDP documents (excluding the Contractor's Proposals) or its correction.

ICSub/NAM

Clause 2.8.1 of ICSub/NAM is all as ICSub/D except that clause 2.8.4 in the list of documents relates to NAM design works.

Is the direction a variation?
Clause 2.9 of ICSub/NAM/C mirrors the above in respect of ICSub/D except 'NAM Design' is substituted for 'SCDP' in its drafting (see clause 2.9.1 and 2.9.2 of ICSub/NAM).

Notification of divergence in statutory requirements

DBSub

Under clause 2.12 of DBSub/C, if the sub-contractor *or the contractor* finds any divergence between the statutory requirements and the following, then the sub-contractor or the contractor, as appropriate, is required to give a written notice to the other specifying the divergence immediately:

- the documents listed at clause 2.10 of DBSub/C;
- any direction requiring a variation issued under clause 3.4 of DBSub/C.

In addition, if the divergence is between the statutory requirements and the Sub-contractor's Proposals or the Contractor's Requirements, the sub-contractor, at the time of issuing the notice of divergence, is to inform the contractor in writing of his proposed amendment for removing the divergence (refer to the '17 day' procedure below.)

The procedure is then as follows:

(1) The '10 day' procedure: the divergence is not between the statutory requirements and the Sub-contractor's Proposals or the Contractor's Requirements:

(a) Within 10 days of finding a divergence, or of receiving a notice from the sub-contractor in respect of a divergence the contractor must issue directions accordingly.
(b) If the direction(s) varies the sub-contract, the direction(s) is treated as direction(s) requiring a variation to be issued.

(2) The '17 day' procedure: the divergence is between the statutory requirements and the Sub-contractor's Proposals or the Contractor's Requirements:

(a) Within 17 days of receipt of the sub-contractor's proposed amendment (where applicable) for removing the divergence the contractor must issue directions accordingly.
(b) The sub-contractor must comply with such directions at no cost.
(c) The only exception is where the divergence is caused by a change to the statutory requirement after the sub-contract base date which necessitates an alteration or modification to the sub-contractor's designed works. This exception is treated as being a direction requiring a variation to the Contractor's Requirements.

MPSub

Clause 11.4 refers. See comments earlier in this chapter under the question of inadequacies/discrepancies in the Contractor's Requirements.

ICSub

Clause 2.10.1 of ICSub/C deals only with any divergence between the statutory requirements and:

- the documents listed at clause 2.8 of ICSub/C; or
- any direction requiring a variation issued under clause 3.4 of ICSub/C.

The sub-contractor or the contractor, as appropriate, is required to give a written notice to the other specifying the divergence immediately. The '10 day' procedure (as per DBSub above) also applies here.

ICSub/D

Clause 2.10.1 of ICSub/D refers. All as clause 2.12 of DBSub/C above, save that the '17 day' procedure relates to divergences between the statutory requirements and SCDP Documents.

ICSub/NAM

Clause 2.10.1 of ICSub/D refers. All as DB/Sub above, except that under the '17 day' procedure:

- The divergences are between the statutory requirements and NAM designed documents.
- The sub-contractor must comply with any directions issued in respect of the above at no cost. Unlike DBSub, ICSub/NAM has no exception.

MWSub/D

Clause 5.2 of MWSub/D refers.

Under clause 5.2.1 of MWSub/D if the sub-contractor finds any divergence between the following then the sub-contractor must immediately give the contractor a written notice specifying the divergence:

- the statutory requirements and the sub-contract documents; or
- the statutory requirements and any contractor's instruction requiring a variation issued under clause 10.1 of MWSub/D.

Clause 5.2.2 is important to note. Except where the sub-contractor is in breach of his obligation under clause 5.2.1 of MWSub/D the sub-contractor is not liable under the sub-contract if the sub-contract works fail to comply with the statutory requirements where this non-compliance arises from the sub-contractor having carried out work in accordance with the sub-contract documents or any instruction of the contractor.

Chapter 5
Time

Time and the adjustment to the period for completion

What are the sub-contractor's (or sub-subcontractor's) obligations in respect of time for commencement and completion?

DBSub

In line with clause 2.3 of DBSub/C, the sub-contractor is required to commence works in accordance with the programme details stated in the sub-contract particulars.

In respect of the construction works (but not the design works), the contractor is to issue a written notice to commence works to the sub-contractor. Although a written notice to commence the design works is not required by the sub-contract, it would obviously be sensible for such a written notice to be provided in any event.

When it comes to the completion of the works, the sub-contractor is required to carry out and complete the works in accordance with the programme details stated in the sub-contract particulars, and reasonably in accordance with the progress of the main contract works or each relevant section, as appropriate.

The above provisions are subject to the clauses of the sub-contract that deal with the adjustment to the period for completion (dealt with later within this chapter).

MPSub

Pursuant to clause 20.1, the contractor is to give the sub-contractor written notice that access to the site will be given to the sub-contractor and shall then provide such access as is reasonably necessary to enable the sub-contractor to execute and complete the sub-contract works. That access may not be exclusive access, and may be subject to any restrictions set out in the Contractor's Requirement documents which are to be identified in the sub-contract particulars. The actual period of the notice to be given is to be stated in the sub-contract particulars, but if no period is stated a default period of seven days applies.

There is also facility within the sub-contract particulars for the period that the sub-contractor may require for pre-site activities prior to commencing works on site to be stated. If no period is inserted in the sub-contract particulars for this matter, the time period is deemed to be nil.

In respect of the completion of the works, the sub-contractor is to carry out and complete the works in accordance with the specific dates or programme requirements set out in the sub-contract particulars (against the entry for clause 20.2.2) and is to proceed with the works regularly and diligently and reasonably in accordance with the progress of the project, so as to achieve practical completion within the period for completion stated in the sub-contract particulars.

The above provisions are subject to the clauses of the sub-contract that deal with extension of time (dealt with later in this chapter).

ICSub

In line with clause 2.2 of ICSub/C, the sub-contractor is required to commence works in accordance with the programme details stated in the sub-contract particulars. The contractor is to issue a written notice to commence works to the sub-contractor.

In respect of the completion of the works, the sub-contractor is to carry out and complete the works in accordance with the programme details stated in the sub-contract particulars, and reasonably in accordance with the progress of the main contract works or each relevant section, as appropriate.

The above provisions are subject to the clauses of the sub-contract that deal with the adjustment to the period for completion (dealt with later within this chapter).

ICSub/D

All as DBSub, except that the provisions are contained under clause 2.2 and not clause 2.3.

ICSub/NAM

All as ICSub/D.

MWSub/D and ShortSub

The sub-contractor's obligations as to time are set out under clause 5.1, and these obligations are to carry out and complete the sub-contract works in accordance with the sub-contract documents, and with due diligence.

This clause clearly has little effect unless a document which sets out the contractor's programme requirements is incorporated in the sub-contract under the second recital.

In line with clause 8.1, the contractor is to issue a written notice to commence works to the sub-contractor, and the sub-contractor is to commence the sub-contract works on site within 14 days of receipt of that written instruction.

Clause 8.2 requires the sub-contractor to proceed with the works regularly and diligently and reasonably in accordance with the progress of the main contract works. It also requires the sub-contractor to achieve practical completion of the sub-contract works within the period for completion (a period inserted under article 3 of the sub-contract).

The above provisions are subject to the clauses of the sub-contract that deal with extension of time (dealt with later within this chapter).

SubSub

The sub-subcontractor's obligations as to time are set out under clause 5.1, and these obligations are to carry out and complete the sub-subcontract works in accordance with the sub-subcontract documents, and with due diligence.

This clause clearly has little effect unless a document which sets out the sub-contractor's programme requirements is incorporated in the sub-subcontract under the third recital.

In line with clause 8.1, the sub-contractor is to issue a written notice to commence works to the sub-subcontractor, and the sub-subcontractor is to commence the sub-subcontract works on site within ten days of receipt of that written instruction.

Clause 8.2 requires the sub-subcontractor to proceed with the works regularly and diligently and reasonably in accordance with the progress of the sub-contract works and the main contract works. It also requires the sub-subcontractor to achieve practical completion of the sub-subcontract works within the period for completion (a period inserted under article 3 of the sub-subcontract).

The above provisions are subject to the clauses of the sub-contract that deal with extension of time.

Is the sub-contractor (or sub-subcontractor) entitled to an adjustment of the period for completion?

DBSub

Yes, this may be an extension *or a reduction* of time (refer to clauses 2.16 to 2.19).

MPSub

Yes, this may be an extension *or a reduction* of time (refer to clause 22.7.1).

ICSub, ICSub/D, ICSub/NAM, MWSub/D, ShortSub and SubSub

Yes, but only an extension of time *not a reduction to time.*

Under what circumstances may a sub-contractor's (or sub-subcontractor's) time period be reduced?

DBSub

A sub-contractor's time period may be reduced where a pre-agreed adjustment of time (i.e. a reduction in this case) applies:

- in respect of an agreed schedule 2 quotation; or
- because a relevant sub-contract omission (i.e. the omission of any work or obligation through a direction for a variation or in respect of a provisional sum for defined work) has the effect of reducing the sub-contractor's time period.

It should be noted that clause 2.18.6.3 makes it clear that no reduction in time as noted above under the second bullet point is to fix a shorter period or periods for completion than that stated in the sub-contract particulars. Therefore any reduction to the sub-contractor's time period can only be made as a result of a relevant sub-contract omission after an extension of time has previously been granted.

Also, no pre-agreed adjustment that extends the sub-contract period shall be subsequently reduced as a result of a relevant sub-contract omission, unless the relevant sub-contract omission relates *directly* to the variation that formed the basis of the pre-agreed adjustment.

MPSub

A sub-contractor's time period may be reduced:

- where acceleration measures have been agreed (as clause 23);
- where cost savings and value improvements apply (as clause 28); or
- where changes reduce the scope of the works (as clause 29).

It should be noted that clause 22.8 makes it clear that no reduction in time as noted above is to fix a shorter period for completion than one that has already been notified, except by agreement between the contractor and the sub-contractor.

ICSub, ICSub/D, ICSub/NAM, MWSub/D, ShortSub and SubSub

Under no circumstances may a sub-contractor's or a sub-subcontractor's time period be reduced.

What events must occur to allow an adjustment of the time for the sub-contractor's (or sub-subcontractor's) period to be considered?

DBSub

An adjustment of the time for the sub-contractor's period will only be considered in the event that:

- a relevant sub-contract event occurs;
- a pre-agreed adjustment applies; or
- a relevant sub-contract omission (i.e. the omission of any work or obligation through a direction for a variation or in respect of a provisional sum for defined work) occurs.

As noted above, a pre-agreed adjustment of time applies in respect of an agreed schedule 2 quotation, and a relevant sub-contract omission is an omission of any work or obligation through a direction for a variation or in respect of a provisional sum for defined work.

Relevant sub-contract events are listed under clause 2.19 of DBSub/C and these are as follows:

- Variations – clause 2.19.1.
- Directions of the contractor given in order to comply with clauses 2.15, 3.10 and 3.11 (excluding an instruction for expenditure of a Provisional Sum for defined work), or 3.16 of the main contract conditions – clause 2.19.2.1.
- Directions of the contractor for the opening up or testing of any work, materials or goods under clause 3.12 or 3.13.3 of the main contract conditions, *unless the inspection or test shows that the works, materials or goods were not in accordance with the sub-contract* – clause 2.19.2.2.
- Directions of the contractor for the opening up or testing of any work, materials or goods under clause 3.10 of the sub-contract conditions, *unless the inspection or test shows that the works, materials or goods were not in accordance with the sub-contract* – clause 2.19.2.3.
- Deferment of the giving of possession of the site or any section under clause 2.4 of the main contract conditions – clause 2.19.3.
- Where there are bills of quantities, the execution of work for which an approximate quantity included in those bills is not a reasonably accurate forecast of the quantity of the work required – clause 2.19.4.
- The suspension by the sub-contractor under clause 4.11 of his obligations under the sub-contract – clause 2.19.5.
- The suspension by the contractor under clause 4.11 of the main contract conditions of his obligations under the main contract – clause 2.19.6.

Clause 3.21 of the sub-contract states that if the contractor under clause 4.11 of the main contract conditions gives the employer a written notice of his intention to suspend the performance of his obligations under the main contract because of non-payment, he is to immediately copy that notice to the sub-contractor, and if he then suspends his obligations under the main contract, he is to immediately notify the sub-contractor.

However, it is most important for the sub-contractor to note that he is not to suspend his performance in respect of the sub-contract works simply because he has been notified by the contractor that the contractor has suspended his performance in respect of the main contract works. In fact, following the notice from the contractor advising that the contractor has suspended his performance in respect of the main contract works, the sub-contractor should (where possible) simply proceed regularly and diligently with his sub-contract works, until and/or unless he receives an express direction from the contractor (under clause 3.22.1 of the sub-contract) to cease the carrying out of the sub-contract works. This cessation is to continue until the contractor directs the sub-contractor to recommence his works (in line with clause 3.22.2).

- Any impediment, prevention or default, whether by act or by omission, by the employer or any of the employer's persons except to the extent caused or contributed to by any default, whether by act or omission, of the sub-contractor or of any of the sub-contractor's persons – clause 2.19.7.
- Any impediment, prevention or default, whether by act or by omission, by the contractor (including when the contractor is acting as the principal contractor) or any of the contractor's persons except to the extent caused or contributed to by any default, whether by act or omission, of the sub-contractor or of any of the sub-contractor's persons – clause 2.19.8.

- The carrying out by a statutory undertaker of work in pursuance of his statutory obligations in relation to the main contract works, or the failure to carry out such work – clause 2.19.9.
- Exceptionally adverse weather conditions – clause 2.19.10. It should be noted that this relevant sub-contract event only relates to 'exceptionally adverse' weather conditions. In this context it should be noted that 'exceptional' may be defined as 'unusual' or 'not typical', whilst 'adverse' may be defined as 'unfavourable' or 'harmful'.

  The sub-contract does not provide any guidance as to how to judge whether the adverse weather conditions encountered are exceptional. A common approach is to compare the actual adverse weather conditions encountered on site against the records of the previous 5, 10, 20 or 30 years' weather conditions maintained by a meteorological centre that is in the same locality as the site.
- Loss or damage occasioned by any of the specified perils (as defined under clause 6.1) – clause 2.19.11.
- Civil commotion or the use or threat of terrorism and/or the activity of the relevant authorities in dealing with such event or threat – clause 2.19.12.
- Strike, lock-out or local combination of workmen affecting any of the trades employed upon the main contract works or any of the trades engaged in the preparation, manufacture or transportation of any of the goods or materials required for the main contract works – clause 2.19.13.
- The exercise after the main contract base date by the United Kingdom Government of any statutory power which directly affects the execution of the main contract works – clause 2.19.14.
- Delay in receipt of any necessary permission or approval of any statutory body which the sub-contractor has taken all practical steps to avoid or reduce – clause 2.19.15.
- *Force majeure* – clause 2.19.16.
  Force majeure is a French law term and this term is generally wider than the common law term act of God. The usual English authority[1] states broadly that it covers all *matters independent of the will of man and which it is not in his power to control.*

MPSub

An adjustment of the time for the sub-contractor's period will only be considered in the event that one of the following items occurs:

- *Force majeure* – clause 22.1.1.
  See definition of *Force majeure* under clause 2.19.16 of DBSub.
- The occurrence of one or more of the specified perils (as defined under clause 1) – clause 22.1.2.
- The exercise after the base date (a date identified in the sub-contract particulars) by the United Kingdom Government of any statutory power which directly affects the execution of the project, other than for alterations to statutory requirements as referred to by clause 10.2 – clause 22.1.3.

---

[1] *Lebeaupin v. Crispin* [1920] 2 KB 714.

- The use or threat of terrorism and/or the activities of the relevant authorities in dealing with such a threat – clause 22.1.4.
- Any Change as defined under clause 1 – clause 22.1.5.
  It should be particularly noted that there will be no adjustment to the period for completion in respect of any matters that are specifically stated by the sub-contract not to give rise to a Change.
- Any interference with the regular progress of the sub-contract works by Others, as defined – clause 22.1.6.
- The valid exercise by the sub-contractor of his rights under section 112 of the Housing Grants, Construction and Regeneration Act 1996 – clause 22.1.7.
  Section 112 of the above Act deals with the sub-contractor's right to suspend his performance (subject to the necessary notices being given) because of non-payment.
- Any other breach or act of prevention by the contractor or by any of his sub-contractors and suppliers or any other person under the control and direction of the contractor, other than the sub-contractor – clause 22.1.8.

ICSub

An adjustment of the time for the sub-contractor's period will only be considered in the event that a relevant sub-contract event occurs.

Relevant sub-contract events are listed under clause 2.13 of ICSub/C, and these are as follows:

- Variations – clause 2.13.1.
- Directions of the contractor given in order to comply with clauses 2.13, 3.12, or 3.13 (excluding an instruction for expenditure of a Provisional Sum for defined work) – clause 2.13.2.1.
- Directions of the contractor for the opening up or testing of any work, materials or goods pursuant to the main contract conditions, *unless the inspection or test shows that the works, materials or goods were not in accordance with the sub-contract* – clause 2.13.2.2.
- Directions of the contractor for the opening up or testing of any work, materials or goods pursuant to the sub-contract conditions, *unless the inspection or test shows that the works, materials or goods were not in accordance with the sub-contract* – clause 2.13.2.3.
- Deferment of the giving of possession of the site or any section under clause 2.5 of the main contract conditions – clause 2.13.3.
- Where there are contract bills or bills of quantities, the execution of work for which an approximate quantity included in those bills is not a reasonably accurate forecast of the quantity of work required – clauses 2.13.4.1 and 2.13.4.2.
- The suspension by the sub-contractor under clause 4.13 of his obligations under the sub-contract – clause 2.13.5.
- The suspension by the contractor under clause 4.11 of the main contract conditions of his obligations under the main contract – clause 2.13.6.
  Clause 3.19 of the sub-contract states that if the contractor under clause 4.11 of the main contract conditions gives the employer a written notice of his intention to suspend the performance of his obligations under the main contract

because of non-payment, he is to immediately copy that notice to the sub-contractor, and if he then suspends his obligations under the main contract, he is to immediately notify the sub-contractor.

However, it is most important for the sub-contractor to note that he is not to suspend his performance in respect of the sub-contract works simply because he has been notified by the contractor that the contractor has suspended his performance in respect of the main contract works. In fact, following the notice from the contractor advising that the contractor has suspended his performance in respect of the main contract works, the sub-contractor should (where possible) simply proceed regularly and diligently with his sub-contract works, until and/or unless he receives an express direction from the contractor (under clause 3.20.1 of the sub-contract) to cease the carrying out of the sub-contract works. This cessation is to continue until the contractor directs the sub-contractor to recommence his works (in line with clause 3.20.2).

- Any impediment, prevention or default, whether by act or by omission, by the employer, the architect, the quantity surveyor or any of the employer's persons except to the extent caused or contributed to by any default, whether by act or omission, of the sub-contractor or of any of the sub-contractor's persons – clause 2.13.7.
- Any impediment, prevention or default, whether by act or by omission, by the contractor (including when the contractor is acting as the principal contractor) or any of the contractor's persons except to the extent caused or contributed to by any default, whether by act or omission, of the sub-contractor or of any of the sub-contractor's persons – clause 2.13.8.
- The carrying out by a statutory undertaker of work in pursuance of his statutory obligations in relation to the main contract works, or the failure to carry out such work – clause 2.13.9.
- Exceptionally adverse weather conditions – clause 2.13.10.

  It should be noted that this relevant sub-contract event only relates to 'exceptionally adverse' weather conditions. In this context it should be noted that 'exceptional' may be defined as 'unusual' or 'not typical', whilst 'adverse' may be defined as 'unfavourable' or 'harmful'.

  The sub-contract does not provide any guidance as to how to judge whether the adverse weather conditions encountered are exceptional. A common approach is to compare the actual adverse weather conditions encountered on site against the records of the previous 5, 10, 20 or 30 years' weather conditions maintained by a meteorological centre that is in the same locality as the site.
- Loss or damage occasioned by any of the specified perils (as defined under clause 6.1) – clause 2.13.11.
- Civil commotion or the use or threat of terrorism and/or the activity of the relevant authorities in dealing with such event or threat – clause 2.13.12.
- Strike, lock-out or local combination of workmen affecting any of the trades employed upon the main contract works or any of the trades engaged in the preparation, manufacture or transportation of any of the goods or materials required for the main contract works – clause 2.13.13.
- The exercise after the main contract base date by the United Kingdom Government of any statutory power which directly affects the execution of the main contract works – clause 2.13.14.

- *Force majeure* – clause 2.13.15.
  Force majeure is a French law term and this term is defined under clause 2.19.16 of DBSub, above.

ICSub/D

All as ICSub.

ICSub/NAM

All as ICSub.

MWSub/D and ShortSub

An adjustment of the time for the sub-contractor's period will only be considered in the event that the ordering of any variation of the sub-contract works or any other reason *beyond the control of the sub-contractor* causes a delay to the completion of the sub-contract works.

SubSub

An adjustment of the time for the sub-subcontractor's period will only be considered in the event that the ordering of any variation of the sub-subcontract works or any other reason *beyond the control of the sub-subcontractor* causes a delay to the completion of the sub-subcontract works.

Is a sub-contractor (or a sub-subcontractor) required to give notice that it is in delay?

DBSub

Clause 2.17.1 of DBSub/C requires the sub-contractor to give written notice to the contractor if and whenever it becomes reasonably apparent that the commencement, progress or completion of the sub-contract works is being or is likely to be delayed.

That notice is to provide the material circumstances, the cause or causes of the delay, and is to identify any event which in his opinion is a relevant sub-contract event.

When giving this notice, or otherwise in writing as soon as possible afterwards, the sub-contractor is to give particulars of the expected effects of the identified event, and is to give an estimate of the expected delay to the completion of the sub-contract works, and is to forthwith notify the contractor in writing should there be any material change in the estimated delay or to any of the particulars provided.

MPSub

Clause 22.2 requires the sub-contractor to give written notice to the contractor when he becomes aware that the progress of the sub-contract works is being or is likely to be delayed due to any cause.

That notice is to detail the cause of the delay, and its anticipated effect upon the progress and completion of the sub-contract works.

Where the sub-contractor considers that the cause of the delay is one of the items identified in clauses 22.1.1 to 22.1.8, he is to provide supporting documentation to demonstrate to the contractor the effect upon the progress and completion of the sub-contract works, and is to revise any documentation provided so that the contractor is aware at all times of the anticipated or actual effect of that cause of delay.

Clause 22.6 requires the sub-contractor to provide documentation to support any further adjustment to the period for completion that he considers fair and reasonable, not later than 28 days after practical completion of the sub-contract works.

ICSub

Clause 2.12.1 of ICSub/C requires the sub-contractor to give written notice to the contractor if and whenever it becomes reasonably apparent that the commencement, progress or completion of the sub-contract works is being or is likely to be delayed.

That notice is to provide the cause of the delay.

Clause 2.12.4.2 notes that the sub-contractor is to provide such information in support of his notice as is reasonably required by the contractor.

ICSub/D

All as ICSub.

ICSub/NAM

All as ICSub.

MWSub/D and ShortSub

Clause 11.1 requires the sub-contractor to give written notice to the contractor if the sub-contractor *is delayed* in completing the sub-contract works within the period for completion by the ordering of any variation of the sub-contract works or for other reasons beyond the control of the sub-contractor. It should be noted that such a notice is only required to be given when the sub-contractor is delayed, rather than simply when it is likely to be delayed.

SubSub

Clause 11.1 requires the sub-subcontractor to give written notice to the sub-contractor if the sub-subcontractor *is delayed* in completing the sub-subcontract works within the period for completion by the ordering of any variation of the sub-subcontract works or for other reasons beyond the control of the sub-subcontractor. It should be noted that such a notice is only required to be given when the sub-subcontractor is delayed, rather than simply when it is likely to be delayed.

What is the recipient to do upon receipt of a sub-contractor's (or a sub-subcontractor's) notice of delay?

DBSub

When the contractor receives a notice of delay, he may request that the sub-contractor provides more information pursuant to clause 2.17.3 of DBSub/C.

Assuming that the contractor is content that he has sufficient information from the sub-contractor then he is to take the following action:

- in line with clause 2.18.1.1 he is to consider whether the event which is stated to be a cause of delay is a Relevant Sub-contract Event;
- then (in line with clause 2.18.1.2) he is to consider whether the said event is likely to delay the completion of the sub-contract works beyond the period or periods for the completion of the sub-contract work.

Following this (except where the sub-contract conditions expressly provide otherwise), the contractor is to:

- give an extension of time to the sub-contractor by fixing a revised period or periods for completion as he then estimates to be fair and reasonable (clause 2.18.1); or
- he is to notify the sub-contractor that he does not consider that the sub-contractor is entitled to an extension of time (clause 2.18.2).

Whichever response the contractor makes, clause 2.18.2 requires him to:

- issue his decision as soon as is reasonably practicable;
- issue his decision within 16 weeks of receipt of the notice or the required particulars;
- in the case where the date from the receipt of the notice or the required particulars, to the date of expiry of the period or periods for the completion of the sub-contract works is less than 16 weeks, then the contractor should endeavour to issue his decision prior to the said expiry date.

When issuing his decision in respect of an extension of time, in line with clause 2.18.3. the contractor is to state:

- The extension of time that he has attributed to each Relevant Sub-contract Event.

- Any *reduction in time* that he has attributed to a Relevant Sub-contract Omission (i.e. the omission of any work or obligation through a direction for a variation or in respect of a provisional sum for defined work, as fully defined under clause 2.16.3).

  In respect of the above possible *reduction in time*, the contractor can only take into account any Relevant Sub-contract Omission that occurred since the last occasion when a new period for completion was fixed by the contractor.

  Also, although the contractor can reduce the extension of time previously given, clause 2.18.6.3 makes it clear that under no circumstance can the period or periods for completion of the sub-contract works be reduced from that stated in the sub-contract particulars (item 5).

Further, the contractor cannot reduce any extension of time previously given by way of a Pre-agreed Adjustment, unless the relevant variation that related directly to the Pre-agreed Adjustment was the subject of the Relevant Sub-contract Omission in question (as clause 2.18.6.4).

In addition to the above, clause 2.18.5 notes that if the expiry of the period or periods for completion of the sub-contract works (or of such works in a section) occurs before the sub-contractor's practical completion, the contractor may (not later than the expiry of 16 weeks after the date of the sub-contractor's practical completion):

- extend the relevant period for completion fixed;
- shorten the relevant period for completion fixed (if he considers that is fair and reasonable having regard to any Relevant Sub-contract Omissions issued after the last occasion on which a new period for completion was fixed); or
- confirm the period or periods for completion previously fixed.

When extending or shortening the relevant period for completion, the contractor is to state:

- the extension of time that he has attributed to each Relevant Sub-contract Event;
- any *reduction in time* that he has attributed to a Relevant Sub-contract Omission.

MPSub

In line with clause 22.4, within 56 days of receipt of a delay notice from the sub-contractor pursuant to clause 22.2 in which the cause of the delay is identified by the sub-contractor as being one of those in clause 22.1.1 to 22.1.8, the contractor is to either:

- notify the sub-contractor of such adjustment to the period for completion as he then calculates to be fair and reasonable; or is to
- notify the sub-contractor why he considers that the period for completion should not be adjusted.

When making an adjustment to the period for completion, the contractor may refer to the documentation provided by the sub-contractor, and may take into account other information available to him.

Also, the contractor can review any adjustment to the period for completion that he grants in the light of further documentation being provided by the sub-contractor, or when the effects of any identified cause of delay becomes more apparent.

Also, in line with clause 22.6, within 56 days of receipt of a delay notice from the sub-contractor pursuant to clause 22.6 the contractor is to either:

- notify the sub-contractor of such adjustment to the period for completion as he then calculates to be fair and reasonable; or is to

- notify the sub-contractor why he considers that the period for completion should not be further adjusted.

ICSub

If the contractor receives a notice of delay, he may request, pursuant to clause 2.12.4.2 of ICSub/C, that the sub-contractor provides no more information than is reasonably necessary to allow him to assess the matter.

Clauses 2.12.1 and 2.12.2 require the contractor, upon receipt of a notice of delay, to estimate the length of delay beyond the period for completion, or beyond any previously revised period for completion, and, as soon as he is able, to make in writing a fair and reasonable extension of such period. It is unclear whether a notice of delay needs to be issued by the sub-contractor before an extension of time is granted by the contractor under clause 2.12.2. That clause implies that if any relevant event referred to in clauses 2.13.1 to 2.13.8 occurs after the period for completion (or revised period for completion), but before practical completion is achieved, the contractor is to estimate the length of any delay and make a fair and reasonable extension of the period for completion, but it does not expressly state that this exercise is to be undertaken only after the sub-contractor has provided a written notice of delay.

Clause 2.12.3 notes that up to 16 weeks after the date of practical completion of the sub-contract works or such works in a section, the contractor may make an extension of time whether upon reviewing a previous decision or otherwise and *whether or not the sub-contractor has given notice* as required by clause 2.12.1.

ICSub/D

All as ICSub.

ICSub/NAM

All as ICSub.

MWSub/D and ShortSub

Clause 11.1 requires the contractor to give a reasonable extension of time if the sub-contractor is delayed in completing the sub-contract works within the period for completion by the ordering of any variation of the sub-contract works or for other reasons beyond the control of the sub-contractor.

SubSub

Clause 11.1 requires the sub-contractor to give a reasonable extension of time if the sub-subcontractor is delayed in completing the sub-subcontract works within the period for completion by the ordering of any variation of the sub-subcontract works or for other reasons beyond the control of the sub-subcontractor.

What happens if a sub-contractor (or a sub-subcontractor) does not issue a notice of delay?

DBSub

Clause 2.18.5 of DBSub/C notes that (irrespective of whether or not a notice of delay has been issued by the sub-contractor) if the expiry of the period or periods for completion of the sub-contract works (or of such works in a section) occurs before the sub-contractor's practical completion, the contractor may, (not later than the expiry of 16 weeks after the date of the sub-contractor's practical completion):

- extend the relevant period for completion fixed;
- shorten the relevant period for completion fixed (if he considers that is fair and reasonable having regard to any Relevant Sub-contract Omissions issued after the last occasion on which a new period for completion was fixed); or
- confirm the period or periods for completion previously fixed.

When extending or shortening the relevant period for completion, the contractor is to state:

- The extension of time that he has attributed to each Relevant Sub-contract Event.
- Any *reduction in time* that he has attributed to a Relevant Sub-contract Omission.

MPSub

The MPSub does not envisage a situation where a contractor would grant an extension of time without first receiving a delay notice from a sub-contractor.

The sub-contractor's delay notice is therefore a condition precedent to the award of an extension of time by the contractor.

If a sub-contractor does not issue a notice of delay he may therefore lose the right to an extension of time, although, because no consequences are expressly provided for in the event that a sub-contractor does not provide a delay notice, it is considered that (in the case of a delay event that the sub-contractor had no liability for) the application of the 'prevention principle' may have effect and this may mean that the sub-contractor would be entitled to an extension of time even if no notice were given.

The prevention principle comes from a generally stated legal principle that a party cannot benefit from its own wrong.

Another point to note is that under clause 20.4, the contractor is to ensure at all times that the sub-contractor is aware of the actual and projected progress of the project, including the date by which the contractor reasonably anticipates completing the project (or section if applicable). Clearly, if the contractor failed to comply with this requirement, the sub-contractor may use this as a reason why he had not submitted a delay notice.

ICSub

Clause 2.12.3 of ICSub/C notes that up to 16 weeks after the date of practical completion of the sub-contract works or such works in a section, the contractor may make an extension of time whether upon reviewing a previous decision or otherwise and *whether or not the sub-contractor has given notice* as required by clause 2.12.1.

ICSub/D

All as ICSub.

ICSub/NAM

All as ICSub.

MWSub/D and ShortSub

Clause 11.1 requires the contractor to give a reasonable extension of time if the sub-contractor is delayed in completing the sub-contract works within the period for completion by the ordering of any variation of the sub-contract works or for other reasons beyond the control of the sub-contractor. Although this requirement is preceded within clause 11.1 by the need for the sub-contractor to provide a notice of delay, it is considered that such a notice is not a condition precedent to the contractor granting an extension of time, and it is therefore considered that clause 11.1 when read as a whole requires the contractor to grant an extension of time (if appropriate) irrespective of any notice being provided by the sub-contractor. Despite this interpretation, it is always safer for the sub-contractor to submit a delay notice in any event.

SubSub

All as the principle outlined under MWSub/D and ShortSub above.

What factors are to be considered when deciding whether or not there is an entitlement to an extension of time?

The most important and indeed difficult task in terms of determining delay is to establish the nexus of cause and effect; the mere happening of an event which happens to be a Relevant Sub-contract Event (for example), confers no entitlement to an extension to the contract period.

The test is whether the event in question actually caused a delay to the completion date (or to a section thereof). In considering this matter, it is generally actual progress which must be delayed, the originally planned/programmed progress is not generally relevant[2].

What needs to be considered are the actual facts on site to determine whether a particular event affected operations on the critical path to the completion date.

What has to be established in determining an extension of time entitlement can briefly be described as follows:

---

[2] *Glenlion v. The Guinness Trust* (1987) 39 BLR 89.

- that an event occurred at all;
- that the event occurred in the manner asserted;
- that the event comes within a category providing an express entitlement to an extension of time within the sub-contract (e.g. that it is a Relevant Sub-contract Event);
- that the required notices have been given;
- that the event outlined has had a particular delaying effect on the critical path to completion.

In certain instances concurrent delays can occur, i.e. two delay events having an impact on the critical path to completion at the same time.

If these two delay events are not the liability of the sub-contractor (or, in fact, if both delay events are the liability of the sub-contractor) then the question of concurrent delays is not a major concern. The difficult issue in respect of concurrent delays occurs when one of the delay events is not the liability of the sub-contractor, but the other delay event is the liability of the sub-contractor. The question then arises as to how the allocation of the delay to completion is allocated between the two concurrent delay events.

The courts have dealt with this issue in a variety of ways and some of the more common approaches used are the 'apportionment' approach, the 'American' approach, the 'chronology of events' approach, the 'dominant cause' approach, the application of the 'but for' test, and the 'Malmaison'[3] approach. Despite all of these approaches, an analysis of the numerous court cases shows that in reality the courts deal with each case on its own merits, and there are no hard and fast rules when it comes to matters of concurrency. It is commonly believed that under JCT contracts, generally if there is a situation where there are two concurrent delay events, one of which a sub-contractor is responsible for (i.e. a culpable delay) and one of which a sub-contractor is not responsible for (i.e. an excusable delay), then the sub-contractor would always be entitled to an extension of time (but not necessarily any loss and/or expense). Whilst this may often be the case, and whilst this appears to be supported by clause 22.7.3 of MPSub, which says that in considering any adjustment to the period for completion the contractor shall 'make a fair and reasonable adjustment to the period for completion notwithstanding that completion of the sub-contract works may also have been delayed due to the concurrent effect of a cause of delay that is not listed in clauses 22.1.1 to 22.1.8', it certainly is not to be considered as being a path that the courts will follow in all situations.

What measures are a sub-contractor (or a sub-subcontractor) required to take prevent delay?

DBSub

Clause 2.18.6.1 of DBSub/C requires the sub-contractor to 'constantly use his best endeavours to prevent delay in the progress of the Sub-contract Works or of such

---

[3] Named after the *Henry Boot Construction (UK) Ltd* v. *Malmaison Hotel (Manchester) Ltd case* [1999] 70 Con LR 32.

works in any Section, however caused, and to prevent their completion being delayed or further delayed beyond the relevant period for completion'.

Clause 2.18.6.2 requires the sub-contractor to 'do all that may reasonably be required to the satisfaction of the Contractor to proceed with the Sub-contract Works'.

The wording of clause 2.18.6.1 and clause 2.18.6.2 is fairly typical JCT wording, and there has been much debate over the years as to what the wording 'constantly use his best endeavours to prevent delay' and 'do all that may reasonably be required' means in effect.

Unfortunately, there appears to be no direct case law available to shed any light on this matter. The wording is generally considered to mean that the sub-contractor would be expected to re-programme the works, re-schedule material deliveries and information request schedules, and keep all parties advised of the situation; and it may even mean that the sub-contractor may need to expend some monies to meet this obligation; but it is not normally considered that the sub-contractor would be expected to expend substantial sums of money in respect of this obligation.

Some commentators consider that the wording does not require a sub-contractor to expend any monies at all, but this is probably taking too extreme a view.

MPSub

Clause 20.3 requires the sub-contractor to 'at all times use his reasonable endeavours to prevent or reduce delay to the progress or to the completion of the sub-contract works and the project'.

As noted above in respect of DBSub, the phrase 'use his reasonable endeavours to prevent delay or reduce delay' are fairly typical words used in JCT contracts and sub-contracts, and the normal interpretation of these words is provided under the commentary for DBSub above.

In respect of the MPSub, the importance of this obligation on the sub-contractor is emphasised under clause 22.7.2 which says that when considering any adjustment to the period for completion the contractor is to have regard to any breach by the sub-contractor of clause 20.3.

ICSub

Clause 2.12.4.1 of ICSub/C requires the sub-contractor to 'constantly use his best endeavours to prevent delay and shall do all that may be reasonably required to the satisfaction of the contractor to proceed with the sub-contract works'.

As noted above in respect of DBSub, the phrases 'constantly use his best endeavours to prevent delay' and 'do all that may reasonably be required' are fairly typical words used in JCT contracts and sub-contracts, and the normal interpretation of these words is provided under the commentary for DBSub above.

ICSub/D

All as ICSub.

ICSub/NAM

All as ICSub.

MWSub/D and ShortSub

Clause 11.2 requires the sub-contractor to 'constantly use his best endeavours to prevent or minimise any delay in the progress of the whole or any part of the sub-contract works'.

As noted above in respect of DBSub, the phrases 'constantly use his best endeavours to prevent delay' are fairly typical words used in JCT contracts and sub-contracts, and the normal interpretation of these words is provided under the commentary for DBSub above.

SubSub

All as MWSub/D and ShortSub above.

Is a sub-contractor entitled to a bonus for early completion?

The only sub-contract that refers to the possibility of a bonus for early completion is the MPSub which says, under clause 24, that 'if the date of Practical Completion is before the expiry of the Period for Completion the Contractor shall be liable to pay the Sub-contractor a bonus calculated at the rate set out in the Sub-contract Particulars for the period from the date of Practical Completion to the expiry of the Period for Completion'.

Naturally, if no rate of bonus was set out in the Sub-contract Particulars, then clause 24 would have no effect.

Practical completion and lateness

Is practical completion defined?

The DBSub, the ICSub, the ICSub/D, the ICSub/NAM, the MWSub/D, the Short-Sub and the SubSub do not define practical completion.

However, the MPSub does define practical completion.

What is practical completion?

Where a sub-contract does not define practical completion (e.g. DBSub, ICSub, ICSub/D, ICSub/NAM, MWSub/D, ShortSub and SubSub), the definition of practical completion may be deduced from various court cases, as follows:

(1) the works can be practically complete notwithstanding that there are latent defects (i.e. defects that are not apparent at the date of practical completion);

(2) a certificate of practical completion may not be issued if there are patent defects (i.e. defects that are apparent at the date of practical completion)[4];

(3) practical completion means the completion of all the construction work that has to be done[5];

(4) however, a contractor would be expected to have discretion to certify practical completion where there are very minor items of work left incomplete, on *de minimis* principles[6].

The above deduced definition does not apply to the MPSub which defines practical completion as being when the sub-contract works are complete for all practical purposes and, in particular when:

(1) The relevant statutory requirements (as defined in MPSub) have been complied with and any necessary consents or approvals obtained.

(2) Neither the existence nor the execution of any minor outstanding works would affect their use.

(3) Any stipulations identified by the 'requirements' (as defined in MPSub, i.e. the documents prepared by the contractor setting out his requirements for the sub-contract works) as being essential for practical completion to take place have been satisfied.

(4) all 'as-built' information and operating and maintenance information required by the sub-contract to be delivered at practical completion has been delivered to the contractor.

What happens when the sub-contractor is of the opinion that he has achieved practical completion?

DBSub

Clause 2.20.1 of DBSub/C requires the sub-contractor to notify the contractor in writing when, in his opinion, the sub-contract works as a whole or such works in a Section are practically complete, and he has complied sufficiently with the requirement to provide as-built drawings, etc., and the requirement to provide information for the Health and Safety File.

MPSub

Clause 20.5 of MPSub requires the sub-contractor to notify the contractor (either in writing or by some other means of communication if agreed between the contractor and the sub-contractor, as clause 5.1 of MPSub) when, in his opinion, practical completion of the sub-contract works has occurred.

---

[4] *Jarvis & Sons* v. *Westminster Corp.* [1970] 1 WLR 637 and *HW Nevill (Sunblest)* v. *William Press* (1981) 20 BLR 78.
[5] *Jarvis & Sons* v. *Westminster Corp.* [1970] 1 WLR 637.
[6] *HW Nevill (Sunblest)* v. *William Press* (1981) 20 BLR 78.

ICSub

Clause 2.14.1 of ICSub/C requires the sub-contractor to notify the contractor in writing when, in his opinion, the sub-contract works as a whole or such works in a Section are practically complete, and he has complied sufficiently with the requirement to provide information for the Health and Safety File.

ICSub/D

Clause 2.14.1 of ICSub/D/C requires the sub-contractor to notify the contractor in writing when, in his opinion, the sub-contract works as a whole or such works in a Section are practically complete, and he has complied sufficiently with the requirement to provide as-built drawings, etc., and the requirement to provide information for the Health and Safety File.

ICSub/NAM

Clause 2.14.1 of ICSub/NAM/C requires the sub-contractor to notify the contractor in writing when, in his opinion, the sub-contract works as a whole or such works in a Section are practically complete, and he has complied sufficiently with the requirement to provide as-built drawings, etc. (where the sub-contractor is responsible for design), and the requirement to provide information for the Health and Safety File.

MWSub/D and ShortSub

The MWSub/D and the ShortSub makes no provision for the sub-contractor to take any action when he is of the opinion that he has achieved practical completion.

SubSub

The SubSub makes no provision for the sub-subcontractor to take any action when he is of the opinion that he has achieved practical completion.

When is practical completion deemed to have been achieved?

DBSub

Clause 2.20.1 of DBSub/C states that if the contractor does not dissent in writing to the sub-contractor's written notice that practical completion has been achieved, within 14 days of receipt of same, practical completion of such work shall be deemed for all the purposes of this sub-contract to have taken place on the date so notified.

If the contractor does dissent, and any such dissension must provide reasons for same, then clause 2.20.2 of DBSub/C states that as soon as the contractor is satisfied that the works are practically complete and he is satisfied that the sub-contractor has complied sufficiently with the requirement to provide as-built drawings, etc., and the requirement to provide information for the Health and Safety File, then he is to notify the sub-contractor in writing of this fact as soon as practicable thereafter, and for all of the purposes of the sub-contract practical completion will be deemed to have taken place on the date notified by the contractor.

Alternatively, the date that practical completion is achieved may be agreed between the parties, or may be determined by the dispute resolution procedures of the sub-contract.

Clause 2.20.2 of DBSub/C makes it clear that the date of practical completion of the sub-contract works shall not, under any circumstances, be any later than the certified date of practical completion of the main contract works or any relevant Section thereof, and adds that the contractor is to notify the sub-contractor in writing of the certified date of practical completion of the main contract works or any relevant Section thereof.

MPSub

Clause 20.5 of MPSub states that if the contractor agrees with the sub-contractor's notice that practical completion has been achieved, he is to issue a statement recording the date of practical completion.

However, if the contractor does not agree that practical completion has been achieved, he is to notify the sub-contractor of the work he requires to be completed before practical completion will occur.

When the sub-contractor considers that the required work has been completed, he is to notify the contractor and when the contractor is satisfied that the works have been completed he is to issue a statement recording the date of practical completion.

Where the contractor does not issue a statement recording the date of practical completion, practical completion is deemed to have occurred on the same date as practical completion of the project as a whole occurs.

ICSub

Clause 2.14.1 of ICSub/C states that if the contractor does not dissent in writing to the sub-contractor's written notice that practical completion has been achieved, within 14 days of receipt of same, practical completion of such work shall be deemed for all the purposes of this sub-contract to have taken place on the date so notified.

If the contractor does dissent, and any such dissension must provide reasons for same, then clause 2.14.2 of ICSub/C states that as soon as the contractor is satisfied that the works are practically complete and he is satisfied that the sub-contractor has complied sufficiently with the requirement to provide information for the Health and Safety File, then he is to notify the sub-contractor in writing of this fact as soon as practicable thereafter, and for all of the purposes of the sub-contract practical completion will be deemed to have taken place on the date notified by the contractor.

Alternatively, the date that practical completion is achieved may be agreed between the parties, or may be determined by the dispute resolution procedures of the sub-contract.

Clause 2.14.2 of ICSub/C makes it clear that the date of practical completion of the sub-contract works shall not, under any circumstances, be any later than the certified date of practical completion of the main contract works or any relevant Section thereof.

ICSub/D

All as ICSub above.

ICSub/NAM

All as ICSub above.

MWSub/D and ShortSub

Clause 8.3 states that the contractor shall determine and notify the sub-contractor in writing of the date when the sub-contract works are practically complete.

SubSub

Clause 8.3 states that the sub-contractor shall determine and notify the sub-subcontractor in writing of the date when the sub-subcontract works are practically complete.

What happens if the sub-contractor fails to complete on time?

DBSub

Clause 2.21 of DBSub/C states that if the sub-contractor fails to complete the sub-contract works within the relevant period for completion, and the contractor gives the requisite notice within a reasonable time after the expiry of that period, the sub-contractor is required to pay the contractor his loss and expense resulting from the failure.

This loss and expense may include:

- the contractor's loss and expense;
- the contractor's other sub-contractors' loss and expense;
- where applicable, any liquidated damages suffered by the contractor under the main contract.

An alternative to the sub-contractor paying for the above is for the sub-contractor to allow the amount of the contractor's loss and expense to be deducted from payments otherwise due to him.

If the contractor follows this latter option then it is important that the withholding notices required by clauses 4.10.3 and 4.12.3 of DBSub/C are issued timeously.

With regard to the above matter, reference should also be made to clause 4.21 of DBSub/C which deals with the recovery by the contractor of any direct loss and/or expense caused to the contractor where the regular progress of the main contract works is materially affected by an act, omission or default of the sub-contractor, or any of the sub-contractor's persons.

MPSub

If the sub-contractor fails to proceed with or complete the sub-contract works in accordance with the sub-contract, clause 21 of MPSub states that the sub-contractor is liable to pay the contractor any loss and/or expense incurred by the contractor.

This loss and expense may include:

- the contractor's loss and expense;
- the contractor's other sub-contractors' loss and expense;
- where applicable, any liquidated damages suffered by the contractor under the main contract.

This clause operates on the basis that, as soon as is reasonably practicable:

- the contractor notifies the sub-contractor that loss and/or expense has been or is likely to be incurred; and
- the contractor notifies the sub-contractor of the loss and/or expense that has been incurred and provides details of that loss and/or expense to the sub-contractor.

ICSub

Clause 2.15 of ICSub/C states that if the sub-contractor fails to complete the sub-contract works within the relevant period for completion, and the contractor gives the requisite notice within a reasonable time after the expiry of that period, the sub-contractor is required to pay the contractor his loss and expense resulting from the failure.

This loss and expense may include:

- the contractor's loss and expense;
- the contractor's other sub-contractors' loss and expense;
- where applicable, any liquidated damages suffered by the contractor under the main contract.

An alternative to the sub-contractor paying for the above is for the sub-contractor to allow the amount of the contractor's loss and expense to be deducted from payments otherwise due to him.

If the contractor follows this latter option then it is important that the withholding notices required by clauses 4.12.3 and 4.14.3 of the sub-contract conditions are issued timeously.

With regard to the above matter, reference should also be made to clause 4.18 of the sub-contract conditions which deals with the recovery by the contractor of any direct loss and/or expense caused to the contractor where the regular progress of the main contract works is materially affected by an act, omission or default of the sub-contractor, or any of the sub-contractor's persons.

ICSub/D

All as ICSub above.

ICSub/NAM

All as ICSub above.

MWSub/D and ShortSub

No specific provision is allowed for the situation where the sub-contractor fails to complete on time.

However, if such a situation arose, and if the contractor incurred costs as a result of the sub-contractor's failure, then the contractor would be entitled to recover damages from the sub-contractor that arose from the sub-contractor's breach of contract.

Clause 12.7 of the sub-contract conditions notes that the contractor is entitled to withhold payment of all or any part of any sums otherwise due where a sum is due from the sub-contractor to the contractor under the sub-contract.

In such a situation, the contractor would need to ensure that the required withholding notice (in line with clause 12.8 of the sub-contract conditions) was issued at the required time (i.e. no later than five days before the final date for payment).

SubSub

All as MWSub/D and ShortSub, except that the SubSub deals with a sub-contract/ sub-subcontractor relationship rather than a contractor/sub-contractor relationship.

Chapter 6
Defects, Design Documents and Warranties

Defects

What is a defect?

A defect is an imperfection, a shortcoming or a failing.

In construction industry terms this may, for example, be paint flaking from walls, a room thermostat not operating correctly, or a water pump not operating at all.

Is a 'defect' defined in the sub-contracts?

None of the sub-contracts in question define what a defect is other than the MPSub which defines a 'defect' under clause 1 as being 'any fault in the Sub-contract Works that arises as a consequence of a failure by the Sub-contractor to comply with his obligations under this Sub-contract, together with the consequences of that fault.'

What is the sub-contractor's liability in respect of defects?

DBSub

Clause 2.22 of the sub-contract conditions requires the sub-contractor to make good at his own cost and in accordance with any direction of the contractor all defects, shrinkages, and other faults in the sub-contract works or in any part of them due to materials or workmanship not in accordance with the sub-contract. This liability also extends to any failure of the sub-contractor to comply with his obligations in respect of the sub-contractor's designed works.

MPSub

Clause 27.1 states that during the rectification period the contractor may instruct the sub-contractor to remedy any defect, and the sub-contractor is to comply with such instruction within a reasonable time and at no cost to the contractor.

ICSub

Clause 2.16 of the sub-contract conditions requires the sub-contractor to make good at his own cost and in accordance with any direction of the contractor all defects, shrinkages, and other faults in the sub-contract works or in any part of them due to materials or workmanship not in accordance with the sub-contract.

ICSub/D

Clause 2.16 of the sub-contract conditions requires the sub-contractor to make good at his own cost and in accordance with any direction of the contractor all defects, shrinkages, and other faults in the sub-contract works or in any part of them due to materials or workmanship not in accordance with the sub-contract. This liability also extends to any failure of the sub-contractor to comply with his obligations in respect of the sub-contractor's designed portion.

ICSub/NAM

Clause 2.16 of the sub-contract conditions requires the sub-contractor to make good at his own cost and in accordance with any direction of the contractor all defects, shrinkages, and other faults in the sub-contract works or in any part of them due to materials or workmanship not in accordance with the sub-contract. This liability also extends to any failure of the sub-contractor to comply with his obligations in respect of the NAM designed works.

MWSub/D and ShortSub

Clause 8.4 notes that the contractor shall notify the sub-contractor of any defects that appear in the sub-contract works during the rectification period (or the defects liability period) for the main contract works, and the sub-contractor shall, at his own expense, make good such defects within a reasonable time from notification.

SubSub

Clause 8.5 notes that the sub-contractor shall notify the sub-subcontractor of any defects that appear in the sub-subcontract works during the rectification period (or the defects liability period) for the main contract works, and the sub-subcontractor shall, at his own expense, make good such defects within a reasonable time from notification.

Over what period of time is the sub-contractor liable for the rectification of defects?

DBSub

Clause 2.22 of the sub-contract conditions does not state the period during which the sub-contractor's liability for the rectification of defects applies.

However, it is submitted that this would be the period noted as the rectification period in the main contract, which, unless specifically stated otherwise, would, by default, be six months. It is quite common (particularly where service installations are involved) for the rectification period to be extended to 12 months.

MPSub

The MPSub defines the rectification period as being the period commencing on practical completion of the sub-contract works and expiring 12 months after the date practical completion of the main contract works occur.

ICSub

Clause 2.16 of the sub-contract conditions does not state the period during which the sub-contractor's liability for the rectification of defects applies.

However, it is submitted that this would be the period noted as the rectification period in the main contract, which, unless specifically stated otherwise, would, by default, be six months. It is quite common (particularly where service installations are involved) for the rectification period to be extended to 12 months.

ICSub/D

All as ICSub above.

ICSub/NAM

All as ICSub above.

MWSub/D and ShortSub

Clause 8.4 notes that the contractor shall notify the sub-contractor of any defects that appear in the sub-contract works during the rectification period (or the defects liability period) for the main contract works, and the sub-contractor shall, at his own expense, make good such defects within a reasonable time from notification.

SubSub

Clause 8.5 notes that the sub-contractor shall notify the sub-subcontractor of any defects that appear in the sub-subcontract works during the rectification period (or the defects liability period) for the main contract works, and the sub-subcontractor shall, at his own expense, make good such defects within a reasonable time from notification.

Does the sub-contractor have any liability for defects after the end of the rectification period?

In the absence of words to the contrary, the sub-contractor's liability for not completing the sub-contract works (including the design works) in accordance with the sub-contract continues for six years from the date of practical completion of the main contract works (in the case of contracts executed under hand) or for 12 years from the date of practical completion of the main contract works (in the case of contracts executed as a deed).

Of course, the sub-contractor's liability for defects which are not related to the sub-contractor not completing the sub-contract works (including the design works) in accordance with the sub-contract ends at the end of the rectification period, and the sub-contractor is not liable for any such defects notified after the end of the rectification period.

However, if a sub-contractor achieves practical completion some considerable time before the contractor achieved practical completion, the sub-contractor's liability for defects could be extended for a period considerably longer than the rectification period (in terms of weeks or months) in respect of the main contract works. This latter point could be particularly significant in the situation where a sub-contractor is relying on manufacturer's warranties to cover for the rectification period of, say, 12 months, when (because of the above situation) the period of liability could considerably exceed 12 months.

What is the latest date that a list of defects can be provided?

Although the sub-contracts in question do not provide any express provision in respect of the latest date that a list of defects can be provided, it is considered that this would be not later than 14 days after the expiry of the rectification period, which is generally consistent with the requirement of the main contract forms.

How often can defects lists be issued and what is the protocol for the clearance of defects?

None of the sub-contracts stipulate how often defects lists may be provided during the rectification period, nor do they provide for any timetable or protocol for the clearance of the defects, etc., that are listed. It is for this reason that contractors may wish to address (in a numbered document) how defects should be prioritised and the timetable is to be complied with, in respect of such prioritised defects.

Under the MPSub, clause 27.2 notes that when all of the instructed defect works have been cleared by the sub-contractor, the contractor is to issue a statement to that effect.

Can a contractor carry out defects for a sub-contractor (or a sub-contractor carry out defects for a sub-subcontractor), and recover the costs incurred?

It is generally considered that one of the purposes of defect rectification clauses is to give the sub-contractor the *right* to remedy defects which come within the remit of the clause. Therefore, if a contractor did not give notice to the sub-contractor (or a sub-contractor did not give notice to a sub-subcontractor, as appropriate) to clear defects, but simply cleared the defects with his own (or other) resources, he might not be able to recover any costs at all or he might not be able to recover more than the amount that it would have cost the sub-contractor to clear the defects[1].

---

[1] *Pearce & High Ltd* v. *Baxter* [1999] BLR 101 and *William Tomkinson & Sons Ltd* v. *Parochial Church Council of St Michael* (1990) 6 Const LJ 319.

However, the sub-contracts being considered do, in certain instances, allow contractors to carry out defects for a sub-contractor (or allow sub-contractors to carry out defects for a sub-subcontractor) in certain instances as further detailed below.

DBSub

Clause 2.35 of the design and build main contract conditions allows the employer to instruct the contractor not to carry out defects, shrinkages or other faults that have been notified to the contractor on a defects list. If such an instruction is issued then the said clause 2.35 adds that an appropriate deduction shall be made in the calculation of the ascertained final sum in respect of the defects, shrinkages or other faults not made good by the contractor.

Although not stated by clause 2.35 of the main contract conditions, it is considered that, in the normal course of events, the 'appropriate deduction' would be based upon an abatement to the contractor's value for the works rather than on the cost of completing the defects, etc., by others.

In respect of the sub-contractor, clause 2.23 of DBSub/C notes that where an 'appropriate deduction' has been made under clause 2.35 of the main contract conditions, then, to the extent that such deduction is attributable to one of the following, the deduction (or an appropriate proportion of the deduction) shall be borne by the sub-contractor and shall be taken into account in the calculation of the final sub-contract sum, or shall be recoverable by the contractor from the sub-contractor as a debt:

- inaccurate setting out by the sub-contractor; or
- defects or other faults in the sub-contract works.

The fact that an appropriate proportion of the deduction is to be taken into account in the calculation of the final sub-contract sum may be considered to strengthen the argument that the deduction is to be an abatement to value rather than a set-off of costs for completion by others.

MPSub

Clause 27.1 of the sub-contract conditions notes that if the sub-contractor fails to clear the defects notified to him by the contractor, the contractor may engage others to carry out the works in accordance with clause 9.3 of the sub-contract conditions. Clause 9.3 requires the contractor to give seven days notice in writing to the sub-contractor before engaging others in such a situation. If the contractor does engage others to clear defects in line with the above procedure, the contractor is entitled to recover the *cost* of engaging others.

In respect of any defects that have not been remedied within a reasonable period of time from the expiry of the rectification period, clause 27.3 of the sub-contract conditions notes that the contractor should issue a statement listing:

- Those defects that the contractor intends to engage others to rectify (together with a proper estimate of the *cost* of undertaking those rectification works).

- Those defects that the contractor does not intend to rectify at all (together with details of the deduction to the sub-contractor's account that the contractor proposes to make – any such deduction, it is submitted, would be based upon an abatement of value rather than being cost based).

Despite the above notes, clause 27.4 of the sub-contract conditions states that the provisions set out in clause 27 generally are without prejudice to any other rights or remedies (e.g. common law rights) that the parties may possess.

ICSub

Clause 2.9 of the intermediate main contract conditions allows the architect/contract administrator to instruct the contractor not to carry out remedial works arising from the contractor's inaccurate setting out, and adds that if such an instruction is issued an appropriate deduction shall be made in the calculation of the ascertained final sum in respect of those errors not amended.

In addition, clause 2.30 of the intermediate main contract conditions allows the architect/contract administrator to instruct the contractor not to carry out defects, shrinkages or other faults that have been notified to the contractor on a defects list. If such an instruction is issued then the said clause 2.30 adds that an appropriate deduction shall be made in the calculation of the ascertained final sum in respect of the defects, shrinkages or other faults not made good.

Although not stated by either clause 2.9 or clause 2.30 of the main contract conditions, it is considered that, in the normal course of events, the 'appropriate deduction' would be based upon an abatement to the contractor's value for the works rather than on the cost of completing the defects, etc., by others.

In respect of the sub-contractor, clause 2.17 of ICSub/C notes that where an 'appropriate deduction' has been made under clause 2.9 or 2.30 of the main contract conditions, then, to the extent that such deduction is attributable to one of the following, the deduction (or an appropriate proportion of the deduction) shall be borne by the sub-contractor and shall be taken into account in the calculation of the final sub-contract sum, or shall be recoverable by the contractor from the sub-contractor as a debt:

- inaccurate setting out by the sub-contractor; or
- defects or other faults in the sub-contract works.

The fact that an appropriate proportion of the deduction is to be taken into account in the calculation of the final sub-contract sum may be considered to strengthen the argument that the deduction is to be an abatement to value rather than a set-off of costs for completion by others.

ICSub/D

All as ICSub above.

ICSub/NAM

All as ICSub above.

MWSub/D and ShortSub

The MWSub/D and the ShortSub do not state that a contractor can carry out defects for a sub-contractor and recover the costs incurred. However, clause 9.4 of the sub-contract conditions states that if a sub-contractor fails to comply within seven days to a written notice from the contractor requiring compliance with an instruction issued, the contractor may employ and pay other persons to carry out the work and all additional costs are recoverable from the sub-contractor.

SubSub

The SubSub does not state that a sub-contractor can carry out defects for a sub-subcontractor and recover the costs incurred. However, clause 9.4 of the sub-sub-contract conditions states that if a sub-subcontractor fails to comply within five days to a written notice from the sub-contractor requiring compliance with an instruction issued, the sub-contractor may employ and pay other persons to carry out the work and all additional costs are recoverable from the sub-subcontractor.

Sub-contractor's design documents

What 'as-built' design documents is a sub-contractor to provide?

DBSub

Clause 2.24 of DBSub/C requires the sub-contractor to provide to the contractor (at no charge), the following documents before practical completion of the sub-contract works (or such works in any relevant Section), which show or describe the sub-contractor's designed works as-built information, and which relate to the maintenance and operation of that portion, including any installations forming part of it.

The documents to be provided are:

- such design documents as may be specified in the sub-contract documents; and/or
- such design documents as the contractor may reasonably require.

The above requirement does not detract from the sub-contractor's obligations under clause 3.20.4 of DBSub/C in respect of the Health and Safety File.

Clearly it is in the interest of the sub-contractor to identifying what design documents are necessary and be in a position to issue this information *prior* to practical completion.

MPSub

MPSub has no express clause similar to clause 2.24 of DBSub/C; however, by reference to the definition of practical completion contained at clause 1 of MPSub, one of the stated requirements for practical completion is as follows: 'All "as-built" information and operating and maintenance information required by this sub-contract to be delivered at Practical Completion has been delivered.'

ICSub

ICSub/C has no equivalent clause to clause 2.24 of DBSub/C. ICSub is not intended for use where the sub-contractor is required to undertake design.

ICSub/D

Clause 2.20 of ICSub/D/C deals with the position regarding design documents but the position is otherwise all as set out under DBSub above.

ICSub/NAM

Clause 2.19 of ICSub/NAM deals with the position regarding NAM design documents but the position is otherwise all as set out under DBSub above.

MWSub/D

Clause 5.1.5 requires the sub-contractor to provide the contractor (as and when necessary, and without charge) with two copies of such drawings or details as are reasonably necessary to enable the contractor to comply with his obligations under the main contract. Therefore if, under the main contract, the contractor was liable to provide as-built drawings, then the sub-contractor would also be required to provide as-built drawings pursuant to clause 5.1.5.

ShortSub

Not applicable. The ShortSub is not intended for use where the sub-contractor is required to undertake design.

SubSub

SubSub has no equivalent clause to clause 2.24 of DBSub, and it is unlikely that the SubSub would be used in the case where a sub-subcontractor had any design input in any event.

Does the sub-contract deal with copyright in the sub-contractor's design documents?

DBSub

Clause 2.25 of DBSub/C deals with the rights/licences conferred upon the contractor and the employer in respect of the copyright in the sub-contractor's design documents.

It should be particularly noted that the listed rights are conditional upon the sub-contractor being paid all monies due under the sub-contract in full (see clause 2.25.2).

The copyright in all of the sub-contractor's design documents remain vested in the sub-contractor.

However, the contractor shall have an irrevocable, royalty-free, non-exclusive licence with the full right to sub-licence the employer, to copy and reproduce the sub-contractor's design documents for any purpose relating to the main contract works, including, without limitation, those listed below:

- the construction of the main contract works;
- the completion of the main contract works;
- the maintenance of the main contract works;
- the letting of the main contract works;
- the sale of the main contract works;
- the promotion of the main contract works;
- the advertisement of the main contract works;
- the reinstatement of the main contract works;
- the refurbishment of the main contract works;
- the repair of the main contract works.

The contractor also has an irrevocable, royalty-free, non-exclusive licence, with the full right to sub-licence the employer, to copy *but not to reproduce* (which must mean using the sub-contractor's design documents as the basis of a similar design issue) the sub-contractor's design documents for any *extension* of the main contract works.

Clause 2.25.3 makes it clear that the sub-contractor is not liable if the employer and/or the contractor uses the sub-contractor's design documents for any purpose other than that they were prepared for.

MPSub

Clause 19 of MPSub deals with copyright, and is similar, but not identical, to DBSub.

Clause 19.1 of MPSub again states that the copyright in all of the sub-contractor's design documents remains vested in the sub-contractor, with an irrevocable, royalty-free, non-exclusive licence granted to the contractor and the employer to use and copy the design documents for any purpose relating to the project, including, without limitation, those listed below:

- the design of the project;
- the execution of the project;
- the completion of the project;
- the maintenance of the project;
- the letting of the project;
- the sale of the project;
- the promotion of the project;
- the advertisement of the project;
- the reinstatement of the project;
- the refurbishment of the project;
- the repair of the project.

Also, as DBSub, the licence permits the contractor and the employer to copy and use the sub-contractor's design documents for the purposes of an extension of the project *but not to reproduce* (which must mean using the sub-contractor's design documents as the basis of a similar design issue) the sub-contractor's design documents for any *extension* of the project.

The employer is also, as DBSub, conferred the right to grant sub-licences.

Clause 19.2 deals with the position where the sub-contractor does not own copyrights in his design documents (i.e. copyright may belong to a third party who prepared the design on his behalf). In such circumstances the sub-contractor must procure from that third party a copyright licence conferring full title guarantee to the contractor and the employer in respect of that design document in the same terms required under clause 19.1.

Clause 19.3 makes clear that the sub-contractor is not liable if the employer and/or the contractor uses the sub-contractor's design documents for any purpose other than that permitted in the licence set out at clause 19.1.

ICSub

Not applicable. ICSub is not intended for use where the sub-contractor is required to undertake design.

ICSub/D

Clause 2.19 of ICSub/D/C is all as DBSub above except that the design documents provided under ICSub/C relate to the sub-contractor's designed portion under this sub-contract.

Again, the requirement at Clause 2.19 of ICSub/D/C does not detract from the sub-contractor's obligations under clause 3.18.4 of ICSub/D/C in respect of the Health and Safety File.

ICSub/NAM

Clause 2.18 of ICSub/NAM is all as clause 2.24 of DBSub above. Indeed, the clauses are identically drafted except that under ICSub/NAM the sub-contractor provides 'NAM design documents' (whereas DBSub uses the term 'design documents') which show or describe the relevant 'NAM designed works' (DBSub uses the term 'designed works').

The above requirement to provide this information does not detract from the sub-contractor's obligations under clause 3.18.4 of ICSub/NAM in respect of the Health and Safety File.

MWSub/D

The MWSub/D does not specifically address this issue. However, as clause 5.1.5 requires the sub-contractor to provide the contractor (as and when necessary, and without charge) with two copies of such drawings, details, calculations and information as are reasonably necessary to enable the contractor to comply with his obligations under the main contract, it may be the case that any copyright obligations that the contractor has under the main contract would be passed onto the sub-contractor through clause 5.1.5.

ShortSub

Not applicable. ShortSub is not intended for use where the sub-contractor is required to undertake design.

SubSub

Not applicable. It is fairly clear that the SubSub is not intended for use where the sub-contractor is required to undertake design.

Collateral warranties

Why do we have collateral warranties?

As noted earlier within this book, it is normally far easier to pursue an action in contract than it is to pursue an action in tort, and it is for this reason that the employer will normally require a collateral warranty to be in place.

A collateral warranty is a separate agreement that establishes a direct contractual link between a sub-contractor (particularly one that has a design liability) and the employer (and/or other interested parties).

Does the sub-contract oblige the sub-contractor to provide a warranty?

DBSub

Clause 2.26 of DBSub/C notes that where part 2 of the main contract particulars (as annexed to the schedule of information) and/or other sub-contract document provides for the giving by the sub-contractor of collateral warranties to the following, then the sub-contractor is to execute and deliver a collateral warranty within 14 days from receipt of a notice from the contractor that identifies the beneficiary and requires execution of a collateral warranty:

- a purchaser;
- a tenant or funder; and/or to
- the employer.

MPSub

MPSub does not expressly require the sub-contractor to provide a warranty.

Further, whilst the main contract (MPF) adopts the the Contracts (Rights of Third Parties) Act 1999 to confer rights on third parties against the main contractor, this does not apply to MPSub. Such third party rights are excluded under MPSub.

However, because the main contract (MPF) is intended for use on large projects, and many developers, funders, purchasers and tenants still want direct remedies against sub-contractors, it remains to be seen if MPSub's omission of collateral warranties and third party rights will result in bespoke non-standard amendments to MPSub being made to introduce such provisions into the sub-contract.

ICSub

Clause 2.18 of ICSub/C mirrors clause 2.26 of DBSub/C (see above).

ICSub/D

Clause 2.18 of ICSub/D/C mirrors clause 2.26 of DBSub/C (see above).

ICSub/NAM

ICSub/NAM/C does not contain an express clause obliging a sub-contractor to enter into a collateral warranty.

However, the notes at the beginning of the Intermediate Named Sub-contract Tender and Agreement records that consideration should be given to using an Intermediate Named Sub-contractor/Employer Agreement (ICSub/NAM/E) in conjunction with ICSub/NAM where the employer requires the named sub-contractor to give:

- undertakings to the employer in respect of the sub-contract works and any related design he is to carry out; and/or
- collateral warranties to purchasers/tenants or any funder in relation to the main contact works.

ICSub/NAM/E is a separate contractual agreement which is outside the scope of this book.

MWSub/D

MWSub/D does not include an express term requiring the sub-contractor to provide a collateral warranty.

ShortSub

ShortSub does not include an express term requiring the sub-contractor to provide a collateral warranty.

This is likely to be because the intention of its use is aimed at low risk packages, i.e. small sub-contract packages, or packages of straightforward content, both of which are fully designed. This form is unsuitable if the sub-contract works are of a complex nature, or if they involve design work.

SubSub

The SubSub does not include an express term requiring the sub-contractor to provide a warranty. Again, as previously noted, it is unlikely that this sub-contract form would be used for sub-subcontractors with a design liability.

What if the sub-contractor wishes to amend the terms of the collateral warranty?

DBSub

Clause 2.26.1 of DBSub/C states that if the sub-contractor requires any (reasonable) amendment to the proposed collateral warranty he is to notify the contractor within seven days of receipt of the proposed collateral warranty, of the proposed amendments. The contractor would then obviously need to seek approval for the proposed changes from the employer.

Pursuant to clause 2.26.1 of DBSub/C the sub-contractor is thereafter to execute and deliver the collateral warranty (with any approved amendments, as applicable) within seven days of being notified of the decision of the employer in respect of the request made for any amendments (irrespective of whether the requested amendments were agreed to or not).

ICSub

Clause 2.18.1 of ICSub/C mirrors clause 2.26.1 of DBSub/C (see above).

ICSub/D

Clause 2.18.1 of ICSub/D/C mirrors clause 2.26.1 of DBSub/C (see above).

ICSub/NAM, MWSub/D, ShortSub and SubSub

Not applicable.

Is the collateral warranty to be executed as a deed?

DBSub

Clause 2.26.2 of DBSub/C makes it clear that where the main contract is executed as a deed (i.e. where a 12-year limitation of liability period applies), the collateral warranties are to be executed as a deed.

Likewise, where the main contract is executed under hand (i.e. where a six-year limitation of liability period applies), the collateral warranties are to be executed under hand.

ICSub

Clause 2.18.2 of ICSub/C mirrors clause 2.26.2 of DBSub/C (see above).

ICSub/D

Clause 2.18.2 of ICSub/D/C mirrors clause 2.26.2 of DBSub/C (see above).

ICSub/NAM, MWSub/D, ShortSub and SubSub

Not applicable.

Does the sub-contract state that any particular collateral warranty is to be used?

DBSub (and ICSub and ICSub/D) do not stipulate any particular form of collateral warranty to use.

However, the DBSub guide notes that, if the collateral warranties are not the standard collateral warranties that the JCT produces (see below), then copies of the required forms of collateral warranties should be annexed to the schedule of information forming part of the DBSub/A, ICSub/A and ICSub/D/A.

Obviously, if the sub-contractor has any queries regarding, or requires any changes to the terms of the proposed collateral warranty (whether this be the standard warranty or an ad hoc warranty), these should be raised by the sub-contractor before the sub-contract is entered into, since, if they are not, it will be more difficult for the sub-contractor to insist on the required changes at a later date.

What provisions are typically found in a collateral warranty?

Collateral warranties normally consist of the following parts:

(1) The parties:

    (a) The warrantor – i.e. the sub-contractor in this case.
    (b) The warrantor's employer – i.e. the contractor in this case.
    (c) The beneficiary – i.e. the purchaser and/or a tenant; and/or the funder; and/or the employer.

(2) Recitals
These set the scene so as to enable anyone reading the document to understand its background. They should briefly describe the project, the roles of the parties and the reason for the warranty being entered into.

(3) The warranty
This is the core and most important part of the warranty. It usually begins with words such as 'The [warrantor] warrants as at and with effect from . . . [e]' and then goes on, by means of a number of sub-clauses, to set out the different obligations of the warrantor.

The most common warranty is that the warrantor has and will continue to perform his obligations under whichever contract the warranty is in respect of.

The normal effect of such warranties is to repeat for the beneficiary (e.g. the employer) the obligations that the warrantor (i.e. the sub-contractor) has to his client (i.e. the contractor).

(4) Supplementary warranties
Matters covered by supplementary conditions, which generally should not be required, include:

    (a) That the warrantor has exercised reasonable skill and care in the selection of materials (a list of deleterious materials is usually included).
    (b) That the works designed by the warrantor will satisfy any performance specification contained in the contract with the warrantor's client.
    (c) That the works will be of sound manufacture and workmanship and (for design and build contracts) will satisfy the requirements of the employer.

(5) Qualification
Collateral warranties are often qualified by a number of provisions. It is common to see a provision that the warrantor shall have no greater liability to the beneficiary, nor a liability of longer duration, than the warrantor would have had

if the beneficiary had been named as the warrantor's client in the principal contract to which the warranty refers. The purpose of such a qualification is to ensure that the warranty does not extend the underlying obligations. This is important to the warrantor as any extension of his liabilities may fall outside the limits of his insurances.

(6) Professional indemnity insurance

Collateral warranties usually contain provisions that require the warrantor to maintain professional indemnity insurance. The level of cover will be specified and it will also be specified that it shall be maintained for six or twelve years from practical completion (normally of the main contract works), provided that the insurance is available at commercially reasonable rates and on reasonable terms. This proviso is sufficient protection to the warrantor and it is not necessary to qualify the requirement by asking it to use best or reasonable endeavours to maintain the insurance. The warrantor will often be required to produce evidence that the insurance is being maintained and renewed.

One issue that can arise in relation to professional indemnity insurance is whether the insurance covers each and every claim or if it is on an aggregate basis for all claims within any year. Most beneficiaries will be advised not to accept aggregate insurance, as there is a danger that there will be no cover for some claims made in the same year as others because of an aggregate limit in value being reached.

It should also be borne in mind that many professional indemnity insurance policies are provided on a 'claims made' basis which means that the applicable insurance is that which covers the period when the claim was made, not that which covers the period when the alleged defect occurred.

In respect of professional indemnity insurance it is in the interest of both the warrantor and the beneficiary that the terms of the warranty fall within the terms of the warrantor's policy. In this regard it should be noted that the purpose of the professional indemnity insurance is to provide a fund of monies upon which the beneficiary can call in the event that a claim needs to be made. If the terms of the warranty do not fall within the terms of the warrantor's policy, that fund of money will not exist.

A reality of the construction industry is that many smaller contractors and professional consultants have very limited assets and, in the absence of satisfactory insurance, would be unable to meet any significant claim under a warranty.

It is, therefore, in the interests of both the warrantor (i.e. the sub-contractor) and the beneficiary that the terms of the warranty fall within the risks usually covered by a professional indemnity insurance policy.

The warrantor obviously does not wish to be faced with claims that he cannot meet (or which, if imposed, would put him out of business), and the beneficiary seeks the comfort of knowing that a source of accessible funds is available. Common sense dictates that it is necessary to strike a balance between these two demands.

(7) Step-in rights

Such provisions are popular where the beneficiary is a funder or freehold purchaser. The purpose of 'step-in rights' is to allow the beneficiary to take over the contract with the warrantor in the event that either the warrantor has a right to terminate that contract or if the beneficiary's own contract has been breached.

(8) Copyright licence

Collateral warranties, as a general rule, contain an irrevocable royalty-free copyright provision in favour of the beneficiary. Such a licence enables the beneficiary to reproduce and use the warrantor's design information. The rights of the beneficiary are normally limited to the use of the information for purposes related to the development.

(9) Contracts (Rights of Third Parties) Act 1999

The Contracts (Rights of Third Parties) Act is almost invariably excluded in most standard and bespoke collateral warranty forms.

There are two other important points to be aware of:

(1) An important factor with collateral warranties, from the warrantor's viewpoint, is that he should have no greater liability to the beneficiary than he has to his client and he has the same defences against a claim from the beneficiary as he would have against his client.
(2) It is also important to note that the terms of a collateral warranty that a party to a contract is required to give to a third party must be known and/or available at the time the principal contract is formed. In the event that this is not the case then there is unlikely to be sufficient certainty over the particular terms requiring the giving of a collateral warranty, making such a requirement difficult to enforce.

Do the JCT produce any standard forms of collateral warranty for use with JCT sub-contracts?

In respect of the DBSub, ICSub and ICSub/D, as noted above, the normal situation would be that the standard collateral warranties that the JCT produces would be used. In which case, many of the above issues (where applicable) would be automatically covered.

The two standard collateral warranties produced by the JCT are:

- the Sub-contractor Collateral Warranty for a Funder (SCWa/F);
- the Sub-contractor Collateral Warranty for a Purchaser or Tenant (SCWa/P&T)

The said collateral warranties are each provided with a three-page Guidance Notes section.

It should be noted that, in respect of the JCT standard collateral warranties, the section of the collateral warranty that relates to professional insurance should be deleted if the sub-contractor has no design liability.

Chapter 7
Control of the Sub-contract Works

Assignment and sub-letting

What is assignment?

A contract contains both benefits and burdens for both parties; one party's benefit is usually the other party's burden.

In law, the general principle is that benefits under a contract may be assigned (that is, legally transferred from the original contracting party to a third party) without the consent of the other party, but burdens under a contract may not. Therefore, without the consent of the contractor, a sub-contractor could not normally assign his liability to complete, as this is a burden, not a benefit[1].

Strictly speaking, if both the benefit and the burden of a contract are legally transferred to a third party, then that is a novation, and not an assignment. A novation is a tripartite agreement by which an existing contract between party X and party Y is legally discharged and a new contract, usually on the same terms as the first contract, is made between party X and (third) party Z.

Can the sub-contractor assign the sub-contract?

DBSub

Clause 3.1 of the sub-contract states that the sub-contractor shall not assign the sub-contract or any rights thereunder without the written consent of the contractor.

Against the above background, the reference in clause 3.1 of the sub-contract that the sub-contractor shall not assign the sub-contract obviously does not, in truth, make a great deal of sense.

However, in the two consolidated appeals of *Linden Gardens* v. *Lenesta Sludge* and *St Martins* v. *Sir Robert McAlpine*[2], Lord Browne-Wilkinson said:

> 'Although it is true that the phrase "assign this contract" is not strictly accurate, lawyers frequently use those words inaccurately to describe an assignment of the benefit of a contract since every lawyer knows that the burden of a contract cannot be assigned.'

---

[1] For this principle, see *Nokes v. Doncaster Amalgamated Collieries Ltd* [1940] AC 1014.
[2] *Linden Gardens v. Lenesta Sludge Disposals* [1993] 3 WLR 408 and *St Martins Property Corporation Ltd v. Sir Robert McAlpine Ltd* [1993] 3 All ER 417.

In the light of the above, what clause 3.1 of the sub-contract does is to make clear by express terms that, irrespective of the general principle of law, no benefit (or right) (in addition to any burden) under the sub-contract may be assigned to any third party without the written consent of the contractor.

MPSub

Assignment is dealt with at clause 38 as follows:

Clause 38.1 of the sub-contract deals with assignment by the sub-contractor and states that the sub-contractor may not assign either the benefit or burden of the sub-contract without the consent of the contractor. Whilst drafted differently to DBSub above, the same comments apply here equally. Note, however, that under the MPSub, consent is not qualified as needing to be in writing.

Clause 38.2 also deals with assignment by the contractor. It provides that the contractor may assign the benefit of the sub-contract at any time without the consent of the sub-contractor. This express clause appears to affirm the contractor's rights under the general law.

ICSub

All as DBSub above.

ICSub/D

All as DBSub above.

ICSub/NAM

All as DBSub above.

MWSub/D

Clause 5.6 states that the sub-contractor shall not make any assignment of the benefit of the sub-contract without the prior written consent of the contractor.

ShortSub

Clause 5.5 of ShortSub is as per MWSub/D above.

SubSub

As ShortSub.

Does the sub-contractor need to get consent to sub-let any works?

Subject to an express term contained in a contract, the general law usually permits a sub-contractor to sub-let his contractual obligations to a third party for their performance (termed vicarious performance) on the strict basis that the sub-contractor still remains fully liable and responsible to the contractor under the sub-contract for all his contractual obligations.

When the above position applies it does not matter who performs the work, provided that they have the relevant skill and expertise to do so.

However, if the obligation to be undertaken is of a personal nature (i.e. the particular skill and expertise of an individual sub-contractor is being relied upon to perform the contractual obligations) then the general law will not permit sub-letting.

DBSub

Clause 3.2 of DBSub/C states that the sub-contractor shall not, without the written consent of the contractor, sub-let the whole or any part of the sub-contract works (clause 3.2.1), or sub-let the design for the sub-contractor's designed portion (clause 3.2.2).

In neither instance is the contractor's consent to be unreasonably delayed or withheld.

It should be noted that these clauses do not say that the contractor's consent is *not* to be delayed or withheld, but that such consent is not to be *unreasonably* delayed or withheld.

What is unreasonable will depend on the particular individual circumstances that exist in relation to the proposed sub-subcontractor, i.e. it will be considered on a case by case basis. There are no hard and fast rules. Typically, some influencing circumstances may include, but not be limited to: the sub-subcontractor's relevant competence and technical expertise; size of the organisation and available resources to do the work efficiently and on time; their financial standing; track record; safety record; general reputation; etc.

By way of example, if a contractor had serious doubts about the ability of a proposed sub-subcontractor (i.e. a proposed sub-contractor to the sub-contractor) to deal with a particular aspect of the sub-contract works, or of the sub-contractor's designed portion, then it may be entirely reasonable for him to delay his consent to the use of the proposed sub-subcontractor until he is provided with documentary evidence (perhaps references, etc.) that he may have requested the sub-contractor to provide to him, before he gives his consent to the use of the proposed sub-subcontractor.

Alternatively, if the contractor had received or was receiving a very poor service from the proposed sub-subcontractor on another project, and therefore had serious doubts that the proposed sub-subcontractor would be capable of carrying out the sub-contract works (or a part thereof) to the required standard, then it may be entirely reasonable for him to withhold his consent to the use of the proposed sub-subcontractor entirely.

Note also that clause 3.2 makes it quite clear that, even if the contractor does give his written consent to the sub-contractor to sub-let the whole or any part of the sub-contract works, the sub-contractor is to remain wholly responsible for carrying out and completing the sub-contract works in all respects in accordance with the sub-contract.

In addition to the foregoing, note that the contractor's consent to any sub-letting of design shall not in any way affect the obligations of the sub-contractor under clause 2.13.1 of DBSub/C or any other provision of DBSub.

MPSub

The sub-contract contains no express provisions regarding sub-letting. The general legal principles noted above would therefore apply, i.e. unless of a personal nature, it does not matter if another third party undertakes the work provided always that the sub-contractor remains responsible to the contractor for such work.

ICSub

Clause 3.2 states that the sub-contractor shall not, without the written consent of the contractor, sub-let the whole or any part of the sub-contract works. It also states that the contractor's consent is not to be unreasonably delayed or withheld.

It should be noted that this clause does not state that the contractor's consent is *not* to be delayed or withheld, merely that such consent is not to be *unreasonably* delayed or withheld.

What is unreasonable will depend on the particular individual circumstances that exist in relation to the proposed sub-subcontractor, i.e. it will be considered on a case by case basis. There are no hard and fast rules. Typically, some influencing circumstances may include, but not be limited to, the sub-subcontractor's relevant competence and technical expertise; size of the organisation and available resources to do the work efficiently and on time; their financial standing; track record; safety record; general reputation; etc.

Note also that clause 3.2 makes it clear that, even if the contractor does give his written consent to the sub-contractor to sub-let the whole or any part of the sub-contract works, the sub-contractor is to remain wholly responsible for carrying out and completing the sub-contract works in all respects in accordance with the sub-contract.

ICSub/D

All as DBSub above.

ICSub/NAM

All as DBSub above, except for the reference to clause 2.13.1 found within clause 3.2 of DBSub/C.

MWSub/D

Clause 5.6 states that the sub-contractor shall not sub-let any of the sub-contract works without the prior written consent of the contractor. It is interesting to note that the clause does not state that the contractor's consent must not be unreasonably withheld or delayed.

ShortSub

Clause 5.5 of ShortSub is as per MWSub/D above.

SubSub

As ShortSub.

Are there any conditions for sub-letting?

DBSub

Clause 3.3 provides only two specific conditions of sub-letting. These are:

- Clause 3.3.1: the sub-subcontractor's employment under the sub-subcontract shall terminate immediately upon the termination (for any reason) of the sub-contractor's employment under the sub-contract.
- Clause 3.3.2: the sub-contractor shall provide for simple interest to be paid to the sub-subcontractor at the level of, and subject to the terms equivalent to those of clauses 4.10.5 and 4.12.4 of DBSub/C in respect of any payment, or any part of it, not properly paid to the sub-subcontractor by the final date for payment. The interest rate stated in the DBSub being a rate 5% per annum above the official dealing rate of the Bank of England current at the date that a payment due under the sub-contract becomes overdue.

Apart from the two qualifications that must be included in any sub-letting arrangement, a sub-contractor is at liberty to agree any other terms (that are enforceable) with his sub-subcontractor as he wishes.

Whilst the JCT has published a sub-subcontract form for use in such a situation (denoted as SubSub) the sub-contractor is not obliged to use that particular form when sub-letting.

MPSub

No specific conditions for sub-letting are stated.

ICSub

All as DBSub above.

ICSub/D

All as DBSub above.

ICSub/NAM

All as DBSub above.

MWSub/D

No specific conditions of sub-letting are stated.

ShortSub

No specific conditions of sub-letting are stated.

SubSub

As ShortSub.

Person-in-charge

What is the obligation regarding a person-in-charge?

DBSub

Clause 3.8 requires the sub-contractor to ensure that at all times during the execution of the sub-contract works he has on site a competent person-in-charge. Clause 3.8 also adds that any directions given by the contractor to that competent person-in-charge will be deemed to have been issued to the sub-contractor.

MPSub

This is dealt with at clause 25.

Clause 25.1 requires both parties (the sub-contractor and the contractor) to have at all times a person appointed to act as their respective representatives with authority to exercise all of his powers and functions under the sub-contract. The appointment of such a representative is to take effect upon (and can also revoked by) notification being given to the other party.

Clause 25.2 states that both the contractor and the employer may appoint advisers in connection with the project, and the sub-contractor may be notified of their appointment and role by the contractor. This clause also states that the sub-contractor must cooperate with such advisers, but immediately qualifies this by stating that such advisers have no authority to act on behalf of the contractor. Sub-contractors need to be alive to this when undertaking such cooperation, and to bear in mind that the adviser has no authority to instruct changes to the sub-contract.

ICSub

Clause 3.7 of ICSub is identical to clause 3.8 of DBSub. Therefore, see DBSub above for comments on this clause.

ICSub/D

All as ICSub above.

ICSub/NAM

All as ICSub above.

MWSub/D, ShortSub and SubSub

There is no provision for a person-in-charge.

Access provided by the sub-contractor

Must the sub-contractor provide access for the contractor, employer and architect/contract administrator?

DBSub

Clause 3.9 obliges the sub-contractor to allow access at all reasonable times to the contractor, the employer and to any person authorised by either the contractor or the employer, to any work which is being prepared for or is to be utilised in the sub-contract works.

The only restrictions to this requirement being those that are necessary to protect any proprietary rights. This latter situation may arise where a product involves particular trade secrets and/or the manufacturer wishes to prevent the manufacturing process from becoming known to third parties.

MPSub

No specific requirements for access are stated.

ICSub

Clause 3.8 obliges the sub-contractor to allow access at all reasonable times to the contractor, to the architect/contract administrator, and to any person authorised by either the contractor or the architect/contract administrator, to any work which is being prepared for or is to be utilised in the sub-contract works.

However, the only restrictions to the sub-contractor allowing access are stated as being those necessary for the sub-contractor to protect any proprietary rights (see comments under DBSub, above).

ICSub/D

All as ICSub above.

ICSub/NAM

All as ICSub above.

MWSub/D

There is no express obligation on the sub-contractor to provide access. This is probably due to the fact that MWSub/D is drafted as a simple, straightforward arrangement intended for a small package of work or one that is of straightforward content with low risk involved.

Notwithstanding this, clause 6 obliges the contractor to provide the sub-contractor with sufficient access to the site and not to hinder or prevent the sub-contractor in performing his obligations under the sub-contract.

ShortSub

As MWSub/D above.

SubSub

There is no express obligation on the sub-subcontractor to provide access.

Notwithstanding this, clause 6 obliges the sub-contractor to provide the sub-subcontractor with sufficient access to the site and not to hinder or prevent the sub-subcontractor in performing his obligations under the sub-subcontract.

Opening up the works and remedial measures

The procedures to be followed in respect of the opening up of works and the remedial measures to works that may be required after the works have been opened up are set out below.

MWSub/D, ShortSub and SubSub

The MWSub/D, ShortSub and SubSub do not contain any express provisions dealing with this issue. Clause 10.1, however, obliges the sub-contractor (sub-subcontractor under SubSub) to comply with any reasonable written instruction and, in practice, this may be how such matters are dealt with (where applicable). Alternatively, the sub-contracts could be amended to incorporate express provisions, or set procedures could possibly be incorporated by way of a sub-contract document inserted at:

- the second recital of MWSub/D;
- the second recital of ShortSub;
- The third recital of SubSub.

Can the contractor issue instructions/directions to open up and test?

DBSub

Clause 3.10 of DBSub/C allows the contractor to issue directions requiring the sub-contractor to:

- open up for inspection any work covered up;
- arrange for or carry out any test of any materials and goods (whether or not already incorporated in the sub-contract works); or
- arrange for or carry out any test of any executed work.

Is the sub-contractor paid for complying with same?
Clause 3.10 of DBSub/C states that the sub-contractor is to be paid the cost of any such opening up or testing directed (including the cost of any resultant making good) except where the opening up or testing:

- is included in the sub-contract sum; or
- shows that the materials, goods or works were not in accordance with the sub-contract requirements.

In the above circumstances such cost are borne by the sub-contractor.

If the work is not defective the sub-contractor may also be entitled to an extension of time (clause 2.19.2.3) and loss and/or expense (clause 4.20.2.3).

MPSub

MPSub clause 26.1 is drafted on a similar basis to clause 3.10 of DBSub/C, namely the contractor's instruction to open up or test is a change unless:

- it is provided for in the contract; or
- it shows that the works, materials or goods are not as per the contract.

ICSub

Clause 3.9 of ICSub/C is all as clause 3.10 of DBSub above.

ICSub/D

Clause 3.9 of ICSub/D/C is all as clause 3.10 of DBSub above.

ICSub/NAM

Clause 3.9 of ICSub/NAM/C is all as clause 3.10 of DBSub above.

What if work is found not to be in accordance with the sub-contract ('non-compliant work')

DBSub

If any work, materials or goods are found not to be in accordance with the sub-contract (termed 'non-compliant work') the contractor, in addition to his other powers, may:

- Under clause 3.11.1 of DBSub/C, issue directions requiring the removal from site, or rectification of all or any of the non-compliant work, with or without similar instructions having been issued under clause 3.13 of the main contract conditions. Note that if a similar instruction under clause 3.13 of the main contract conditions:

  — Has not been issued:
    The contractor must, before issuing the direction, consult with the sub-contractor and give due regard to the Sub-contract Code of Practice set out in schedule 1 to DBSub/C. The Sub-contract Code of Practice is dealt with later in this chapter.
  — Has been issued:
    The contractor's direction issued under this clause must state this.

- Under clause 3.11.2 of DBSub/C, after consultation with the sub-contractor, issue directions requiring a variation as is reasonably necessary in consequence of a clause 3.11.1 direction. Clause 3.11.2 states that provided such directions are reasonably necessary the sub-contractor will *not* be entitled to payment, or an extension of time, as a consequence of the direction.

- Having due regard to the Sub-contract Code of Practice set out in schedule 1 to the DBSub/C, issue such directions under clause 3.10 of DBSub/C to open up for inspection or to test (as is reasonable in all of the circumstances) to establish to the contractor's reasonable satisfaction the likelihood or extent (as appropriate to the circumstances) of any further similar non-compliance (refer to clause 3.11.3 of DBSub/C). Provided such directions are reasonable, the sub-contractor will not be paid for the opening up, testing, or subsequent making good irrespective of whether the opening up and/or testing shows the work to be compliant or non-compliant. If the works are found to be compliant, the sub-contractor may, however, be entitled to an extension of time (i.e. if works directed delay the completion date of the sub-contract works or any Section) as per clause 2.19.2.3.

It should be noted that the Sub-contract Code of Practice is enclosed at schedule 1 of DBSub.

MPSub

Where works, materials or goods are not in accordance with the sub-contract, clause 26 of MPSub provides that the contractor may:

- Instruct that they be removed from the site, either totally or partially (clause 26.2.1 of MPSub).
- Under clause 26.2 of MPSub, instruct, following consultation with the sub-contractor, their use for the sub-contract works on the basis that:
  — the sub-contractor becomes liable to pay the contractor an appropriate amount (calculated in accordance with the prices and principles set out in the pricing document); and
  — the sub-contractor has no entitlement to an extension of time or loss and/or expense.
- Instruct, after consultation with the sub-contractor, such further works as are necessary as a consequence of the removal or use of the non-conforming work, materials or goods (clause 26.2.3 of MPSub).
- In order to establish to the contractor's reasonable satisfaction that other similar work, materials or goods are in accordance with the sub-contract, under clause 26.2.4 of MPSub the contractor may instruct such further opening up, testing or inspection as is reasonable.

Clause 26.3 of MPSub provides that no instruction issued under 26.2 will constitute a change.

ICSub

ICSub deals with this slightly differently.

Obligation on sub-contractor to notify
Clause 3.10.1 of ICSub/C states that if during the execution of the sub-contract works the sub-contractor discovers any work, materials or goods that are not in accordance with the sub-contract, he must 'forthwith' write to the contractor

advising the immediate action he proposes (at no cost to the contractor) to establish whether any similar problems exist in any work already executed or materials or goods already supplied (whether or not incorporated in the sub-contract works).

If the contractor (see clauses 3.10.1 to 3.10.3):

- has not received the sub-contractor's above statement within five days of discovery;
- is not satisfied with the sub-contractor's proposed action; or
- cannot wait for the sub-contractor's written proposals due to safety consider-ations or statutory obligations;

he can issue directions requiring the sub-contractor (who must 'forthwith comply' at no cost to the contractor) to open up for inspection any work covered up or test any materials, goods or executed works in order to establish that there is no similar failure, including making good thereafter.

What if the sub-contractor objects to a direction under clause 3.10.1
The contractor's direction under clause 3.10.1 as to the nature or the extent of the opening up for inspection or testing should be reasonable. What is reasonable will depend on the particular circumstances of each case.

The sub-contractor's right to object is dealt with at clause 3.10.2 and clause 3.10.3 of ICSub/C:

- The sub-contractor's objection must be in writing (with reasons) and issued within seven days of the contractor's direction.
- Irrespective of his objection the sub-contractor must comply with the direction.
- If within seven days from receiving the sub-contractor's objection the contractor does not respond in writing, either withdrawing or modifying his direction to remove the sub-contractor's objection, then any dispute or difference can be referred to a person appointed under the sub-contract dispute resolution procedures.
- This person will subsequently decide whether the contractor's direction was reasonable and if not, will decide upon payment, if any, or extension of time, if any, due to the sub-contractor in respect of compliance with the direction.

Clause 3.11.1 gives the contractor the authority to issue directions specifically requiring the removal from site of any work, materials or goods that are not in accordance with the sub-contract.

ICSub/D

All as ICSub above.

ICSub/NAM

All as ICSub above.

What about instructions or deductions under the main contract?

DBSub

Clause 3.13.2 of the DB main contract
If any work, materials or goods are not in accordance with the contract then the employer may consult with the contractor and, with the agreement of the employer, may allow all or any of such work, materials or goods to remain; and an appropriate deduction would then be made to the ascertained final (main contract) sum.

Clause 3.12 of DBSub/C
If under clause 3.13.2 of the main contract conditions:

- The consultation relates (in whole or in part) to the sub-contract works the contractor must immediately consult with the sub-contractor (clause 3.12.1 of DBSub/C). This appears to imply that this should be before the employer makes his decision.
- When the employer allows all or any non-compliant work to remain, the contractor must notify the sub-contractor in writing and (where applicable) an appropriate deduction would then be made to the final sub-contract sum or, if this was not possible, then the amount of such deduction is recoverable by the contractor from the sub-contractor as a debt (clause 3.12.2 of DBSub/C).

Clause 3.12.3 of DBSub/C requires the sub-contractor to comply with any contractor's directions requiring a variation as are reasonably necessary in consequence of the employer allowing all or any non-compliant work to remain. Like clause 3.11.2 of DBSub/C (discussed earlier in this chapter) provided such directions are reasonably necessary the sub-contractor is neither paid for the variation nor entitled to an extension of time.

MPSub

MPSub does not refer to any instructions or deductions under the main contract under clause 26 of MPSub. As discussed earlier, clause 26.2 deals with non-compliant work being allowed to remain, subject to an appropriate deduction.

ICSub

ICSub does not have an equivalent clause to clause 3.12 of DBSub and ICSub has no express provisions permitting non-compliant work to remain.

ICSub/D

All as ICSub above.

ICSub/NAM

All as ICSub above.

What about workmanship issues?

DBSub

Clause 3.13 of DBSub/C deals with the situation where 'workmanship' rather than simply 'work' is not in accordance with the sub-contract.

Accordingly, where there has been a failure to carry out work:

- in a proper and workmanlike manner; and/or
- in accordance with the Construction Phase Plan (previously known as the Health and Safety Plan);

the contractor, in addition to his other powers, may, after consultation with the sub-contractor, issue such directions (whether requiring a variation or otherwise) as are in consequence reasonably necessary. Where the failure to comply is the subject of an instruction issued by the employer under clause 3.14 of the main contract conditions, the contractor must inform the sub-contractor of this fact when issuing his directions (clause 3.13.2 of DBSub/C).

Provided such directions are reasonably necessary, the sub-contractor is not entitled to payment or an extension of time or loss and expense in respect of the variation.

MPSub

Whilst MPSub contains no express clause solely addressing workmanship like clause 3.13 of DBSub/C above, such issues would, it is submitted, be addressed within clause 26 of MPSub (dealt with earlier within this chapter).

ICSub

If there is a failure to carry out work in a proper and workmanlike manner, clause 3.11.2 of ICSub/C permits the sub-contractor to issue such directions (whether requiring a variation or otherwise) as are in consequence reasonably necessary.

The sub-contractor is to comply with these directions without any cost to the contractor.

ICSub/D

All as ICSub above.

ICSub/NAM

All as ICSub above.

What about non-compliant work by others?

DBSub

Under clause 3.14.1 of DBSub/C the sub-contractor must comply with the contractor's directions to take down, re-execute, re-fix and/or re-supply any sub-contract work that has been properly executed, or materials or goods properly fixed or supplied under the sub-contract where necessitated due the removal or rectification of non-compliant work by others (i.e. by the contractor or *another* sub-contractor). Such directions, however, must be issued before the date of practical completion of the sub-contract works or of such works in any relevant Section.

Where such directions are issued then:

- in line with clause 3.14.1 of DBSub/C, a copy of the direction must be sent to the sub-contractor whose non-compliant work or failure to comply gave rise to the direction being issued; and
- under clause 3.14.2 of DBSub/C the sub-contractor is entitled to:
  - payment for such work directed on the basis of a fair valuation;
  - an extension of time (assuming that the works in question caused a delay to the completion date of the sub-contract works or such works in any Section); and
  - to recover any consequential (and proven) loss and expense.

MPSub

MPSub does not have an equivalent clause to clause 3.14 of DBSub/C.

ICSub

Clause 3.12 of ICSub/C contains the same provisions as clause 3.14 of DBSub/C above.

ICSub/D

All as ICSub above.

ICSub/NAM

All as ICSub above.

Indemnity by the sub-contractor

DBSub

In the situation where there is non-compliant work by the sub-contractor or there is a failure by him to carry out work in accordance with the sub-contract, clause 3.15 of DBSub/C requires the sub-contractor to indemnify the contractor in respect of any liability and any costs the contractor incurs as a direct result of such non-compliance and/or such failure by the sub-contractor to carry out work in accordance with the sub-contract.

MPSub

MPSub does not have an express equivalent clause to clause 3.15 of DBSub/C.

ICSub

Clause 3.13 of ICSub/C contains the same provisions as clause 3.15 of DBSub/C above.

ICSub/D

All as ICSub above.

ICSub/NAM

All as ICSub above.

Schedule 1 – the Sub-contract Code of Practice

DBSub

Only DBSub refers to the Sub-contract Code of Practice and this is referred to when directions in respect of clause 3.11.1 and 3.11.3 of DBSub/C are being considered by the contractor. Although the contractor is not bound to follow the code of practice, if a dispute arose, his case would be greatly weakened unless he could show that the terms of the code had been fully considered before his directions were issued. The purpose of the code is to ensure that directions under clause 3.11.1 and 3.11.3 of DBSub/C are not issued unreasonably.

A review of the sub-contract code of practice is outside the scope of this book. However, if such a review is required, reference should be made to Peter Barnes' book titled *The JCT 05 Standard Building Sub-contract* published by Blackwell Publishing.

Attendance and site conduct

What attendances are to be provided?

DBSub

By the sub-contractor?
Clause 3.16.1 of DBSub/C states that all items of attendance, other than those listed either:

- in item 6 of the sub-contract particulars (DBSub/A); and/or
- in a numbered document referred to in sub-contract particulars (item 16 of DBSub/A);

are required to be provided by the sub-contractor.

By the contractor?
The sub-contract particulars at item 6.1 of DBSub/A provides a basic list of attendance items that will be provided by the contractor free of charge. This list is as follows:

- Provision and erection of all necessary scaffolding and scaffold boards for work over 3.30 metres.
  Below this height is the responsibility of the sub-contractor.

- Provision of reasonable measures to prevent access by unauthorised persons.
  This is a new matter introduced in April 2007 by amendment 1 to DBSub. The corresponding footnote to this item advises that this item is to be amended as appropriate following due consideration being given to the Work at Height Regulations 2005.

- Space for temporary site accommodation.
  This does not, however, cover the accommodation itself – see clause 3.17 of DBSub/C.

- Single phase supply of electricity at 240v for temporary site accommodation.
  The actual connection from that point of supply to the temporary site accommodation is to be carried out by (or at the expense of) the sub-contractor (see clause 3.17 of DBSub/C).

- Single phase supply of electricity at 110v for tools and temporary lighting.
  All the contractor is required to do is provide a point of supply of 110 volt electricity for the sub-contractor to connect to. Distribution of the electricity to the area of work, etc., is the responsibility of the sub-contractor.
  It is preferable for the location of the points of electricity supply to be agreed in advance and for that agreement to be recorded on a numbered document referred to in sub-contract particulars item 16 of DBSub/A.

- Water supplied at the points identified in the sub-contract documents for temporary site accommodation and for the carrying out of the sub-contract works.
  The comments under the previous item apply equally here.

- Use of mess rooms.
  That is usually a site canteen.

- Use of sanitary accommodation.
  That is site toilets.

- Use of welfare facilities.
  For example, drying rooms, shower rooms, and/or first aid rooms.

- All reasonable non-exclusive use of hoisting facilities that the contractor has on site at the time the sub-contract works are being carried out.
  A sub-contractor should particularly note that:

  — he only has reasonable and *non-exclusive* use of hoisting facilities;
  — he has no control over the type of hoisting facilities that will be provided by the contractor.
    For this reason it is usually sensible to agree with the contractor:
    — a hoisting/lifting register on a daily or weekly basis permitting the time periods for any hoisting required by the sub-contractor so that it can be planned in advance;
    — the hoisting facilities that will be provided (agreed in advance and recorded on a numbered document within sub-contract particulars item 16 of DBSub/A).

- The benefit of all reasonable watching to be provided by the contractor under the main contract.

The fact that watching may be provided by the contractor under the main contract does not absolve the sub-contractor from his liability for the loss or damage of his unfixed site materials.

This list at item 6.1 of the sub-contract particulars is to be amended (as appropriate) and added to (if appropriate).

If there is insufficient room on DBSub/A for further attendance items to be listed, further sheets should be used and should be annexed to the DBSub/A as a numbered document.

It is important that the list of attendance items under item 6.1 is carefully considered and correctly amended, because all attendances (other than those finally listed) are to be provided by the sub-contractor.

What about other attendance items not listed in DBSub/A?

In addition to the attendance items listed above, the following are some of the more common attendance items that the contractor and the sub-contractor may wish to discuss and/or agree are to be provided by the contractor; with any such agreement being recorded on a numbered document referred to in sub-contract particulars item 16 of DBSub/A. This is not intended to be a conclusive list, and the particular items of attendance will vary according to the particular circumstances.

- Responsibility for setting out of the sub-contract works.
  Commonly the contractor provides main grid lines and site datum levels from which the sub-contractor sets out the sub-contract works.

- Unloading/distribution of materials.
  These are normally the responsibility of the sub-contractor. However, in certain circumstances (for example, materials being delivered before a sub-contractor has a presence on site) it may be agreed that the contractor will unload (but will accept no liability for) those particular materials or goods, or distribute the materials to (or near to) the area of work.

- Space for storage of materials on site.

- Covered/secure store.

- Steps/trestles/towers.
  For the avoidance of any possible future disagreement it is wise for the sub-contractor and the contractor to agree which party will provide, erect and remove, steps, trestles and towers (for work that is not more than 3.30 metres high) that are necessary for the execution of the sub-contract works. If this matter is not discussed then the default position is that it is the sub-contractor's liability to provide, erect and remove the said steps, trestles and towers.

- Personal Protective Equipment (PPE).
  This is normally the responsibility of the sub-contractor to ensure that his operatives have adequate and suitable PPE. However, if very special PPE is required, the contractor may agree to provide this.

- Protection of sub-contractor's goods, materials or works that are not fully and finally incorporated into the main contract works.
  This is normally the responsibility of the sub-contractor. Accordingly, if the contractor agrees to provide any protection materials or carry out any protection work, this needs to be confirmed as an additional attendance item.

- Responsibility for the control of or removal of water.
  This problem can often be overlooked, but should be discussed and agreed to avoid any future disagreements. This can be a particular problem in situations such as:

  — diamond drilling (where water is used to cool the cutting tool, etc.), particularly where the diamond drilling is carried out in an otherwise finished area of the works;
  — grit or sand blasting (with a water base), where the excess water needs to be removed (without the grit or sand blocking up the drain runs);
  — temporary roofs, where consideration is often not given to the discharge of rainwater into gutters, rainwater down-pipes and drains.

- Materials, fuel, power and/or water for commissioning and testing.
  Item 6.1 of the sub-contract particulars specifically notes that materials, fuel, power and/or water for commissioning and testing are excluded as an attendance item that a contractor would normally be expected to provide to a sub-contractor free of charge.
  The sub-contractor may therefore need to allow for the necessary materials, fuel, power and/or water for commissioning and testing within his tender.
  A point that is sometimes overlooked is that three-phase electricity is sometimes required for testing and commissioning (e.g. for lift installations, etc.) *before* the permanent supply has been connected. This factor can have major programme implications.

MPSub

MPSub deals with this in a virtually identical manner to DBSub above.

Clause 14.1 of MPSub states that the contractor shall provide free of charge all items of attendance identified in the sub-contract particulars. Any attendances not identified in the sub-contract particulars are the responsibility of the sub-contractor.

The sub-contract particulars at the clause 14.1 entry provides a basic list of common possible attendance items (similar to DBSub's discussed above) which is to be amended and/or added to (as appropriate). If there is insufficient room, MPSub states that further sheets should be used to record such attendances, being signed or initialled by the parties and annexed to MPSub.

Unlike DBSub, however, MPSub's list has an additional note column enabling the parties to record against each corresponding attendance item any particular agreement or limitations in respect of any listed item of attendance or the duration that a particular item of attendance will be provided for.

ICSub

All as DBSub/C above (except that clause 3.14 applies rather than clause 3.16 of DBSub).

ICSub/D

All as ICSub above.

ICSub/NAM

All as ICSub above.

MWSub/D

Clause 5.5 deals with attendances, and says that except for any attendances specifically identified in the sub-contract documents (which the contractor provides free of charge) everything necessary to enable the sub-contractor to execute the sub-contract works is to be provided by the sub-contractor.

Accordingly, if there are any attendances to be provided by the contractor these must be recorded under the second recital under the entry 'further documents forming part of this sub-contract', and may include scaffolding, use of on site facilities, water, electricity and any other applicable items.

ShortSub

Clause 5.4 of ShortSub is as per clause 5.5 of MWSub/D above.

SubSub

SubSub is similar to ShortSub. All attendances are provided by the sub-subcontractor except for any specific attendances recorded in the sub-subcontract documents as being provided, free of charge, by 'others'. The third recital is where further sub-subcontract documents are to be recorded.

Who can use the scaffolding on site?

DBSub

As noted above, one of the free issue items of attendance provided by the contractor at item 6.1 of the sub-contract particulars is the provision and erection of all necessary scaffolding and scaffold boards for work over 3.30 metres. Obviously this does not apply to work undertaken below this height, which (unless agreed otherwise) is the responsibility of the sub-contractor.

In addition to this 'free' attendance item, clause 3.16.2 of DBSub/C notes that the contractor, the sub-contractor, the contractor's persons and the sub-contractor's persons in common with all others having a like right shall for the purposes of the main contract works (and only for the purposes of the main contract works) be entitled to use any erected scaffolding belonging to or provided by the contractor or the sub-contractor while it remains erected.

Clause 3.16.2 relates to:

- Both scaffolding belonging to or provided by the contractor and scaffolding belonging to or provided by the sub-contractor.
- Scaffolding for work both over 3.30 metres high and less than 3.30 metres high.
- The use of the scaffolding in question while it remains erected.

  This presumably means that scaffold may be used whilst it remains erected for its primary purpose, but that it will not remain erected thereafter for any secondary purpose.

Also, the scaffolding would (by implication) need to be used without alteration (e.g. amending lift heights, strengthening for loadings, etc.).

Clause 3.16.2 of DBSub/C also adds that the use of this scaffold by either party is on the express condition that no warranty or other liability on the part of the contractor or the contractor's persons or of the sub-contractor or the sub-contractor's persons, as appropriate, shall be created or implied under this sub-contract in regard to the fitness, condition or suitability of such scaffolding.

This is a very significant point since it is the responsibility of both the contractor and the sub-contractor to ensure that their operatives on site are working from safe scaffolding.

With this point in mind, it may be the case that the sub-contractor (or the contractor, as appropriate) before using the scaffolding referred to under clause 3.16.2 of DBSub/C would need to satisfy himself, at least, that the scaffolding:

- Was erected, altered, and dismantled (as appropriate) by competent persons.
- Had been regularly inspected.
- Was secured to the building or structure (if appropriate) in enough places to prevent collapse.
- Was strong enough to allow for the work that was proposed to be carried out from it, and for the materials that were proposed to be stored on it.

MPSub

MPSub has no equivalent clause to clause 3.16.2 of DBSub/C.

ICSub

All as DBSub/C above (but clause 3.14.2 of ICSub/C applies).

ICSub/D

All as ICSub above.

ICSub/NAM

All as ICSub above.

MWSub/D

MWSub/D has no equivalent clause to clause 3.16.2 of DBSub/C.

ShortSub

As MWSub/D above.

SubSub

As SubSub above.

What does the sub-contract say about possible misuse of scaffolding, etc.?

DBSub

Clause 3.16.3 of DBSub provides that the contractor and the sub-contractor shall not, and shall respectively ensure that the contractor's persons and the sub-contractor's persons do not, wrongfully use or interfere with the plant, ways, scaffolding, temporary works, appliances or other property belonging to or provided by others; also that no party should infringe upon any statutory requirements.

However, clause 3.16.3 also makes it clear that the above points do not affect the rights of the contractor or of the sub-contractor to carry out their respective statutory duties and contractual obligations under the sub-contract or the main contract.

Any misuse of scaffolding or other property belonging to or provided by the contractor is covered by the indemnity given by the sub-contractor to the contractor under clause 2.5.2 of DBSub/C.

MPSub

MPSub has no equivalent clause to clause 3.16.3 of DBSub/C.

ICSub

All as DBSub/C above (but clause 3.14.3 of ICSub/C applies).

ICSub/D

All as ICSub above.

ICSub/NAM

All as ICSub above.

MWSub/D

MWSub/D has no equivalent clause to clause 3.16.3 of DBSub/C.

ShortSub

ShortSub has no equivalent clause to clause 3.16.3 of DBSub/C.

SubSub

As ShortSub above.

Temporary buildings

DBSub

Site accommodation would normally include for site offices, workshops, stores and sheds, etc. (but not for mess rooms, sanitary accommodation and welfare facilities which are normally provided by the contractor).

Clause 3.17 of DBSub/C makes it clear that, subject to clause 3.16.1, it is for the sub-contractor (at his own expense) to provide, erect, maintain, move and subsequently remove all temporary site accommodation that he requires.

It may also be necessary for the sub-contractor to move his temporary site accommodation to a different location on site during the course of the works.

Who determines the location of sub-contractor's site accommodation?
Clause 3.17 states that this is determined by the contractor, but also gives the sub-contractor the right to make a reasonable objection if, for example, the sub-contractor considers his site accommodation will be placed in an unsuitable location on site.

Normally, space on site is fairly limited, and it is therefore usual for the contractor to agree with the sub-contractor upon the number of pieces of individual site accommodation that will be permitted, the size of the various items of the site accommodation, and any stacking arrangements that will be necessary.

Clause 3.17 of DBSub/C states that for the purposes of the provision of, the erection of, the maintenance of, the moving of and/or the removal of temporary site accommodation by the sub-contractor, the contractor agrees to give all reasonable facilities.

What are reasonable facilities?
It is submitted that in the context of the above, all reasonable facilities simply means that the contractor will allow the sub-contractor, or give the opportunity to the sub-contractor to provide, erect, maintain, move, and/or remove temporary site accommodation; it does not mean that the contractor will provide any equipment or resources (e.g. cranage or labour) to the sub-contractor for any of the above listed operations.

Joint Fire Code
Item 6.3 of the sub-contract particulars may have an impact on the construction of the temporary accommodation where the Joint Fire Code applies, and this matter is dealt with in more detail under Chapter 2 of this book.

MPSub

Clause 14.1 of MPSub states that the sub-contractor is to provide all accommodation unless the sub-contract particulars state that these are provided by the contractor.

ICSub

All as DBSub above (but clause 3.15 of ICSub/C applies).

ICSub/D

All as ICSub above.

ICSub/NAM

All as ICSub above.

MWSub/D

MWSub/D has no equivalent clause to clause 3.17 of DBSub/C. See comments earlier regarding the provision of attendances under this sub-contract.

ShortSub

As MWSub/D above.

SubSub

As ShortSub above.

Site clearance – how is the sub-contractor's waste dealt with?

DBSub

In respect of item 6.4 of the sub-contract particulars:

- If no entry is made, the normal (default) position is that all rubbish resulting from the carrying out of the sub-contract works is to be disposed of off-site by the sub-contractor.
- Item 6.4. is only completed if there are any specific requirements for the manner in which rubbish is to be disposed of, for example that the sub-contractor is to dispose of his rubbish to a central skip to be provided by the contractor; or may be that certain waste must be disposed of in a particular manner.

To the extent that the sub-contractor has liability for site clearance (as set out above) clause 3.18 of DBSub/C states that the sub-contractor shall:

- clear away all rubbish resulting from his carrying out of the sub-contract works so as to keep access to the sub-contract works clear at all times; and
- upon practical completion of the sub-contract works or of such works in each Section, clear up and leave those works clean and tidy to the reasonable satisfaction of the contractor together with all areas made available to him (and used by him) for the purposes of carrying out the sub-contract works.

MPSub

Sub-contract particulars (clause 14.2 entry) refers and provides two options for dealing with the sub-contractor's rubbish as follows:

- option 1: taken to a central location for the subsequent disposal by the contractor; or
- option 2: removal and disposal off-site by the sub-contractor.

If this entry is not completed option 1 applies in default.

Clauses 14.2 and 14.3 deal with rubbish clearance by the sub-contractor who is to:

- regularly, in the manner required in the sub-contract particulars, clear away all rubbish resulting from the execution of the sub-contract works to keep the site, at all times, free from the sub-contractor's rubbish (clause 14.2); and
- upon practical completion of the sub-contract works clear up and leave the sub-contract works clean and tidy to the reasonable satisfaction of the contractor together with all areas made available to him (and used by him) for the purposes of carrying out the sub-contract works (clause 14.3).

ICSub

All as DBSub above (but clause 3.16 of ICSub/C applies).

ICSub/D

All as ICSub above.

ICSub/NAM

All as ICSub above – with the exception that the requirements for rubbish clearance are entered in IT11.4 of ICSub/NAM/IT (which is identical to item 6.4 of the sub-contract particulars in DBSub/A).

MWSub/D

MWSub/D has no equivalent clause to clause 3.18 of DBSub/C.

ShortSub

As MWSub/D above.

SubSub

As ShortSub above.

Health and safety and CDM

Health and safety

DBSub

Clause 3.19 of DBSub/C requires the sub-contractor to comply at no cost to the contractor with:

- all health and safety legislation relevant to the sub-contract works and the manner in which they are being carried out;
- all reasonable directions of the contractor to the extent necessary for compliance by the contractor and the sub-contractor with such legislation as it affects the sub-contract works; and

- within the time reasonably required, any written request by the contractor for information reasonably necessary to demonstrate compliance by the sub-contractor with clause 3.19.

MPSub

Clause 15 of MPSub covers health and safety:

- Clause 15.1 obliges the sub-contractor to comply with all health and safety legislation relevant to the sub-contract works, and further comply with all the contractor's instructions issued to ensure such compliance by both the sub-contractor and the contractor. Any such instructions do not constitute a change, unless they cause the sub-contractor to exceed his legal obligations.
- Subject to receiving the contractor's reasonable request, the sub-contractor must provide to the contractor information demonstrating his compliance with clause 15.1 (clause 15.2).
- Clause 15.3: whether or not the Joint Fire Code applies to the project is identified by an entry in the sub-contract particulars. Where it applies both parties must comply with its requirements. Additionally, and without delay, the parties must implement any remedial measures necessary following notification by an insurer of their non-compliance with the Joint Fire Code. Such remedial measures do not constitute a change.

ICSub

Clause 3.17 of ICSub/C mirrors clause 3.19 of DBSub/C above.

ICSub/D

All as ICSub above.

ICSub/NAM

All as ICSub above.

MWSub/D

Clause 5.1 of MWSub/D places several obligations upon the sub-contractor, one of which is that he shall carry out and complete the sub-contract works in compliance, *inter alia*, with the Construction Phase Plan (where applicable) and the statutory requirements (including the CDM Regulations). Clause 5.1.4 also states that in relation to the sub-contractor's designed portion the sub-contractor must comply with regulations 11, 12 and 18 of the CDM Regulations.

Clause 6 also obliges the contractor to comply with the CDM Regulations.

MWSub/D's guidance notes further state that the contractor will need to agree with the sub-contractor how health and safety issues will be dealt with, acknowledging that this may involve compliance with the CDM Regulations and the development of, for example, method statements, risk assessments, etc.

It also clarifies the extent to which the CDM Regulations apply (which is discussed later below).

ShortSub

Clause 5.6 states that the sub-contractor shall, amongst other matters, comply with any relevant statutory requirements (including the CDM Regulations). Clause 6 also states that the contractor shall comply with the CDM Regulations.

Like MWSub/D, ShortSub's guidance notes also state the contractor will need to agree with the sub-contractor how health and safety issues will be dealt with, acknowledging that this may involve compliance with the CDM Regulations and the development of, for example, method statements, risk assessments, etc.

SubSub

As ShortSub above (except that 5.6 deals with the sub-subcontractor's obligations in this respect and clause 6 deals with the sub-contractor's).

What is the relevant health and safety legislation?

Generally, the laws which govern health and safety are not industry specific, but they do relate to many construction activities (including design, where appropriate). There are several Acts of Parliament and regulations that apply.

The fundamental Act governing health and safety in construction is the Health and Safety at Work Act 1974. This Act has over sixty regulations, but the principal regulations of this Act which affect design and construction are:

- Health and Safety (First Aid) Regulations 1981
- Noise at Work Regulations 1989
- Electricity at Work Regulations 1989
- The Health and Safety Information for Employees Regulations 1989
- Manual Handling Operations Regulations 1992
- Personal Protective Equipment Regulations 1992
- Gas Safety (Installation and Use) Regulations 1994
- Reporting of Injuries, Diseases and Dangerous Occurrences Regulations 1995 (known as RIDDOR)
- Provision and Use of Work Equipment Regulations 1998 (known as PUWER 98)
- The Lifting Operations and Lifting Equipment Regulations 1998 (known as LOLER 98)
- The Control of Major Accident Hazards Regulations 1999 (known as COMAH)
- Management of Health and Safety at Work Regulations 1999
- The Chemicals (Hazards Information and Packaging for Supply) Regulation 2002 (known as CHIP)
- The Control of Substances Hazardous to Health Regulations 2002 (known as COSHH)
- The Work at Height Regulations 2005
- The Construction (Design and Management) Regulations 2007 (known as the CDM Regulations)

Some of the more important elements of the above Acts and Regulations are discussed below.

What was the intention of the Health and Safety at Work Act 1974?

The intention of this Act was to make further provision for securing the health, safety and welfare of persons at work, for protecting others against risks to health or safety in connection with the activities of persons at work, for controlling the keeping and use and preventing the unlawful acquisition, possession and use of dangerous substances, and for controlling certain emissions into the atmosphere.

The first part of section 2 of the Act contains a general statement of the duties of employers (which would include contractors and sub-contractors) to their own employees while at work and is qualified in subsection (2) which instances particular obligations to:

(1) Provide and maintain plant and systems of work that are safe and without risks to health. Plant covers any machinery, equipment or appliances including portable power tools and hand tools.
(2) Ensure that the use, handling, storage and transport of articles and substances is safe and without risk.
(3) Provide such information, instruction, training and supervision to ensure that employees can carry out their jobs safely.
(4) Ensure that any workshop under his control is safe and healthy and that proper means of access and egress are maintained, particularly in respect of high standards of housekeeping, cleanliness, disposal of rubbish and the stacking of goods in the proper place.
(5) Keep the workplace environment safe and healthy so that the atmosphere is such as not to give rise to poisoning, gassing or the encouragement of the development of diseases. Adequate welfare facilities should be provided.

In respect of the above, and in all of the following regulations, an employer would be a contractor and/or a sub-contractor, if and as appropriate.

Further duties are placed on an employer by:

- Section 2(3) of the Act; which requires the employer to prepare and keep up to date a written safety policy supported by information on the organisation and arrangements for carrying out the policy. The safety policy has to be brought to the notice of employees; however, where there are five or less employees this section does not apply.

   The safety policy should consist of three parts:

   — A general statement of intent.
   — Details of the organisation (people and their duties).
   — Details of the practical arrangements (i.e. systems and procedures), e.g.

      — safety training;
      — safe systems of work;
      — environmental control;
      — safe place of work;
      — safe plant and equipment;
      — noise control;
      — use of toxic materials;
      — utilisation of safety committee(s) and safety representatives;

— fire safety and prevention;
— medical facilities and welfare;
— maintenance of records;
— accident reporting and investigation;
— emergency procedures.

- Section 2(6) of the Act, which requires an employer to consult with any safety representatives appointed by recognised trade unions to enlist their cooperation in establishing and maintaining high standards of safety.
- Section 2(7) of the Act, which requires an employer to establish a safety committee if requested by two or more safety representatives.

The employer is not allowed by section 9 of the Act to charge any employee for anything done or provided to meet statutory requirements.

The employees' duties are laid down in sections 7 and 8 of the Act, and these sections state that, whilst at work, every employee must take care of the health and safety of himself and of other persons who may be affected by his acts or omissions.

Also, employees should cooperate with their employer to meet legal obligations and they must not, either intentionally or recklessly, interfere with or misuse anything, whether plant equipment or methods of work, provided by their employer to meet the obligations under this or any other related Act.

Health and Safety (First Aid) Regulations 1981

These regulations deal with the requirements for first aid.

Noise at Work Regulations 1989

These regulations require employers to take action to protect employees from hearing damage.

Electricity at Work Regulations 1989

These regulations require people in control of electrical systems to ensure that they are safe to use and are maintained in a safe condition.

The Health and Safety Information for Employees Regulations 1989

These regulations require (amongst other things) that employers display a poster telling employees what they need to know about health and safety.

Manual Handling Operations Regulations 1992

These regulations place an obligation on employers to carry out an assessment of manual handling activities undertaken and to reduce any identified risks. Manual handling includes lifting, pulling, pushing, carrying, lowering and turning.

Once the risks have been identified, the appropriate control measures must be identified and used.

Personal Protective Equipment Regulations 1992

These regulations place an obligation on employers to assess and review the provision and suitability of personal protective equipment at work; as noted above personal protective equipment is usually simply referred to as 'PPE'.

An employer must assess the need to provide PPE, but PPE should be provided only when other control measures have been examined and either implemented or dismissed as not being reasonably practicable.

It is the responsibility of an employer to provide PPE and the employer should not charge its employees for the PPE issued.

Gas Safety (Installation and Use) Regulations 1994

These regulations cover safe installation, maintenance and use of gas systems and appliances in domestic and commercial premises.

Reporting of Injuries, Diseases and Dangerous Occurrences Regulations 1995 (known as RIDDOR)

These regulations require that certain accidents must be reported to the Health and Safety Executive (usually referred to as the HSE).

These accidents are:

- An accident that is fatal or involves a major injury such as a fracture, amputation, or loss of sight.
- A work-related accident which results in more than three days off work.
- An accident on or related to the site that results in a member of the public being killed or sent to hospital.

Provision and Use of Work Equipment Regulations 1998 (known as PUWER 98)

These regulations place an obligation on all employers to carry out an assessment of work equipment being used within a business and to reduce risks, so far as is reasonably practicable.

The Lifting Operations and Lifting Equipment Regulations 1998 (known as LOLER 98)

These regulations require that lifting equipment provided for use at work is strong and stable enough for the particular use, that it is positioned and installed to minimise any risks, that it is used safely and is subject to ongoing inspection.

The Control of Major Accident Hazards Regulations 1999 (known as COMAH)

These regulations require those who manufacture, store or transport dangerous chemicals or explosives in certain quantities to notify the relevant authority.

Management of Health and Safety at Work Regulations 1999

Regulation 3 of the Management of Health and Safety at Work Regulations 1999, places a legal duty on employers to carry out a risk assessment.

What is a risk assessment?

This is simply a careful examination of what, in the works to be carried out, could cause harm to people (including employees, other operatives and the public).

There are five basic steps to producing a risk assessment:

- Look for the hazards.
- Decide who might be harmed, and how they might be harmed.
- Evaluate the risks arising from the hazards identified and decide whether existing precautions are adequate or if more should be done.
- Record the findings.
- Review the assessment from time to time and revise it if necessary.

What is a method statement?

This is a method of control, usually used after a risk assessment of an operation is carried out.

The method statement is used to control the operation and to ensure that all concerned are aware of the hazards associated with the work and the safety precautions to be taken.

Method statements should be 'project specific' and one of the main criticisms of sub-contractors (in particular) is that they attempt to use 'generic' method statements (rather than 'project specific' method statements).

The Chemicals (Hazards Information and Packaging for Supply) Regulation 2002 (known as CHIP)

This regulation requires suppliers to classify, label and package dangerous chemicals and provide safety data sheets for them.

The Control of Substances Hazardous to Health Regulations 2002 (known as COSHH)

These regulations require employers to control exposure to hazardous substances to prevent illness; both employees and others who may be exposed must be protected.

These hazardous substances include things such as adhesives, paints, cleaning agents, fumes from soldering and welding, biological agents and carcinogenic-inducing chemicals.

To comply with COSHH, there are eight basic steps to follow:

- Assess the risks.
- Decide what precautions are needed.
- Prevent or adequately control exposure.
- Ensure that control measures are used and maintained.
- Monitor the exposure.
- Carry out appropriate health surveillance.
- Prepare plans and procedures to deal with accidents, incidents and emergencies.
- Ensure employees are properly informed, trained and supervised.

The Work at Height Regulations 2005, as amended by
the Work at Height (Amendment) Regulations 2007

- These regulations came into force on the 6 April 2005.
- Statistics record that nearly 25% of deaths at work each year are caused by falls from height.
- These regulations apply to all work at height where there is a risk of a fall liable to cause personal injury and are aimed at preventing deaths and injuries caused each year by falls at work by placing duties on employers, the self-employed and any person who controls the work of others.

The Construction (Design and Management) Regulations 2007
(known as the CDM Regulations)

These regulations are dealt with in more detail in the next section of this chapter.

The CDM Regulations (i.e. the Construction (Design and Management) Regulations 2007)

Obviously, all of the sub-contracts considered in this book address CDM 2007.

DBSub

Clause 3.20 of DBSub/C deals with CDM Regulations as they apply to the sub-contractor.

To reflect their importance, clause 3.20 of DBSub/C includes an express cross-undertaking by the parties to comply with CDM 2007 in relation to the main contract works and the site. Additionally, clause 3.20.1 to 3.20.4 includes further provisions applicable if the project has 'notifiable' status under the CDM Regulations.

The main contract particulars annexed to the schedule of information in DBSub/A will identify whether or not the project is notifiable.

MPSub

MPSub at clause 7.2 includes a simple statement specifically addressing the CDM Regulations, namely, the sub-contractor warrants both his competence and that he will allocate the necessary resources to fulfil the roles under CDM 2007 of contractor and (if required by the sub-contractor) designer.

ICSub

Clause 3.18 virtually mirrors clause 3.20 of DBSub/C above. Like DBSub, it includes an express cross-undertaking by the parties to comply with CDM 2007 in relation to the main contract works and the site, and additional provisions (clause 3.18.1 to 3.18.4) applicable to 'notifiable' projects.

ICSub/D

As ICSub above.

ICSub/NAM

As ICSub above.

MWSub/D

As noted earlier, clause 5.1 of MWSub/D places several obligations upon the sub-contractor, one of which is that he shall carry out and complete the sub-contract works in compliance, *inter alia*, with the Construction Phase Plan (where applicable) and the statutory requirements (including the CDM Regulations).

Clause 5.1.4 also states that in relation to the sub-contractor's designed portion, the sub-contractor must comply with regulations 11, 12 and 18 of the CDM Regulations.

Clause 6 also states that the contractor must comply with the CDM Regulations.

MWSub/D's guidance notes further state that the contractor will need to agree with the sub-contractor how health and safety issues will be dealt with, acknowledging that this may involve compliance with the CDM Regulations and the development of, for example, method statements, risk assessments, etc.

ShortSub

Clause 5.6 states that the sub-contractor must, amongst other matters, comply with any relevant statutory requirements (including the CDM Regulations). Clause 6 also states that the contractor must comply with the CDM Regulations.

ShortSub's guidance notes contain a similar entry to MWSub/D discussed above.

SubSub

As ShortSub above.

Does the sub-contractor get paid for complying with CDM 2007?

No. Clause 3.20.3 of DBSub and clause 3.18.3 of ICSub, ICSub/D and ICSub/NAM state that the sub-contractor shall comply at no cost to the employer or the contractor with all reasonable requirements of the Principal Contractor relating to compliance by the sub-contractor with the CDM Regulations; and further notes that no extension of time shall be given for such compliance.

MPSub, MWSub/D, ShortSub and SubSub are silent, but it is submitted that the effect would be the same.

What happened to the CDM 1994 Regulations?

The CDM Regulations 2007 (CDM 2007) are effective from 6 April 2007 and apply to all projects ongoing or post 6 April 2007. They replaced the CDM Regulations 1994 and the Construction (Health, Safety and Welfare) Regulations 1996.

Accompanying the CDM 2007 Regulations is an approved code of practice, entitled 'Managing Health and Safety in Construction', but commonly referred to as 'ACop'.

What is the purpose of the CDM 2007 Regulations?

The CDM Regulations require that health and safety is taken into account and managed throughout all stages of a project, from conception, design and planning, through to site work and subsequent to maintenance and repair of the building.

Accordingly, the CDM 2007 Regulations place health, safety and welfare duties upon everyone who takes part in the construction process – the employer, the designers, the contractors and the sub-contractors.

Despite certain similarities, the CDM 2007 Regulations are a more coherent and wider set of duties than its predecessor (the CDM 1994 Regulations), with increased emphasis on improving coordination and cooperation between the various parties, encouraging team working, reducing risk in the construction process and avoiding unnecessary bureaucracy.

All duty holders will need to review their current health and safety procedures to ensure compliance. For example, the role of the Planning Supervisor is replaced by a CDM Coordinator, and the Health and Safety Plan is replaced by a Construction Phase Plan.

What are the sub-contractors' duties under the CDM Regulations?

A sub-contractor's duties are the same as the contractors. Additionally, if the sub-contractor designs he assumes the obligations CDM 2007 places on designers.

When do the CDM Regulations apply?

The CDM Regulations apply to most common building, civil engineering and engineering construction work, and are defined widely in the regulations at section 2 (interpretation), including, for example, not only construction, but commissioning, renovation, repair, redecoration, demolition, etc.

The CDM Regulations apply to all design work carried out for construction purposes (including demolition and dismantling).

Domestic projects

CDM 2007 does not apply to a domestic client (i.e. construction works at the client's own home). They do, however, remain applicable to all other relevant parties involved in a domestic project, e.g. the contractor.

When must a project be notified to the Health and Safety Executive (HSE)?

Under CDM 2007, all projects fall into one of two categories: notifiable or non-notifiable. A project must be notified to the HSE if it lasts longer than 30 days or

500 man days of work. Note that the 'five or more workers' requirements of CDM 1997 has been omitted.

The regulations set out the relevant parties' obligations as follows:

- Part 2 (regulations 4 to 13): duties that apply to all projects (notifiable or not) in respect of clients, designers, contractors and sub-contractors. This covers the following matters:

  — competence (regulation 4);
  — cooperation (regulation 5);
  — coordination (regulation 6);
  — general principles of prevention (regulation 7);
  — election by clients (regulation 8);
  — client's duty in relation to arrangements for managing projects (regulation 9);
  — client's duty in relation to information (regulation 10);
  — duties of designers (regulation 11);
  — designs prepared or modified outside Great Britain (regulation 12);
  — duties of contractors (regulation 13).

- Part 3 (regulations 14 to 24): additional duties applicable to notifiable projects. This covers the following matters:

  — appointments by the client where a project is notifiable (regulation 14);
  — client's duty in relation to information where a project is notifiable (regulation 15);
  — the client's duty in relation to the Health and Safety File (regulation 17);
  — additional duties of designers (regulation 18);
  — additional duties of contractors (regulation 19);
  — general duties of CDM Coordinators (regulation 20);
  — notification of project by CDM Coordinator (regulation 21);
  — duties of the Principal Contractor (regulation 22);
  — the Principal Contractor's duty in relation to the construction phase plan (regulation 23);
  — the Principal Contractor's duty in relation to cooperation and consultation with workers (regulation 24).

What are the employer's duties under CDM 2007?

Under CDM 2007 employers (termed 'clients') can no longer delegate their duties to an agent (as occurred under CDM 1994).

The employer's (excluding domestic clients) duties are described below.

Part 2 – all construction projects

Regulations 4 to 10 refer. He is, amongst other matters, to:

- Take reasonable steps to ensure that the CDM Coordinator, Principal Contractor, project designers and any contractors they appoint directly are competent and adequately resourced.
- Ensure suitable project management arrangements for health and safety are in place (including the allocation of sufficient and other resources). Take reasonable

steps to ensure that such arrangements are maintained and reviewed throughout the duration of the project.
- Provide all pre-construction information in the employer's possession to designers and contractors, including notification to the contractor of the minimum notice that will be given for the contractor's commencement of the works.

Part 3 – notifiable projects

Regulations 14, 15, 16 and 17 refer. He is, amongst other matters, to:

- Appoint a CDM Coordinator and a Principal Contractor for each project.
- Provide the CDM Coordinator with pre-construction information, plus all health and safety information relating to the project in the employer's possession (or reasonably obtainable) as is likely to be needed for the Health and Safety File.
- Ensure that the construction phase does not start until a suitable Construction Phase Plan has been prepared by the Principal Contractor and suitable welfare facilities are in place.
- After the Construction Phase Plan has been issued, take reasonable steps to provide access to the Health and Safety File and update this document, as necessary, with new information. In practical terms the client must ensure that the Health and Safety File is reviewed and/or updated as necessary so that it is accurate at the completion of the construction work. This completed file must be kept available for others, i.e. a new owner or for future construction work.

The reader is referred to the CDM 2007 Regulations for a full description of the duties placed upon clients.

What are the designer's duties under CDM 2007?

The term 'designer' includes everyone preparing drawings and specifications for the project, including variations and temporary works designs, and also includes producing detailed design work to satisfy a performance specification. Where appropriate, designers include architects, consulting engineers (e.g. structural engineers, building services engineers, etc.), contractors and sub-contractors.

The designer's duties (which, in certain circumstances, could include a sub-contractor) are set out at regulations 4, 5, 6, 11, 12 (part 2) and regulation 18 (part 3). A designer, amongst other matters, has responsibilities as outlined below.

Part 2 – all construction projects

The designer must:

- Ensure that the employer is aware of his duties under CDM 2007 before commencing work.
- Avoid foreseeable risks to the health and safety of any person when preparing or modifying the design by eliminating hazards where feasible; or reducing risks from those hazards that cannot be eliminated. The regulations oblige designers to give priority to collective measures in achieving this.

- Take reasonable steps to provide adequate information about all aspects of the design, construction or maintenance of the structure as will adequately assist the employer, other designers, and the contractor in discharging their duties under the regulations.

Part 3 – notifiable projects

The designer must:

- Ensure a CDM Coordinator is appointed before commencing work.
- Provide adequate information about all aspects of the design, construction or maintenance of the structure to the CDM Coordinator.
- Cooperate with the CDM Coordinator and any other designer involved in the project.
- Provide design information for inclusion in the Construction Phase Plan and the Health and Safety File.

The reader is referred to the CDM 2007 Regulations for a full description of the duties placed upon clients.

What are the CDM Coordinator's duties under CDM 2007?

Part 3 – notifiable projects

The role of the CDM Coordinator is set out at regulations 20 and 21 (see also regulations 4 and 7 in Part 2). He is, amongst other matters, to:

- Advise and assist the employer in complying with his CDM 2007 duties.
- Ensure the coordination and cooperation of health and safety between the relevant parties during the design and planning phase of the project.
- Collect and distribute information, such as pre-construction information.
- Liaise with the principal contractor regarding the health and safety file, the information the principal contractor needs to prepare the Construction Phase Plan and the ongoing design.
- Take all reasonable steps to ensure that designers comply with their duties under 11 and 18(2) of the regulations.
- Give advice about health and safety competence and resources needed for the project.
- Ensure notice of the project is given to the Health and Safety Executive.
- Prepare and update the Health and Safety File and pass the file to the employer on completion of the project.

The CDM Regulations do not require CDM Coordinators to visit the site or to assess the performance of the Principal Contractor once construction work has begun.

The identity of the CDM Coordinator is normally provided in the main contract particulars as annexed to the schedule of information attached to the sub-contract, e.g. DBSub/A, ICSub/A, etc.

The reader is referred to the CDM 2007 Regulations for a full description of the duties placed upon CDM Coordinators.

What are the Principal Contractor's duties under CDM 2007?

Part 3 – notifiable projects

The Principal Contractor's full list of duties are listed at regulations 22, 23 and 24 (see also regulations 4 to 7 of part 2 and regulations 25 to 44 of part 4 'duties relating to health and safety on construction sites' which is not considered below).
He is, amongst other matters, to:

- Plan, manage and monitor health and safety during the construction phase of the project, in liaison with other contractors and ensure necessary cooperation and coordination between the relevant parties.
- Develop a Construction Phase Plan before work starts on site. Thereafter review and revise the Construction Phase Plan from time to time as appropriate. The plan may need to be developed during the construction phase to take account of changing conditions on site as work progresses or as the design changes. The Principal Contractor's Construction Phase Plan should take account of general issues, specific hazards, risk control measures and the general principles of risk assessment.
- Provide the CDM Coordinator with all information he requires for inclusion in the Health and Safety File.
- Ensure suitable welfare facilities are provided from the start and maintained throughout the construction phase.
- Ensure all contractors are provided with relevant health and safety information, and that all workers have site inductions and (as necessary) further information and training.

It should be noted that whilst the contractor would normally be the Principal Contractor, this may not always be the case.

The identity of the Principal Contractor will be provided in the main contract particulars as annexed to the schedule of information to the sub-contract.

The reader is referred to the CDM 2007 Regulations for a full description of the duties placed upon the Principal Contractor.

What are the contractor's duties under CDM 2007?

Part 2 – all construction projects

The contractor's duties are listed at regulations 4 to 7 and 13 of part 2. The contractor is, amongst other matters, to:

- Ensure that the employer is aware of his duties under CDM 2007 before commencing work.
- Plan, manage and monitor his work and that by others on his behalf to ensure, as far as is reasonably practicable, that his works are executed without risks to health and safety.

- Ensure that they and anyone they employ or engage (e.g. sub-contractors, workers) are competent.
- Ensure any sub-contractor used by the contractor is notified of the minimum amount of time (CDM planning period) that will be allowed for his preparation and planning prior to commencement on site.
- Ensure his workers (whether employed or self-employed and undertaking work under the contractor's control) are provided with necessary information and training to permit work to be carried out safely.
- Cooperate and coordinate their work with others as required.
- Ensure compliance with the relevant requirements in part 4 of the regulations.
- Ensure the provisions of adequate welfare facilities in compliance with schedule 2 of the regulations.

Part 3 – notifiable projects

The contractor's additional duties are listed at regulation 19. The contractor is, amongst other matters, to:

- Ensure a CDM Coordinator is appointed and the HSE notified before commencing work.
- Cooperate with the CDM Coordinator.
- Cooperate with the Principal Contractor, including providing information which may affect the health and safety of those executing the construction work or any person affected by it.
- Comply with the Principal Contractor's reasonable directions, site rules and the Construction Phase Plan.
- Notify the Principal Contractor of any matters that may require a review, alteration or addition of the construction phase plan and/or problems with the plan or risks identified during their work that have significant implications for the management of the project.
- Issue the Principal Contractor with information for the Health and Safety File.
- Notify the Principal Contractor of any death, accidents or dangerous occurrences.
- Ensure a CDM Coordinator is appointed before commencing work.
- Provide adequate information about all aspects of the design, construction or maintenance of the structure to the CDM Coordinator.

As noted earlier, the sub-contractor's duties are the same as the contractors. The CDM 2007 Regulations place extensive duties on contractors regarding health and safety and welfare. The reader is referred to the CDM 2007 Regulations for a full description of the duties placed upon contractors.

Will I have sufficient time to plan and prepare before construction work begins?

Regulation 13(3) provides that every contractor is to notify his sub-contractors (and so on down the contractual chain) of the minimum amount of time available to the sub-contractor for planning and preparation before he begins construction work.

In response, the JCT sub-contracts now include an entry in the contract particulars for inserting the sub-contractor's minimum CDM planning period.

How does the role of the Principal Contractor affect the sub-contractor?

Sub-contractors are required to help the Principal Contractor to achieve safe and healthy site conditions. They should cooperate with other sub-contractors working on the site and provide health and safety information (including risk assessments) to the Principal Contractor.

How does health and safety competence affect sub-letting?

The CDM Regulations require that when sub-letting, a contractor should satisfy himself that those who are to do the work are:

- competent in relevant health and safety issues; and
- intend to allocate adequate resources, including time, equipment and properly trained workers to do the job safely and without risks to health.

It is quite common for sub-contractors to be asked to complete a health and safety competence questionnaire before their tender is considered for acceptance.

Additionally, to reinforce the importance of CDM 2007, express terms are also usually included in the sub-subcontract. For example, clause 3.3.2.1 of the DBSub/C makes it a requirement that a sub-subcontract will include an express undertaking to comply with CDM 2007 in relation to the main contract works at the site.

What is pre-construction information?

The pre-construction information is provided by the employer (regulation 10 of CDM 2007) to designers, and contractors who are, or may be, appointed. Regulation 10(2) specifies that the following information is to be included:

- any information about and/or affecting the site and/or construction work;
- any information concerning the proposed use of the structure as a workplace;
- the minimum amount of time before the construction phase allowed to contractors for planning and preparation for construction work;
- any information in any existing Health and Safety File.

Its purpose is to ensure that information relevant to health and safety is passed on to those who need it so that significant risks during the work can be anticipated and planned for.

What is the Construction Phase Plan?

The Construction Phase Plan (previously known as the Health and Safety Plan) is developed by the Principal Contractor to address key issues of health, safety and welfare relevant to the project.

Issues which need to be considered for inclusion in the plan include:

- How health and safety will be managed during the construction phase.
- What high risk activities will require risk assessments and method statements to be produced?
- Information about welfare arrangements.
- Information on necessary levels of health and safety training for those working on the project and arrangements for project-specific awareness training and refresher training such as toolbox talks.

  A toolbox talk is a method of passing health and safety information on to the site operatives at site level. The toolbox talk is normally given by the site safety representative to a group of operatives on site, and is normally in relation to a particular health and safety issue (for example, the safe use of electric power tools, or the need for leading edge protection, etc.).
- Arrangements for monitoring compliance with health and safety law.

The Construction Phase Plan should be developed before construction work actually starts, and should then be reviewed as necessary to account for changing project circumstances.

Clause 3.20.2 of DBSub/C obliges the contractor to ensure that the sub-contractor is supplied forthwith with a copy of any development of the Construction Phase Plan by the Principal Contractor.

What is the Health and Safety File?

The Health and Safety File is a record of information for the employer or the end user.

The CDM Coordinator ensures that it is produced at the end of the project and that it is then passed to the employer. The file gives details of health and safety risks that will have to be managed during maintenance, repair, renovation or demolition.

Contractors (and sub-contractors) should pass information that may affect the maintenance, repair, renovation or demolition of the finished building to the CDM Coordinator for inclusion within the file.

The employer should make the file available to those who will work on any future design, construction, maintenance or demolition of the structure (and this may include the contractor or the sub-contractor when they are carrying out defect rectification work).

A Health and Safety File will often incorporate operating and maintenance manuals and as-built drawings (prepared by the contractor and the sub-contractors), as well as general details of the construction methods and materials used.

For example, clause 3.20.4 of DBSub/C requires the sub-contractor to provide to the contractor, within the time reasonably required by the contractor, such information in respect of the sub-contract works as is reasonably necessary to enable the contractor to comply with clause 3.18.4 of the main contract conditions which relates to the information reasonably required by the CDM Coordinator for inclusion in the Health and Safety File.

Suspension of main contract by contractor

Generally

All the JCT main contracts to which the sub-contracts in this book refer contain express rights for the main contractor to suspend his performance where the employer has failed to make payment in the specified manner. Such rights are of course subject to the contractor's strict compliance with the notice requirements and procedure stated in the main contract.

Naturally, suspension of the main contract by the contractor will raise questions by the sub-contractor as to what action, if any, he should take under the sub-contract.

In such circumstances the sub-contractor should refer to the provisions of the sub-contract itself for the answer.

However, as a general principle, unless the contract provides otherwise (which the unamended JCT sub-contracts do not):

(1) There is no automatic right for a sub-contractor to suspend his performance under the sub-contract in circumstances where the contractor has suspended the main contract works.
(2) The sub-contractor should not stop carrying out the sub-contract works unless he is instructed or directed by the contactor to do so.

Of course, the above possibility may cause great difficulties in practice, for example in respect of health and safety considerations, and this is a matter that would need to be considered by a sub-contractor on a project-by-project basis.

Does the sub-contract provide for the sub-contractor to be notified of the contractor's proposed suspension/suspension of his performance under the main contract?

DBSub

The DBSub deals with this at clause 3.21.

Position where contractor intends to suspend
Clause 3.21 states that if the contractor under clause 4.14 of the main contract conditions gives the employer a written notice of his intention to suspend the performance of his obligations under the main contract because of non-payment by the employer, he is to immediately copy that notice to the sub-contractor.

Position where contractor actually suspends main contract
In circumstances where the contractor actually suspends his obligations under the main contract, clause 3.21 obliges the contractor to immediately notify the sub-contractor of this occurence.

MPSub

The MPSub does not deal with this issue.

ICSub

Clause 3.19 of ICSub/C refers and is drafted identically to clause 3.21 of DBSub. Refer, therefore, to the comments made at DBSub/C above.

ICSub/D

All as ICSub above.

ICSub/NAM

All as ICSub above.

MWSub/D

MWSub/D does not deal with this issue.

ShortSub

The ShortSub does not deal with this issue.

SubSub

The SubSub does not deal with this issue.

What should the sub-contractor do when notified that the contractor has suspended his performance under the main contract?

DBSub

The sub-contractor *must not* suspend the performance of the sub-contract works simply because he has received notice from the contractor that he has suspended the performance of the main contract works.

Instead, the sub-contractor should, where possible (and noting the above comments regarding health and safety considerations), continue to proceed regularly and diligently with his sub-contract works, until and/or unless he receives an express direction from the contractor to cease the carrying out of the sub-contract works.

If the sub-contractor were to suspend his performance in the absence of an express direction from the contractor, this may be found to be a wrongful suspension by the sub-contractor, and this may amount to a repudiation by the sub-contractor, which, if accepted by the contractor, may lead to the sub-contractor suffering all of the contractor's damages arising from that repudiation.

MPSub

The MPSub does not deal with this issue.

ICSub

All as comments for DBSub above.

ICSub/D

All as comments for DBSub above.

ICSub/NAM

All as comments for DBSub above.

MWSub/D

MWSub/D does not deal with this issue.

ShortSub

The ShortSub does not deal with this issue.

SubSub

The SubSub does not deal with this issue.

In what circumstances should the sub-contractor cease the sub-contract works when the main contract is suspended by the contractor?

Health and safety issues – all sub-contracts

Surprisingly, all of the sub-contracts discussed in this book fail to address a real practical issue.

What happens if the suspension of the main contract renders the execution of the sub-contract works unsafe, i.e. withdrawal of health and safety supervisor, scaffolding, welfare provisions, etc., but the contractor has not directed the sub-contractor to cease the carrying out of the sub-contract work?

Obviously, each case will vary on its own particular facts, but given the potential ramifications and statutory penalties where this applies common sense dictates that the sub-contractor must immediately raise any health and safety concerns with the contractor and request his urgent response and action.

The provisions included in the relevant sub-contracts for directions by the contractor are considered below.

DBSub

Clause 3.22 deals with directions from the contractor to the sub-contractor in two parts:

(1) Direction to cease sub-contract works, and further directions in respect of such cessation (clause 3.22.1 refers).
(2) Direction to recommence the sub-contract works, and further directions in respect of such recommencement (clause 3.22.2 refers).

This is discussed below.

Clause 3.22.1: contractor's direction to cease sub-contract works, and further directions in respect of such cessation
Following suspension of the main contract by the contractor, clause 3.22.1 states that the contractor 'may' direct the sub-contractor to cease the carrying out of the sub-contract works and after that may issue to the sub-contractor such further directions as necessary.

The word 'may', above, denotes that the contractor is not obliged to issue a direction to the sub-contractor to cease the carrying out of the sub-contract works; it is entirely at the contractor's discretion to issue (or not issue) this direction.

As noted earlier within this chapter, there are potentially serious ramifications if the sub-contractor suspends without first receiving a contractor's direction under clause 3.22.1.

The sub-contractor should therefore ensure that he has first received a clear written direction to suspend works before suspending his works. Unless and until such a direction is received the sub-contractor must proceed regularly and diligently with his sub-contract works.

If such a direction is issued, the contractor may issue further directions as may be necessary following such cessation. Although the DBSub/C does not elaborate upon what those 'further directions' may be, it would be reasonable to assume that they would include for things such as a requirement that unfinished works are adequately protected, and that the site working area is left secure.

Clause 3.22.2: direction to recommence the sub-contract works, and further directions in respect of such recommencement (clause 3.22.2 refers)
This clause only applies in circumstances where the contractor has issued a direction to the sub-contractor to cease the sub-contract works pursuant to clause 3.22.1.

Clause 3.22.2 of DBSub/C states that, if the contractor resumes the performance of his obligations under the main contract, the contractor shall, if he has directed the sub-contractor to cease the carrying out of the sub-contract works, direct the sub-contractor to recommence those works and may issue further directions in regard to such recommencement.

Again, although the DBSub/C does not elaborate upon what those 'further directions' may be, it would be reasonable to assume that they would allow for such things as the removal of protection from unfinished works, and the cleaning down of work left standing during the period of suspension to allow other trades to proceed.

The sub-contractor's position is safeguarded from any suspension due to the above in respect of:

- Time – by the inclusion of clause 2.19.6 of DBSub/C as a Relevant Sub-Contract Event to be considered by the contractor when giving the sub-contractor an extension of time to the completion date or to a sectional completion date of the sub-contract works.
- Loss and expense – by the inclusion of clause 4.20.4 of DBSub/C as a Relevant Sub-contract Matter to be considered by the contractor when agreeing the amount of direct loss and expense due to the sub-contractor.

MPSub

Unlike DBSub/C no specific provisions are included.

However, arguably, the contractor may be permitted to instruct the sub-contractor to cease the sub-contract works (although not expressly stated) by virtue of the wide-ranging right to issue instructions at clause 9.1 of MPSub.

Clause 9.1 records that the sub-contractor is to comply with all written instructions issued by the contractor in connection with the design, execution and completion of the sub-contract works, except to the extent that the terms of the sub-contract restrict the contractor's right to issue any particular instruction.

It is not clear how the sub-contractor's position would be safeguarded from any suspension by the contractor. However, this may be covered within MPSub's drafting as follows:

- time – by the inclusion of clause 22.1.6 and/or 22.1.8;
- loss and expense – by the inclusion of clause 30.2.1 and/or 30.2.2.

ICSub

Clause 3.20 of ICSub/C refers and is drafted identically to clause 3.22 of DBSub/C. Refer, therefore, to the comments made at DBSub above.

ICSub/D

All as ICSub above.

ICSub/NAM

All as ICSub above.

MWSub/D

MWSub/D does not provide for the sub-contractor to be notified that the contractor has suspended, or intends to suspend, the main contract works.

ShortSub

As MWSub/D above.

SubSub

The SubSub does not deal with this issue at all.

Other provisions – strikes

What happens when the main contract works are affected by strikes, etc.?

DBSub

Clause 3.23 of DBSub/C deals with strikes.

This clause says that if the main contract works are affected by a strike, lock-out or local combination of workmen affecting any of the trades employed upon them or any of the trades engaged in the preparation, manufacture or transportation of any of the goods or materials required for the main contract works:

- neither the contractor nor the sub-contractor shall be entitled to make any claim upon the other for any loss and expense resulting from such action;
- the contractor shall take all reasonable steps to keep the site open and available for the use of the sub-contractor; and
- the sub-contractor shall take all reasonable practical steps to continue with the sub-contract works.

Clause 3.23 further notes that nothing in clause 3.23 shall affect any other right of the contractor or the sub-contractor under the sub-contract if such action occurs.

Although not elaborated upon, this caveat would probably, for example, protect the parties' respective position in respect of the provisions for termination under the sub-contract. Clause 2.19.13 of DBSub/C also safeguards the position of the sub-contractor by specifically including a relevant event for such matters.

MPSub

The MPSub does not specifically deal with the contractual position in circumstances where the main contract works are affected by a strike, lock-out or local combination of workmen affecting any of the trades employed upon them or any of the trades engaged in the preparation, manufacture or transportation of any of the goods or materials required for the main contract works.

ICSub

Clause 3.21 of ICSub/C deals with strikes. This clause is worded identically to clause 3.23 of DBSub/C. Accordingly, refer to the comments made at DBSub above.

ICSub/D

All as ICSub above.

ICSub/NAM

All as ICSub above.

MWSub/D

MWSub/D does not specifically deal with the contractual position in circumstances where the main contract works are affected by a strike, lock-out or local combination of workmen affecting any of the trades employed upon them or any of the trades engaged in the preparation, manufacture or transportation of any of the goods or materials required for the main contract works.

ShortSub

All as MWSub/D above.

SubSub

As ShortSub above.

Benefits under the main contract

DBSub

Subject to the sub-contractor providing an appropriate indemnity and security for costs to accord with the reasonable requirements of the contractor, clause 3.24 of DBSub/C grants the sub-contractor a right to require the contractor to take certain actions under the main contract which may be relevant in the case of contentious opinions and instructions by the employer under the main contract (provided that those actions are relevant to the sub-contract works and are not inconsistent with any other express terms of the sub-contract).

Any such action as outlined above is to be undertaken at the expense of the sub-contractor.

MPSub

The MPSub does not have an express provision dealing with this.

ICSub

Clause 3.22 also grants the sub-contractor a right to require the contractor to take certain lawful actions under the main contract (subject to the sub-contractor providing an appropriate indemnity and security for costs as stated), which may be relevant in the case of contentious opinions and instructions by the architect/ contract administrator under the main contract (provided that those actions are relevant to the sub-contract works and are not inconsistent with any other express terms of the sub-contract).

Any such action as outlined above is to be undertaken at the expense of the sub-contractor.

ICSub/D

As ICSub above.

ICSub/NAM

As ICSub above.

MWSub/D

MWSub/D does not have an express provision dealing with this.

ShortSub

All as MWSub/D above.

SubSub

The SubSub does not have an express provision for dealing with this.

Certificates/statements or notices under the main contract

For obvious reasons the sub-contractor needs to be informed by the contractor of certain key main contract certificates or statements or notices (the precise terminology relates to the main contract/sub-contract relationship) which have a direct bearing on the sub-contract.

How does the sub-contractor receive this information?

DBSub

Clause 3.25 deals with certain key statements and notices issued under the main contract, and how the sub-contractor goes about obtaining such information from the contractor.

Clause 3.25 obliges the contractor, upon receipt of a written request from the sub-contractor, to notify the sub-contractor of the dates that the following documents are issued under the main contract.

(1) The fixing or confirmation of the completion date for the main contract works or any section.
    Comment: the completion date for the main contract works is relevant in respect of any liquidated damages which the contractor may be required to pay under the provisions of the main contract, which in certain circumstances the sub-contractor may be liable for.

(2) Any written statement issued under clause 2.30 of the main contract conditions.
    Comment: the date given in a written statement issued under clause 2.30 of the main contract conditions is deemed to be practical completion of that portion of the main contract works.

(3) The practical completion certificate or each sectional completion certificate for the main contract works.
    Comment: the practical completion certificate date or each sectional completion certificate date in respect of the main contract works are the deemed latest dates for the practical completion and/or the sectional completion of the sub-contract works.

(4) Each notice of completion of making good (defects).
    Comment: the notice of completion of making good for the main contract works or relevant section establishes the latest possible date that triggers the release of the balance of any retention to the sub-contractor.

(5) The final account and final statement or, if applicable, the employer's final account and employer's final statement.
Comment: the final certificate under the main contract triggers the release of the final payment to be made by the contractor to the sub-contractor.

MPSub

MPSub does not require a written request from the sub-contractor for the release of information:

- Clause 8.3 obliges the contractor to notify the sub-contractor of the date of practical completion of the main contract.
- Clause 20.4 also obliges the contractor, at all times, to ensure that the sub-contractor is aware of the actual and projected progress of the project, including the date by which the contractor reasonably anticipates completing the project (i.e. the main contract works) or, if applicable, any sections.

ICSub

Clause 3.23 deals with certain key certificates issued to the contractor under the main contract, and how the sub-contractor goes about obtaining such information from the contractor.

Clause 3.23 is triggered by the sub-contractor's written request for the information listed at clause 3.23.

Following the contractor's receipt of the sub-contractor's written request, the contractor must notify the sub-contractor of the dates that the following documents are issued under the main contract.

(1) The fixing or confirmation of the completion date for the main contract works or any section.
Comment: the completion date for the main contract works is relevant in respect of any liquidated damages which the contractor may be required to pay under the provisions of the main contract, which in certain circumstances the sub-contractor may be liable for.

(2) Any written statement issued under clause 2.25 of the main contract conditions.
Comment: the date given in a written statement issued under clause 2.25 of the main contract conditions (i.e. partial possession by the employer) is deemed to be practical completion of that portion of the main contract works.

(3) The practical completion certificate or each sectional completion certificate for the main contract works.
Comment: the practical completion certificate date or each sectional completion certificate date in respect of the main contract works are the deemed latest dates for the practical completion and/or the sectional completion of the sub-contract works.

(4) Each certificate of making good (defects).
Comment: each certificate of making good (defects) provides the latest date for triggering the release of the balance of any retention to the sub-contractor.

(5) The final certificate.
    Comment: the final certificate under the main contract triggers the release of
    the final payment to be made by the contractor to the sub-contractor (clause
    4.14.2 of ICSub/C refers).

ICSub/D

All as ICSub above.

ICSub/NAM

All as ICSub above.

MWSub/D

Clause 12.4 of MWSub/D states that the contractor shall, as soon as reasonably
practicable, give written notice to the sub-contractor of:

(1) The date of practical completion of the main contract.
    Comment: this starts the time running for the main contract defects rectification
    period (e.g. 12 months from practical completion).

(2) The date of the expiry of the rectification period under the main contract.
    Comment: provided no defects exist in the sub-contract works, the sub-
    contractor's final payment is due seven days after the expiry of the rectifica-
    tion period under the main contract. If there are defects in the sub-contract
    works on this date, payment falls due seven days after the sub-contractor has
    completed all defects in the sub-contract works (clause 12.5 refers).

ShortSub

All as MWSub/D above. Note, however, that the term 'rectification period' in
MWSub/D is referred to as the 'defects liability period' in ShortSub.

SubSub

The SubSub does not provide any mechanism for the sub-subcontractor to receive
information in respect of the dates of certain key certificates, statements and notices
that have been issued under the main contract.

Chapter 8
Payment

Payment generally

When is the first interim payment due?

DBSub

Clause 4.9.1 notes that the first interim payment is due on the date when the contractor's application for interim payment under the main contract immediately following the commencement of the sub-contract works is to be made. Therefore, in such a case, if a sub-contractor commenced works on 20 June (for example) and the main contractor's application for payment was to be made on 21 June, then the sub-contractor's first interim payment would be due on 21 June.

In the case where no dates for the making of contractor's applications for interim payments are stated in the main contract particulars, then the first interim payment to the sub-contractor shall become due not later than one month after the date of commencement of the sub-contract works on site.

It should be noted that the first option above appears to relate to the commencement of the sub-contract works generally, whilst the default position under the second option relates to the commencement of the sub-contract works *on site*.

The date that a payment becomes due (often referred to as the payment due date) is not the date when the sub-contractor receives payment, but is the date from which the final date of payment is established. Also, the payment due date is not necessarily the date when the sub-contractor's work is valued, since clause 4.13 makes it clear that the valuation date could be not more than seven days before the payment due date.

MPSub

After the sub-contractor commences work, the first interim payment is due on the day of the month inserted in the sub-contract particulars against clause 31.2. Therefore if the '28th day of the month' was inserted in the sub-contract particulars, and a sub-contractor commenced work on 20 June (as the example used for DBSub, above), the first interim payment to the sub-contractor would become due on 28 June.

If no day of the month is inserted in the sub-contract particulars against clause 31.2, the default position is that the '25th day of the month' applies.

ICSub

Clause 4.9.1 notes that the first interim payment shall be due on the date for issue of the interim certificate under the main contract immediately following the commencement of the sub-contract works. Therefore, in such a case, if a sub-contractor commenced works on 20 June (for example) and the date for issue of the next interim certificate under the main contract was 21 June, then the sub-contractor's first interim payment would be due on 21 June.

In the case where no dates for the issuing of interim certificates under the main contract are stated in the main contract particulars, then the first interim payment to the sub-contractor shall become due not later than one month after the date of commencement of the sub-contract works on site.

It should be noted that the first option above appears to relate to the commencement of the sub-contract works generally, whilst the default position under the second option relates to the commencement of the sub-contract works *on site.*

The date that a payment becomes due (often referred to as the payment due date) is not the date when the sub-contractor receives payment, but is the date from which the final date of payment is established.

ICSub/D

All as ICSub.

ICSub/NAM

Clause 4.9.1 notes that the first interim payment shall be due on the date specified in article 3 of the sub-contract.

In the case where no such dates are included in article 3, then the first interim payment to the sub-contractor shall become due not later than one month after the date of commencement of the sub-contract works on site.

It should be noted that the first option above appears to relate to the commencement of the sub-contract works generally, whilst the default position under the second option relates to the commencement of the sub-contract works *on site.*

The date that a payment becomes due (often referred to as the payment due date) is not the date when the sub-contractor receives payment, but is the date from which the final date of payment is established.

MWSub/D and ShortSub

The first interim payment becomes due not later than one month after commencement of the sub-contract works on site.

SubSub

The first interim payment becomes due not later than one month after commencement of the sub-subcontract works on site.

When are subsequent interim payments due?

DBSub

The payment due dates for subsequent interim payments are on the same date in each month following the date of the payment due date for the first interim payment (refer to clause 4.9.2). If any of the dates so calculated falls on a non-business day (i.e. a Saturday, Sunday or a public holiday) then the payment due date for that month shall be the nearest business day to the date in question in that month. It should be noted that the wording of clause 4.9.2 refers simply to the 'nearest business day' rather than to the next business day. Therefore, for example, if the payment due date for any particular month fell on a Saturday, then it would be the case that the nearest business day would be the Friday before rather than the Monday following that date.

In line with clause 4.9.2, interim payments continue up to and including the month following the date of practical completion of the sub-contract works as a whole. Therefore, if practical completion of the sub-contract works was achieved in May (for example) then the last 'monthly' interim payment would be the appropriate date in the following month (i.e. June).

That is not to say that interim payments will not continue after that date, but they will not necessarily continue on a monthly basis.

Clause 4.9.2 notes that after the interim payment in the month following practical completion of the sub-contract works, interim payments are only issued as and when further amounts are ascertained as due and payable.

When these interim payments are due, the payment due date will be on the same date of the month (or, as appropriate, the nearest business day to that date) as the pre-practical completion interim payment dates.

Clause 4.9.2 also notes that where the first interim payment becomes due on a date which does not recur in a subsequent month, then the interim payment for that subsequent month shall be due on the last day of that month. Therefore, for example, if the first interim payment became due on 31 October, then (as there are not 31 days in November) the next interim payment would become due on 30 November (assuming of course that the 30 November is a business day; if it is not then the date applicable would be the nearest business day to the 30 November).

MPSub

Prior to practical completion of the sub-contract works, interim payments are due on the day of the month inserted in the sub-contract particulars against clause 31.2. If no day of the month is inserted in the sub-contract particulars against clause 31.2, the default position is that the '25th day of the month' applies.

After practical completion of the sub-contract works, interim payment advices are to be issued at intervals of not less than one month, but the contractor is not obliged to issue a payment advice where the amount identified as due to either party is less than the amount stated in the sub-contract particulars (against clause 31.2). If no value is inserted in the sub-contract particulars against clause 31.2, the default position is that the amount of £5,000.00 applies.

Payments become due on the date of issue of the payment advice.

ICSub

The payment due dates for subsequent interim payments are on the same date in each month following the date of the payment due date for the first interim payment (refer to clause 4.9.2). If any of the dates so calculated falls on a non-business day (i.e. a Saturday, Sunday or a public holiday) then the payment due date for that month shall be the nearest business day to the date in question in that month. It should be noted that the wording of clause 4.9.2 refers simply to the 'nearest business day' rather than to the next business day. Therefore, for example, if the payment due date for any particular month fell on a Saturday, then it would be the case that the nearest business day would be the Friday before rather than the Monday following that date.

In line with clause 4.9.2, interim payments *only* continue up to and including the month following the date of practical completion of the sub-contract works as a whole. Therefore, if practical completion of the sub-contract works was achieved in May (for example) then the last 'monthly' interim payment would be the appropriate date in the following month (i.e. June). After the above payment, the only other payment that would be made would be the final payment.

Clause 4.9.2 also notes that where the first interim payment becomes due on a date which does not recur in a subsequent month, then the interim payment for that subsequent month shall be due on the last day of that month. Therefore, for example, if the first interim payment became due on 31 October, then (as there are not 31 days in November) the next interim payment would become due on 30 November (assuming of course that the 30 November is a business day; if it is not then the date applicable would be the nearest business day to the 30 November).

ICSub/D

All as ICSub.

ICSub/NAM

All as ICSub.

MWSub/D and ShortSub

Subsequent interim payments are due at monthly intervals after the first interim payment.

A further interim becomes due 17 days after the practical completion of the sub-contract works.

SubSub

Subsequent interim payments are due at monthly intervals after the first interim payment.

Can a sub-contractor make a payment application, and what is its effect?

DBSub

Clause 4.13 states that not later than seven days before an interim payment is due the sub-contractor may submit an application setting out what the sub-contractor considers to be the amount of the gross valuation.

If the sub-contractor does submit such an application, and if the contractor agrees with the gross valuation in the sub-contractor's application, the contractor can adopt that valuation for the purposes of determining the gross valuation due in the appropriate interim payment.

It must be noted that the conditions do not require the sub-contractor to submit an application for payment *at all*, the conditions simply state that the sub-contractor *may* submit an application for payment.

If the sub-contractor did not submit an application for payment, *the contractor would still be liable to make payments to the sub-contractor in line with the agreed payment terms.*

MPSub

Under clause 31.1, a sub-contractor is required (not later than seven days before the sub-contractor considers that a payment advice should be issued by the contractor) to submit detailed application for payment to the contractor.

Although it is considered that a contractor has an obligation to issue payment advices even if a sub-contractor does not issue an application for payment, it should be noted that clause 31.2 only requires the contractor to provide payment advices in 'similar detail' to the applications for payment made by a sub-contractor.

ICSub

The ICSub makes no provision for a sub-contractor to make a payment application.

ICSub/D

The ICSub/D makes no provision for a sub-contractor to make a payment application.

ICSub/NAM

The ICSub/NAM makes no provision for a sub-contractor to make a payment application.

MWSub/D and ShortSub

The MWSub/D and the ShortSub makes no provision for a sub-contractor to make a payment application.

SubSub

The SubSub makes no provision for a sub-subcontractor to make a payment application.

What is the amount due, in respect of interim payments?

DBSub

In line with clause 4.10.1, the amount due in interim payments is based either on:

- the value of any stage payments agreed between the contractor and the sub-contractor;
  or, if no stage payments have been agreed:
- the contractor's gross valuation as referred to in clause 4.13.

Irrespective of which of the above options is used, the following deductions are to be made:

(1) any amount which may be deducted and retained as retention in respect of the sub-contract works in accordance with clause 4.15; and
(2) the total amount previously due as interim payments under the sub-contract.

MPSub

In line with clause 31.5, the amount due in interim payments is based on:

(1) The proportion of the sub-contract sum to which the sub-contractor is entitled, calculated in the manner set out in the pricing document.
(2) The value of any changes executed by the sub-contractor.
(3) The amount of any reductions due to work executed in line with design documents that have not been marked 'A Action' or 'B Action'.
(4) Any amount that either party is liable to pay the other in accordance with the provisions of the sub-contract.

Payments previously made should then be deducted to determine the amount due.

ICSub

In line with clause 4.10, the amount due in interim payments is based either on:

- the value of any stage payments agreed between the contractor and the sub-contractor;
  or, if no stage payments have been agreed:
- the contractor's gross valuation as referred to in clause 4.10.1 and 4.10.2, less the total of the amounts referred to in clause 4.10.3.

Irrespective of which of the above options is used, the following adjustments are also to be made:

(1) an adjustment for any retention applicable; and
(2) the total amount previously due as interim payments under the sub-contract.

ICSub/D

All as ICSub.

ICSub/NAM

All as ICSub.

MWSub/D and ShortSub

The amount due in respect of interim payments is the value of the work properly carried out, together with any applicable loss and/or expense, less the total amount due in any previous payments.

SubSub

All as MWSub/D and ShortSub.

What 'discount' can be deducted by the main contractor?

None of the sub-contracts under consideration in this book allow for trade or cash discount.

How much is due for payment in interim payments?

DBSub

The gross valuation to be made by the contractor is the total of the amounts referred to in clauses 4.13.1 to 4.13.6, less the total of the amounts referred to in clause 4.13.7, applied up to and including a date not more than seven days before the date when the interim payment becomes due.

This valuation would therefore include:

- The total of the sub-contract work on site properly executed by the sub-contractor including any variation work executed, and, where applicable, with the adjustment for fluctuations under option C. Where a priced activity schedule is included in the numbered documents, the value is to be based upon the percentage that is completed of each activity in that activity schedule.
- The total value of the materials and goods delivered to or adjacent to the main contract works for incorporation into the main contract works by the sub-contractor.
- The total value of any off-site materials (if those materials are listed items under the sub-contract, and assuming that the pre-conditions for the payment of off-site materials have been met).
- The total value of payments made or costs incurred by the sub-contractor under clauses 2.18 or 2.20 of the main contract conditions (i.e. in respect of fees or charges legally demandable, and in respect of royalties and patent rights).
- Any loss and expense payments, and the cost of the restoration of the sub-contract works and the replacement or repair of any sub-contract site materials that are lost or damaged, and the removal and disposal of debris which under insurance clause 6.7.4 is to be treated as a variation.
- Any amount payable to or deductible from the sub-contractor under fluctuations options A or B, if applicable.
- Any amount deductible from the sub-contractor in respect of inaccurate setting out by the sub-contractor or to defects or other faults in the sub-contract works, and for any non-compliant work carried out by the sub-contractor which is allowed to remain in place.

MPSub

In line with clause 31.5, the amount due in interim payments is based on:

- The proportion of the sub-contract sum to which the sub-contractor is entitled, calculated in the manner set out in the pricing document.
- The value of any changes executed by the sub-contractor.
- The amount of any reductions due to work executed in line with design documents that have not been marked 'A Action' or 'B Action'.
- Any amount that either party is liable to pay the other in accordance with the provisions of the sub-contract.

Payments previously made should then be deducted to determine the amount due.

ICSub

The valuation to be made by the contractor is the total of:

(1) The relevant percentage as stated in the sub-contract particulars for works executed, etc., taking into account that a retention value may need to be held. The default percentage is 95% if practical completion of the sub-contract works has not been achieved, and 97.5% after practical completion of the sub-contract works has been achieved. These said percentages can of course be entered into the sub-contract particulars at a different level to the default percentages noted above. The applicable percentage is then applied to:

    (a) The value of work properly executed (including the value of variations, but excluding any value related to works carried out under insurance clause 6.7.4 which is otherwise treated as a variation). Where there is an activity schedule then the valuation will be based upon that activity schedule, and the valuation of each activity will be based upon the percentage of the applicable activity complete at the applicable valuation date.

    (b) The value of materials and goods delivered to or adjacent to the main contract works for incorporation therein by the sub-contractor (provided that those materials and goods have been reasonably, properly and not prematurely delivered, and provided that the materials and goods are adequately protected against weather, etc.).

    (c) The value of any listed items (i.e. off-site materials and goods).

(2) In addition to the above, the full amount (i.e. 100%) of any applicable:

    (a) Fees and charges (as clause 2.4 of the sub-contract conditions).

    (b) Inspection and testing (as clauses 3.9 and 3.10.3 of the sub-contract conditions).

    (c) Fluctuations (as clause 4.15 of the sub-contract conditions).

    (d) Loss and expense (as clause 4.16 of the sub-contract conditions).

    (e) Insurance premiums (as clause 6.6.3 of the sub-contract conditions).

    (f) Restoration, etc., of loss or damage (as clause 6.7.4 of the sub-contract conditions).

(3) Also, the ascertained amount of any deduction to be made for:

(a) Deductions under the main contract conditions (as clause 2.17 of the main contract conditions).
(b) Non-compliance with directions (as clause 3.6 of the sub-contract conditions).
(c) Fluctuations (as clause 4.15 of the sub-contract conditions).

ICSub/D

All as ICSub.

ICSub/NAM

All as ICSub.

MWSub/D and ShortSub

The amount due for payment in respect of interim payments is 95% (or such other percentage as may be set out in the sub-contract documents) of the value of the work properly carried out, together with any applicable loss and/or expense, less the total amount due in any previous payments.

After practical completion of the sub-contract works, the amount due for payment increases to 97.5% (or such other percentage as may be set out in the sub-contract documents).

The value of the works properly executed is determined by reference to the rates and prices in the pricing document or by reference to the sub-contract sum if there are no such rates and prices.

SubSub

The amount due for payment in respect of interim payments is the value of the work properly carried out, together with any applicable loss and/or expense, less the total amount due in any previous payments.

The value of the works properly executed is determined by reference to the rates and prices in the pricing document or by reference to the sub-contract sum if there are no such rates and prices.

Who owns the sub-contractor's unfixed materials on site?

DBSub

It should be noted that even if payment is not made for any materials or goods delivered to or adjacent to the main contract works (which are intended for the main contract works) by the sub-contractor, these materials or goods shall not be removed from site unless the contractor consents in writing to such removal. This is confirmed under clause 2.15.1, which also makes it clear that such consent shall not be unreasonably delayed or withheld.

Clause 2.15.2 notes that when the value of any of the sub-contractor's unfixed materials or goods on site have been included in any interim payment under which the amount properly due to the contractor has been paid to him by the employer, those materials or goods shall be and shall become the property of the employer and the sub-contractor shall not deny that they are and have become the property of the employer. Therefore in such a scenario, the sub-contractor's unfixed materials and goods would become the property of the employer even though the sub-contractor may not have been paid for them.

Further, clause 2.15.3 notes that if the contractor pays the sub-contractor for any unfixed materials or goods on site, in advance of a value for such materials and goods being included in any interim payment (under the main contract), such materials or goods shall be and shall become the property of the contractor upon payment being made by him to the sub-contractor.

The above clauses are designed to protect the employer and the contractor from right of title claims by the sub-contractor in respect of unfixed materials and goods on site.

The above concern only exists whilst the materials or goods are unfixed. Materials and goods which are to be incorporated into the works in almost all construction projects are bound at some stage to become 'attached to the soil', and thus become 'fixtures' to the land. Once the materials and goods become fixtures they become the property of the freeholder (normally the employer) in any event.

MPSub

The sub-contract is silent as to which party owns the sub-contractor's unfixed materials on site. Therefore, the normal position would be that the sub-contractor retains ownership until after payment for the materials has been made to him, or until the materials become fixtures to the property or are incorporated into the main contract works.

ICSub

It should be noted that even if payment is not made for any materials or goods delivered to or adjacent to the main contract works (which are intended for the main contract works) by the sub-contractor, these materials or goods shall not be removed from site unless the contractor with the agreement of the architect/contractor administrator consents in writing to such removal. This is confirmed under clause 2.11.1, which also makes it clear that such consent shall not be unreasonably delayed or withheld.

Clause 2.11.2 notes that when the value of any of the sub-contractor's unfixed materials or goods on site have been included in any interim payment certificate (under the main contract) under which the amount properly due to the contractor has been paid to him by the employer, those materials or goods shall be and shall become the property of the employer and the sub-contractor shall not deny that they are and have become the property of the employer. Therefore in such a scenario, the sub-contractor's unfixed materials and goods would become the property of the employer even though the sub-contractor may not have been paid for them.

Further, clause 2.11.3 notes that if the contractor pays the sub-contractor for any unfixed materials or goods on site, in advance of a value for such materials and goods being included in any interim payment certificate (under the main contract), such materials or goods shall be and shall become the property of the contractor upon payment being made by him to the sub-contractor.

The above clauses are designed to protect the employer and the contractor from right of title claims by the sub-contractor in respect of unfixed materials and goods on site.

The above concern only exists whilst the materials or goods are unfixed. Materials and goods which are to be incorporated into the works in almost all construction projects are bound at some stage to become 'attached to the soil', and thus become 'fixtures' to the land. Once the materials and goods become fixtures they become the property of the freeholder (normally the employer) in any event.

ICSub/D

All as ICSub.

ICSub/NAM

All as ICSub.

MWSub/D

All as MPSub.

ShortSub

All as MPSub.

SubSub

All as MPSub.

What are the pre-conditions for the payment of off-site materials?

DBSub

The first point is that payment for off-site materials will not be made unless the materials are listed items under the sub-contract particulars.

However, even if the materials or goods in question are listed items, the sub-contractor is only entitled to be paid if the following conditions have been fulfilled:

(1) The listed items are in accordance with the sub-contract. This effectively means that the item must be in line with the specification requirements. However, a listed item could also not be in accordance with the sub-contract if, because of a variation, the work requiring the listed item had been omitted or altered.
(2) The sub-contractor has provided the contractor with reasonable proof that the property in such listed items is vested in the sub-contractor.

(3) The listed items are insured for their full value against loss or damage under a policy of insurance protecting the rights of the employer, the contractor and the sub-contractor. The insurance needs to be in force from the period when the property (right of title) of the materials or goods passes to the sub-contractor up until the time that those materials or goods are delivered to, or adjacent to, the main contract works.

(4) At the premises where the listed items have been manufactured or assembled or are stored, the materials or goods are set apart or have been clearly and visibly marked (individually or in sets) by letters or figures or by reference to a pre-determined code, and there is in relation to such items clear identification of the contractor and the employer to whose order they are held, and their destination as the main contract works.

(5) That, if required in the sub-contract particulars, a bond in favour of the contractor or the employer (as the contractor directs) is provided from a surety approved by the contractor, in the amount set out in the sub-contract particulars and in the terms set out in part 1 of schedule 3 to the sub-contract. The form of bond in question is what is known as an 'on demand' bond. An 'on demand' bond is a bond that can be called upon without any condition (i.e. purely 'on demand') and the sub-contractor needs to be aware of this fact before entering into the bond.

MPSub

The MPSub does not envisage that payment will be made for off-site materials.

ICSub

All as DBSub, except that the terms of the bond are set out under schedule 1 of the sub-contract and not under part 1 of schedule 3 of the sub-contract.

ICSub/D

All as ICSub.

ICSub/NAM

All as ICSub.

MWSub/D, ShortSub and SubSub

These sub-contracts do not envisage that payment will be made for off-site materials.

When is a payment notice to be given in respect of interim payments?

DBSub

A written payment notice is to be given not later than five days after the date on which an interim payment becomes due. That notice is to specify the amount of the payment which is proposed to be made in respect of the sub-contract works, to what the amount of the payment relates and the basis on which that amount was calculated.

MPSub

Clause 31.3 notes that a statement setting out the amount of the payment proposed to be made and the basis on which that amount was calculated (i.e. the payment notice) should be set out in, or should be attached to each payment advice.

ICSub

All as DBSub.

ICSub/D

All as DBSub.

ICSub/NAM

All as DBSub.

MWSub/D, ShortSub and SubSub

A payment notice is to be given not later than five days after the date on which a payment becomes due. That notice is to specify the amount of the payment to be made and the basis on which that amount was calculated.

What is the effect of not giving a payment notice in respect of interim payments?

Many sub-contractors are of the view that if a payment notice is not given in respect of an application submitted by a sub-contractor, the sub-contractor will automatically be entitled to receive payment in line with the application submitted. This is not the case.

A series of court cases have shown that the lack of a payment notice has very little practical effect. All the sub-contractor is entitled to be paid, irrespective of whether or not a payment notice is given or not, is the amount that is actually due.

The amount that is actually due is not determined by the value claimed by the sub-contractor within a payment application submitted, nor, indeed, by the value indicated in any payment notice issued by the contractor, but is based on the correct gross valuation of the works.

Of course, if a payment notice was not issued, and if there was a dispute regarding the amount paid in respect of a particular interim payment, and if that dispute was referred to adjudication, then the adjudicator may take into account the lack of a payment notice (as a background issue), but this should not have any impact upon his decision as to the correct amount due.

When is the final date for payment in respect of interim payments?

DBSub

The final date for payment of interim payments is 21 days after the date on which those payments become due (as clause 4.9.3).

MPSub

Other than in the case of termination, the final date for payment of interim payments is 21 days after the date on which those payments become due.

ICSub

All as DBSub.

ICSub/D

All as DBSub.

ICSub/NAM

All as DBSub.

MWSub/D and ShortSub

The final date for payment of interim payments is 28 days after the date on which those payments become due.

SubSub

The final date for payment of interim payments is 31 days after the date on which those payments become due.

When is a withholding notice to be given in respect of interim payments?

DBSub

A written withholding notice is to be given not later than five days before the final date for payment of an interim payment. The withholding notice is to specify the amount proposed to be withheld and/or deducted from the amount notified in the payment notice, the ground or grounds for such withholding and/or deduction, and the amount of withholding and/or deduction attributable to each ground.

    A contractor may give the sub-contractor a withholding notice at the same time as he gives a payment notice. If a withholding notice is combined with a payment notice in this way, then for that 'combined' notice to be effective it would need to be issued not later than five days after the applicable payment due date, and the 'combined' notice would need to satisfy the requirement of both a payment notice and a withholding notice.

MPSub

A written withholding notice is to be given not later than seven days before the final date for payment of an interim payment. The withholding notice is to specify the amount proposed to be withheld and/or deducted, the ground or grounds for such withholding and/or deduction, and the amount of withholding and/or deduction attributable to each ground.

ICSub

All as DBSub.

ICSub/D

All as DBSub.

ICSub/NAM

All as DBSub.

MWSub/D and ShortSub

A written withholding notice is to be given not later than five days before the final date for payment of an interim payment. The withholding notice is to specify the amount proposed to be withheld and/or deducted from the amount notified in the payment notice, the ground or grounds for such withholding and/or deduction, and the amount of withholding and/or deduction attributable to each ground.

Clause 12.7 notes that the contractor is entitled to withhold payment of all or part of any sums otherwise due where a sum is due from the sub-contractor to the contractor under the sub-contract. Of course, any such withholding made can only be made if a compliant withholding notice has been issued.

SubSub

A written withholding notice is to be given not later than five days before the final date for payment of an interim payment. The withholding notice is to specify the amount proposed to be withheld and/or deducted from the amount notified in the payment notice, the ground or grounds for such withholding and/or deduction, and the amount of withholding and/or deduction attributable to each ground.

Clause 12.4 notes that the sub-contractor is entitled to withhold payment of all or part of any sums otherwise due where a sum is due from the sub-subcontractor to the sub-contractor under the sub-subcontract. Of course, any such withholding made can only be made if a compliant withholding notice has been issued.

What is the effect of not giving a withholding notice in respect of an interim payment?

Unlike the courts' view of payment notices, it is now quite clear following a series of court cases that if a timeous withholding notice is not issued, no withholding or deduction will be permitted within an adjudication action. This provision is strictly operated, and if a withholding notice is not issued within the timescale detailed within the sub-contract (i.e. no later than five days before the final date for payment, of the payment in question), the proposed withholding and/or deduction will almost certainly be disallowed and the merits of the proposed withholding and/or deduction will not be considered at all.

DBSub

Under DBSub, clause 4.10.4 makes the importance of a withholding notice quite clear by stating:

> 'Subject to any notice given under clause 4.10.3, the Contractor shall no later than the final date for payment pay the amount specified in his notice given under clause 4.10.2 or, in the absence of a notice under clause 4.10.2, the amount calculated in accordance with clause 4.10.1.'

MPSub

Under MPSub, clause 32.1 notes that, provided that an effective notice of withholding has been given, the contractor may withhold from any interim payment that is due to the sub-contractor, any amount that the sub-contractor is liable to pay the contractor in accordance with the terms of the sub-contract and/or any sums owed to the contractor by the sub-contractor as a consequence of any breach of the sub-contract.

ICSub

Under ICSub, clause 4.12.3 makes the importance of a withholding notice quite clear by stating:

> 'Subject to any notice given under clause 4.12.2, the Contractor shall no later than the final date for payment pay the amount specified in his notice given under clause 4.12.1 or, in the absence of a notice under clause 4.12.1, the amount calculated in accordance with clause 4.10.'

ICSub/D

All as ICSub.

ICSub/NAM

All as ICSub.

MWSub/D, ShortSub and SubSub

Unlike the courts' view of payment notices, it is now quite clear following a series of court cases that if a timeous withholding notice is not issued, no withholding or deduction will be permitted within an adjudication action.

When is the final payment due?

DBSub

The final payment is due not later than seven days after either the contractor's final account and final statement under the main contract or the employer's final account and final statement under the main contract become conclusive as to balance due between the employer and the contractor under the main contract.

MPSub

The final payment is due upon the date of issue of the statement issued by the contractor under either clause 27.2 or 27.3 of the sub-contract. The statements issued under clauses 27.2 or 27.3 relate to the clearance of the sub-contractor's defects.

ICSub

The final payment is due not later than seven days after the date of issue (not the date of receipt) of the final certificate under the main contract.

The above position is effectively a 'pay when certified' situation which many commentators consider is so closely akin to 'pay when paid' that it should be outlawed by way of the Housing Grants, Construction and Regeneration Act 1996, but which the courts appear to consider is a legitimate mechanism.

In this situation the sub-contractor's final payment (under the sub-contract) is directly linked to the date of issue of the final certificate (under the main contract), something that the sub-contractor has absolutely no control over.

ICSub/D

All as ICSub.

ICSub/NAM

All as ICSub.

MWSub/D and ShortSub

The final payment is due either seven days after the expiry of the defects liability period under the main contract (if no defects in the sub-contract works exist at that time) or seven days after the date of completion of the defects if defects in the sub-contract works do exist at that time.

SubSub

Under the SubSub there is no 'final payment' as such. The final payment is in reality simply the last interim payment.

When is a payment notice to be given in respect of the final payment?

DBSub

The payment notice in respect of the final payment is to be given not later than five days after the date on which the final payment becomes due. That notice is to be sent by special or recorded delivery. The payment notice is to notify the sub-contractor of the amount of the final payment to be made to the sub-contractor, to what the amount of the payment relates, and the basis on which such an amount was calculated.

MPSub

Clause 31.3 notes that a statement setting out the amount of the payment proposed to be made and the basis on which that amount was calculated (i.e. the payment notice) should be set out in, or should be attached to each payment advice.

ICSub

All as DBSub.

ICSub/D

All as DBSub.

ICSub/NAM

All as DBSub.

MWSub/D, ShortSub and SubSub

A payment notice is to be given not later than five days after the date on which the final payment becomes due. That notice is to specify the amount of the payment to be made and the basis on which that amount was calculated.

What is the effect of not giving a final payment notice in respect of the final payment?

DBSub

Many sub-contractors may be of the view that if a final payment notice is not given in respect of an application (or a final account) submitted by a sub-contractor, the sub-contractor will automatically be entitled to receive payment in line with the application (or final account) submitted.

As noted above in respect of interim payments, this is not the case, all the sub-contractor is entitled to, irrespective of whether or not a final payment notice is issued, is the amount that is actually due.

The amount that is actually due is not determined by the value claimed by the sub-contractor, nor, indeed, by the value indicated in any final payment notice submitted by the contractor, but is based on the correct gross valuation of the works, less any previous payments made.

Of course, if a final payment notice was not issued, and if there was a dispute regarding the amount paid that was referred to adjudication, then the adjudicator may take into account the lack of a final payment notice (again as a background issue only), but this should not have any impact upon his decision as to the correct amount due.

The correct amount due should be determined simply by applying the provisions of the contract to arrive at the final sub-contract sum and then deduct any previous payments made to arrive at the amount due.

However, the final payment notice must be sent by special or recorded delivery, whereas a payment notice for an interim payment does not need to be sent in such a way, and it appears clear that *this is because of the effect that the final payment notice has regarding the finality of the matters referred to in clause 1.9 of the sub-contract conditions* (refer to Chapter 3 of this book).

MPSub

Clause 31.7 notes that the final payment advice is final and binding upon the parties in relation to amounts due from the contractor to the sub-contractor under or in connection with the sub-contract, including any sums due to the sub-contractor as a consequence of claims for breach of contract, breach of statutory duty, negligence or otherwise, unless within 28 days of the final payment advice being issued the sub-contractor disputes any aspect of it by reference to adjudication or litigation.

ICSub

All as DBSub.

ICSub/D

All as DBSub.

ICSub/NAM

All as DBSub.

MWSub/D, ShortSub and SubSub

A series of court cases have shown that the lack of a payment notice has very little practical effect. All the sub-contractor is entitled to be paid, irrespective of whether or not a payment notice is given, is the amount that is actually due.

The amount that is actually due is not determined by the value claimed by the sub-contractor within a payment application submitted, nor, indeed, by the value indicated in any payment notice issued by the contractor, but is based on the correct gross valuation of the works, less only the total amount due in previous interim payments.

What methods are used to calculate the final sub-contract sum?

DBSub

The sub-contract can be placed on the basis of either a sub-contract sum or a sub-contract tender sum.

When the sub-contract sum is used this means that the final sub-contract sum will be calculated on an adjustment basis and when the sub-contract tender sum is used this means that final sub-contract sum will be calculated on a remeasurement basis.

The adjustment basis is generally referred to as a lump sum contract.

In this context, a lump sum contract is a contract to complete specified works for a lump sum price, but payment would still normally be paid if the specified works were not entirely complete. A lump sum contract is different to an 'entire' contract, which rarely applies in any event, where payment is only made if the contract works are *entirely* complete.

Under the adjustment basis, the quality and quantity of work included in the sub-contract sum is that set out in the bills of quantities and the sub-contract design documents, or where there are no bills of quantities, the quality and quantity of work is that included in the sub-contract documents taken together.

If the contractor provides quantities and these are contained in the sub-contract documents, then the quantity of work contained in the sub-contract sum shall be in line with those quantities.

When the adjustment basis applies, the sub-contract sum shall not be altered in any way other than in accordance with the express provisions of the conditions (e.g. by the issue of a variation instruction). This approach is in line with the traditional approach of a lump sum contract, with additions and/or omissions to that lump sum price for variations.

Importantly, it must be noted that, other than for any errors or inadequacies in the bills of quantities, any error, whether an arithmetic error or otherwise, shall be deemed to have been accepted by the parties.

Therefore if there is an arithmetical or other error in the computation of a bill of quantities item, a bill of quantities page total, a bill of quantities section total, or a bill of quantities summary page, these errors are deemed to have been accepted by the parties and no subsequent adjustment will be made for such errors.

Under the adjustment basis, the final sub-contract sum is calculated by making the following adjustments to the sub-contract sum:

(1) The amount stated (either an addition or an omission) by the contractor in his acceptance of any schedule 2 quotation.
(2) The amount (either an addition or an omission) of any variation to any schedule 2 quotation.
(3) The deduction of all Provisional Sums and the addition of the amount of the valuation of the works executed by or the disbursements made by the sub-contractor in accordance with the directions of the contractor as to the expenditure of those Provisional Sums.
(4) The deduction of the value of all work described as provisional included in the sub-contract documents, and the addition of the amount of the valuation of the works executed by or the disbursements made by the sub-contractor

in accordance with the directions of the contractor as to the expenditure against all work described as provisional included in the sub-contract documents.

(5) The deduction of the value of all approximate quantities included in any bill of quantities, and the addition of the amount of the valuation of the works executed by or the disbursements made by the sub-contractor in accordance with the directions of the contractor as to the expenditure against all approximate quantities described in any bill of quantities or in the Contractor's Requirements.

(6) The adjustment for the amount due (whether this be an addition or an omission) of each variation valued in line with the valuation rules, together with the adjustment (whether this be an addition or an omission) of the amount included in the sub-contract documents for any other work which has suffered a substantial change in its conditions due to the variation in question.

(7) The deduction in respect of deductions made under the main contract conditions, and in respect of non-compliant work.

(8) The addition of any payment due to the sub-contractor as a result of payments made or costs incurred by the sub-contractor in relation to fees or charges legally demandable, and royalties and patent rights under the main contract conditions.

(9) The amount of any valuation in respect of non-compliant work by others.

(10) The amount ascertained in respect of the sub-contractor's loss and expense.

(11) Any adjustment (whether this be an addition or an omission) due to the fluctuation option applicable.

(12) The adjustment (whether this be an addition or an omission) of any other amount. Adjustments of any other amount may, for example, include costs incurred by the contractor associated with the non-compliance of directions by the sub-contractor; or associated with insurance taken out on behalf of the sub-contractor.

The alternative to the above is the remeasurement basis.

Under the remeasurement basis, the quality and quantity of work included in the sub-contract tender sum is that set out in the bills of quantities and the sub-contractor's design documents, or where there are no bills of quantities, the quality and quantity of work will be that included in the sub-contract documents taken together. If the contractor provides quantities and these are contained in the sub-contract documents, then the quantity of work contained in the sub-contract tender sum shall be in line with those quantities.

When the remeasurement basis applies, the sub-contract works shall be subject to complete remeasurement.

When the remeasurement basis applies, the final sub-contract sum is based upon a valuation of a complete remeasurement of the works rather than being based on additions to or omissions from the sub-contract tender sum.

This amount is then adjusted by:

(1) The amount stated (either an addition or an omission) by the contractor in his acceptance of any schedule 2 quotation.

(2) The amount (either an addition or an omission) of any variation to any schedule 2 quotation.

(3) The addition of any payment due to the sub-contractor as a result of payments made or costs incurred by the sub-contractor in respect of fees or charges legally demandable and royalties and patent rights payments due under the main contract conditions.
(4) The amount of any valuation in respect of non-compliant work by others.
(5) The deduction of any applicable deductions under the main contract conditions, and any applicable deductions in respect of non-compliant work.
(6) The amount ascertained in respect of the sub-contractor's loss and expense.
(7) Any adjustment (whether this be an addition or an omission) due to the fluctuation option applicable.
(8) The adjustment (whether this be an addition or an omission) of any other amount, which may include for costs incurred by the contractor associated with the non-compliance of directions by the sub-contractor or associated with insurance taken out on behalf of the sub-contractor.

MPSub

In line with clause 31.6, the amount due in the final payments is based on:

- The sub-contract sum.
- The value of any changes executed by the sub-contractor.
- The amount of any reductions due to work executed in line with design documents that have not been marked 'A Action' or 'B Action'.
- Any amount that either party is liable to pay the other in accordance with the provisions of the sub-contract.
- The appropriate deduction recorded by any statement issued under the provisions of clause 27.3 in respect of defects that the contractor does not intend to rectify.

Payments previously made should then be deducted to determine the amount due.

ICSub

The sub-contract can be placed on the basis of either a sub-contract sum (article 3A) or a sub-contract tender sum (article 3B).

When the sub-contract sum is used this means that the final sub-contract sum shall be calculated on an adjustment basis and when the sub-contract tender sum is used this means that final sub-contract sum shall be calculated on a remeasurement basis.

The adjustment basis is generally referred to as a lump sum contract.

In this context, a lump sum contract is a contract to complete specified works for a lump sum price, but payment would still normally be paid if the specified works were not entirely complete. A lump sum contract is different to an 'entire' contract, which rarely applies in any event, where payment is only made if the contract works are *entirely* complete.

Under the adjustment basis, the quality and quantity of work included in the sub-contract sum is that set out in the bills of quantities, or where there are no bills of quantities, the quality and quantity of work is that included in the sub-contract documents taken together.

If the contractor provides quantities and these are contained in the sub-contract documents, then the quantity of work contained in the sub-contract sum shall be in line with those quantities. If there are no such quantities, but work stated or shown on any drawings included in the sub-contract documents is inconsistent with any description of that work in the specification/work schedules, then that stated or shown on those drawings shall be deemed to prevail.

When the adjustment basis applies, the sub-contract sum shall not be altered in any way other than in accordance with the express provisions of the conditions (e.g. by the issue of a variation instruction). This approach is in line with the traditional approach of a lump sum contract, with additions and/or omissions for variations.

Importantly, it must be noted that, other than for any errors or inadequacies in the bills of quantities, any error, whether an arithmetic error or otherwise, shall be deemed to have been accepted by the parties.

Therefore, if there is an arithmetical or other error in the computation of a bill of quantities item, a bill of quantities page total, a bill of quantities section total, or a bill of quantities summary page, these errors are deemed to have been accepted by the parties and no subsequent adjustment will be made for such errors.

Under the adjustment basis, the final sub-contract sum is calculated by making the following adjustments to the sub-contract sum:

(1) The deduction of all Provisional Sums and the addition of the amount of the valuation of the works executed by or the disbursements made by the sub-contractor in accordance with the directions of the contractor as to the expenditure of those Provisional Sums.

(2) The deduction of the value of all work described as provisional included in the sub-contract documents, and the addition of the amount of the valuation of the works executed by or the disbursements made by the sub-contractor in accordance with the directions of the contractor as to the expenditure against all work described as provisional included in the sub-contract documents.

(3) The deduction of the value of all approximate quantities included in any bill of quantities, and the addition of the amount of the valuation of the works executed by or the disbursements made by the sub-contractor in accordance with the directions of the contractor as to the expenditure against all approximate quantities described in any bill of quantities or in the Contractor's Requirements.

(4) The adjustment for the amount due (whether this be an addition or an omission) of each variation valued in line with the valuation rules, together with the adjustment (whether this be an addition or an omission) of the amount included in the sub-contract documents for any other work which has suffered a substantial change in its conditions due to the variation in question.

(5) The deduction in respect of deductions made under the main contract conditions, and in respect of non-compliant work.

(6) The addition of any payment due to the sub-contractor as a result of payments made or costs incurred by the sub-contractor in relation to fees or charges legally demandable, and royalties and patent rights under the main contract conditions.

(7) The amount of any valuation in respect of non-compliant work by others.

(8) The amount ascertained in respect of the sub-contractor's loss and expense.

(9) Any adjustment (whether this be an addition or an omission) due to the fluctuation option applicable.

(10) The adjustment (whether this be an addition or an omission) of any other amount. Adjustments of any other amount may, for example, include costs incurred by the contractor associated with the non-compliance of directions by the sub-contractor; or associated with insurance taken out on behalf of the sub-contractor.

The alternative to the above is the remeasurement basis.

Under the remeasurement basis, the quality and quantity of work included in the sub-contract tender sum is that set out in the bills of quantities (and the sub-contractor's design portion documents), or where there are no bills of quantities, the quality and quantity of work will be that included in the sub-contract documents taken together. If the contractor provides quantities and these are contained in the sub-contract documents, then the quantity of work contained in the sub-contract tender sum shall be in line with those quantities.

When the remeasurement basis applies, the sub-contract works shall be subject to complete remeasurement.

When the remeasurement basis applies, the final sub-contract sum is based upon a valuation of a complete remeasurement of the works rather than being based on additions to or omissions from the sub-contract tender sum.

This amount is then adjusted by:

(1) The addition of any payment due to the sub-contractor as a result of payments made or costs incurred by the sub-contractor in respect of fees or charges legally demandable and royalties and patent rights payments due under the main contract conditions.

(2) The amount of any valuation in respect of non-compliant work by others.

(3) The deduction of any applicable deductions under the main contract conditions, and any applicable deductions in respect of non-compliant work.

(4) The amount ascertained in respect of the sub-contractor's loss and expense.

(5) Any adjustment (whether this be an addition or an omission) due to the fluctuation option applicable.

(6) The adjustment (whether this be an addition or an omission) of any other amount, which may include for costs incurred by the contractor associated with the non-compliance of directions by the sub-contractor or associated with insurance taken out on behalf of the sub-contractor.

ICSub/D

All as ICSub.

ICSub/NAM

All as ICSub.

MWSub/D and ShortSub

The gross amount due in the final payment is simply the value of the work (including variations) properly carried out, determined in accordance with the rates and prices specified in the pricing document (if one exists) or by reference to the sub-contract sum if there are no rates and prices, together with any direct loss and/or expense ascertained under clause 10.3.

SubSub

The gross amount due in the final payment is simply the value of the work (including variations) properly carried out, determined in accordance with the rates and prices specified in the pricing document (if one exists) or by reference to the sub-subcontract sum if there are no rates and prices, together with any direct loss and/or expense ascertained under clause 10.3.

When is the sub-contractor to provide information to the contractor to enable the contractor to calculate the final sub-contract sum?

DBSub

The sub-contractor is to send to the contractor all documents necessary for the purposes of calculating the final sub-contract sum not later than two months after practical completion of the sub-contract works.

Because of the nature of building works the 'documents necessary' needs to be judged on a project by project basis and cannot be more precisely defined, and therefore there may be disputes regarding what documents are actually necessary.

MPSub

Under clause 31.1, a sub-contractor is required (not later than seven days before the sub-contractor considers that a payment advice should be issued by the contractor) to submit detailed application for payment to the contractor.

Although it is considered that a contractor has an obligation to issue the final payment advice even if a sub-contractor does not issue an application for payment, it should be noted that clause 31.2 only requires the contractor to provide payment advices in 'similar detail' to the applications for payment made by a sub-contractor.

ICSub

The sub-contractor is to send to the contractor all documents necessary for the purposes of calculating the final sub-contract sum not later than four months after practical completion of the sub-contract works.

Because of the nature of building works the 'documents necessary' needs to be judged on a project by project basis and cannot be more precisely defined, and, therefore, there may be disputes regarding what documents are actually necessary.

ICSub/D

All as ICSub.

ICSub/NAM

All as ICSub.

MWSub/D and ShortSub

These sub-contracts do not require the sub-contractor to provide information to the contractor to enable the contractor to calculate the final sub-contract sum.

SubSub

The SubSub does not require the sub-subcontractor to provide information to the sub-contractor to enable the sub-contractor to calculate the final sub-subcontract sum.

When is the contractor to prepare and send to the sub-contractor a statement of the calculation of the final sub-contract sum?

DBSub

The contractor is to prepare and send to the sub-contractor a statement of the calculation of the final sub-contract sum not later than eight months after receipt by the contractor from the sub-contractor of the 'documents necessary' to enable the contractor to calculate the final sub-contract sum.

If the contractor does not receive the documents necessary for the purposes of calculating the final sub-contract sum within two months of the practical completion date of the sub-contract works, the contractor is to prepare and send to the sub-contractor a statement of the calculation of the final sub-contract sum based on the information that is in his possession.

MPSub

The contractor is to prepare and send to the sub-contractor a statement of the calculation of the final sub-contract sum with the final payment advice.

ICSub

The contractor is to prepare and send to the sub-contractor a statement of the calculation of the final sub-contract sum:

(1) not later than eight months after receipt by the contractor from the sub-contractor of the 'documents necessary' to enable the contractor to calculate the final sub-contract sum; and
(2) before the architect/contract administrator issues the final certificate under the main contract.

Therefore, in effect, a long-stop period of eight months after receipt by the contractor of the 'documents necessary' to enable the contractor to calculate the final sub-contract sum is set for the contractor to send to the sub-contractor a statement of the calculation of the final sub-contract sum.

This long-stop period may of course be much shorter, and *must* be shorter if the period between the date when the sub-contractor sends to the contractor all documents necessary for the purpose of calculating the final sub-contract sum and the date that the architect/contract administrator issues the final certificate under the main contract conditions is less than eight months.

ICSub/D

All as ICSub.

ICSub/NAM

All as ICSub.

MWSub/D and ShortSub

These sub-contracts do not require the contractor to prepare and send to the sub-contractor a statement of the calculation of the final sub-contract sum.

SubSub

The SubSub does not require the sub-contractor to prepare and send to the sub-subcontractor a statement of the calculation of the final sub-subcontract sum.

What is the effect of the statement of the calculation of the final sub-contract sum?

DBSub

Clause 4.6.4 makes it clear that if nothing in the statement of the calculation of the final sub-contract sum sent by the contractor to the sub-contractor is disputed by the sub-contractor, giving grounds for so disputing, *within one month* of the submission of the statement to the sub-contractor, the said statement will be deemed to have taken into account all adjustments required by the conditions of the sub-contract.

MPSub

As the statement of the calculation of the final sub-contract sum forms part of the final payment advice, the effect of the statement is the same as the effect of the final payment advice.

ICSub

There appears to be no contractual effect of the statement of the calculation of the final sub-contract sum.

ICSub/D

All as ICSub.

ICSub/NAM

All as ICSub.

MWSub/D, ShortSub and SubSub

There is no statement of the calculation of the final sub-contract sum (or sub-subcontract sum) under these sub-contracts and therefore this question is not applicable.

When is the final date for payment in respect of the final payment?

DBSub

The final date for payment of the final payment is 28 days after the date that it becomes due.

MPSub

Other than in the case of termination, the final date for payment of interim payments is 21 days after the date on which those payments become due.

ICSub, ICSub/D, ICSub/NAM, MWSub/D and ShortSub

All as DBSub.

SubSub

The final date for payment of the final payment (i.e. the last interim payment) is 31 days after the date that it becomes due.

When is a withholding notice to be given in respect of the final payment?

DBSub

A written withholding notice is to be given not later than five days before the final date for payment of the final payment. The withholding notice is to specify the amount proposed to be withheld and/or deducted from the amount notified in the payment notice, the ground or grounds for such withholding and/or deduction, and the amount of withholding and/or deduction attributable to each ground.

A contractor may give the sub-contractor a withholding notice at the same time as he gives a payment notice. If a withholding notice is combined with a payment notice in this way, then for that 'combined' notice to be effective it would need to be issued not later than five days after the applicable payment due date, and the 'combined' notice would need to satisfy the requirement of both a payment notice and a withholding notice.

MPSub

A written withholding notice is to be given not later than seven days before the final date for payment of an interim payment. The withholding notice is to specify the amount proposed to be withheld and/or deducted, the ground or grounds for such withholding and/or deduction, and the amount of withholding and/or deduction attributable to each ground.

ICSub, ICSub/D, ICSub/NAM, MWSub/D, ShortSub and SubSub

All as DBSub.

What is the effect of not giving a withholding notice in respect of the final payment?

Unlike the courts' view of payment notices, it is now quite clear following a series of court cases that if a timeous withholding notice is not issued, no withholding or deduction will be permitted within an adjudication action. This provision is strictly operated, and if a withholding notice is not issued within the timescale detailed within the sub-contract (i.e. no later than five days before the final date for payment, of the payment in question), the proposed withholding and/or deduction will almost certainly be disallowed and the merits of the proposed withholding and/or deduction will not be considered at all.

DBSub

Clause 4.12.3 of the sub-contract states:

> 'If no such notice of withholding and/or deduction is given, the Contractor shall not later than the final date for payment pay the amount notified to the Sub-contractor under clause 4.12.2 or (in the absence of any notice under clause 4.12.2) the amount calculated in accordance with clause 4.12.1.'

MPSub

Clause 32.1 notes that, provided that an effective notice of withholding has been given, the contractor may withhold from any interim payment that is due to the sub-contractor, any amount that the sub-contractor is liable to pay the contractor in accordance with the terms of the sub-contract and/or any sums owed to the contractor by the sub-contractor as a consequence of any breach of the sub-contract.

ICSub, ICSub/D and ICSub/NAM

All as DBSub, but clause 4.14.3 applies rather than clause 4.12.3.

MWSub/D, ShortSub and SubSub

Unlike the courts' view of payment notices, it is now quite clear following a series of court cases that if a timeous withholding notice is not issued, no withholding or deduction will be permitted within an adjudication action. This provision is strictly operated, and if a withholding notice is not issued within the timescale detailed within the sub-contract (i.e. no later than five days before the final date for payment, of the payment in question), the proposed withholding and/or deduction will almost certainly be disallowed and the merits of the proposed withholding and/or deduction will not be considered at all.

Is value added tax (VAT) payable?

DBSub

Clause 4.7.1 of the sub-contract conditions makes it clear that both the sub-contract sum and the sub-contract tender sum are exclusive of VAT, and adds that in relation to any payment made to the sub-contractor under the sub-contract, the contractor shall pay the amount of VAT properly chargeable in respect of that payment.

If, after the sub-contract base date, the supply of any goods and services to the contractor by the sub-contractor becomes exempt from VAT and the VAT therefore cannot be recovered by the sub-contractor, the contractor is to pay the sub-contractor an amount equal to the amount of the VAT input tax on the supply to the sub-contractor of the goods and services which contribute to the appropriate part of the sub-contract works but which, because of the exemption, the sub-contractor cannot recover.

This rule only applies where the supply of any goods and services to the contractor by the sub-contractor becomes exempt from VAT *after* the sub-contract base date. The sub-contract base date is given in the sub-contract particulars.

MPSub

Clause 34.1 of the sub-contract conditions makes it clear that all amounts within the sub-contract exclude VAT. Clause 34.2 notes that any VAT properly chargeable in line with legislation is to be paid by the parties, providing that any necessary documentation reasonably necessary to permit such a payment to be properly made is provided by the party receiving the VAT.

ICSub

All as DBSub.

ICSub/D

All as DBSub.

ICSub/NAM

All as DBSub.

MWSub/D, ShortSub and SubSub

Article 1 of the sub-contract conditions makes it clear that the sub-contract sum excludes VAT. Any VAT properly chargeable is to be paid by the contractor to the sub-contractor.

Does the Construction Industry Scheme (CIS) apply to payments made?

DBSub

Clause 4.8 notes that the obligation of the contractor to make any payment under the sub-contract is subject to the provisions of the Construction Industry Scheme (CIS).

MPSub

Clause 31.8 notes that the obligation of the contractor to make any payment under the sub-contract is subject to the provisions of the Construction Industry Scheme (CIS). That clause also expressly states that if the sub-contractor fails to provide those vouchers required under CIS, then the contractor is not obliged to make any further payment to the sub-contractor until such time as the necessary vouchers are provided by the sub-contractor.

ICSub

All as DBSub.

ICSub/D

All as DBSub.

ICSub/NAM

All as DBSub.

MWSub/D, ShortSub and SubSub

Although these sub-contracts do not specifically refer to the Construction Industry Scheme, it is clear that under that scheme *all* payments made from contractors to sub-contractors (or from sub-contractors to sub-subcontractors) must take account of the sub-contractor's (or sub-subcontractor's) tax status as determined by HM Revenue & Customs. Therefore the CIS will apply to payments made.

Retention and retention bonds

The JCT sub-contracts covered in this book deal with retention either by way of a retention percentage deducted or retained, or by way of a retention bond (or both). However, the SubSub does not include either provision.

What is retention?

Retention is a contractual provision permitting the contractor to withhold a sum of money against the value of works executed by the sub-contractor.

Retention is withheld in respect of latent defects in the sub-contractor's works, i.e. defects that are not apparent at practical completion of the sub-contract. Any patent defects (i.e. obvious defects) should be dealt with by way of a reduction in the valuation of the works at the appropriate time.

The specific rules and provisions governing the withholding and release of retention are stated in the contract, along with the retention percentage to be applied to the sub-contractor's interim payments.

The sub-contract will usually provide for the retention to be released by the contractor to the sub-contractor in two parts (normally two halves). The first part is normally released at the practical completion stage. The second part (sometimes referred to as the final release of retention) is normally released when all defects have been cleared at the end of the rectification period (previously known as the defects liability period).

What are the problems associated with retention?

The withholding of retention in the construction industry has generated its own particular difficulties and problems.

Late release of retention

Considerable delay in the release of retention (particularly the second part of retention) is commonplace. Sub-contractors need to make a careful record of when such retentions are due to be released and, at the appropriate time, take action as necessary to obtain its release.

Final release of retention triggered by event under the main contract

It is not unusual for the contractor's standard terms or amendments to a standard sub-contract to seek to make the sub-contractor's final retention release subject to the certificate of making good (defects) being issued under the main contract works.

This may, on a large project, be several years after the practical completion of the sub-contract works by the sub-contractor, particularly early sub-contract packages, e.g. piling, groundworks, etc.

Example

A brickwork sub-contractor achieves practical completion of his sub-contract works on 1 August 2006. The brickwork sub-contract value is £1 million, and the final release of retention (at say 1.5%) equals £15,000.00.

The brickwork sub-contractor subsequently cleared all of his defects at the end of the rectification/defects liability period, and therefore, in the normal course of events, would expect to be paid the retention monies held.

However, the sub-contract is amended to provide that the sub-contractor's final retention release is subject to the certificate of making good (defects) being issued under the main contract.

The certificate of making good (defects) under the main contract works is not expected to be issued until December 2009.

Clearly, the effect of such amendments is that sub-contractors can be kept from their retention monies sometimes for many months (even years) after they have cleared all of their defects, which can create real problems in terms of their cash flow.

When tendering, sub-contractors therefore need to identify any such amendments affecting the release of retention and their implications and to take such action as necessary when preparing their tender.

Indeed, it is not unknown for sub-contractors of smaller sub-contract package sizes to effectively build into their price an assumption that the final retention release will not be made, since making this allowance in its tender is more cost effective than spending disproportionate costs in pursuing a relatively small amount of money.

Insolvency of the main contractor

One of the biggest risks faced by the sub-contractor is the insolvency of the contractor.

Unless the contract provided requires that the retention money be held by the contractor in trust (which rarely occurs, and which is not catered for in the standard JCT sub-contracts), where retention is held by a contractor against a sub-contractor and the contractor becomes insolvent and ceases trading, the likelihood is that the sub-contractor would not recover any of the retention monies at all (particularly as the sub-contractor would only be added to the list of other unsecured creditors).

How is the retention percentage deducted or retained?

DBSub

In line with clause 4.15.1 of DBSub, the retention which may be deducted is to be calculated as follows:

(1) Where the sub-contract works or such works in any Section have not reached practical completion, the retention which the contractor may deduct and retain shall be the percentage stated under item 8 of the sub-contract particulars applied to the part of the gross valuation (appropriate at that time) covered by clauses 4.13.1 to 4.13.3 inclusive (refer to clause 4.15.1.1 of DBSub/C). The retention percentage may be inserted by the parties to the sub-contract under item 8 of the sub-contract particulars; however, if no rate is inserted the default percentage is 3% (which is the same as the default position under the main contract).
(2) Where the sub-contract works or such works in any Section have reached practical completion, the retention which the contractor may deduct and retain shall be one half of the amount calculated under item 1 above (refer to clause 4.15.1.2 of DBSub/C).

MPSub

The MPSub does not contain any retention provisions. The sub-contract guide published by the JCT states that if additional security is required by the contractor a bond or parent company guarantee should be obtained prior to entering into the sub-contract.

ICSub, ICSub/D and ICSub/NAM

Whilst the ICSub/C and ICSub/D/C do not specifically use the term 'retention', the retention provisions in these sub-contracts are far simpler than the more detailed retention rules contained in DBSub/C.

The retention percentages (pre and post-practical completion) may be inserted by the parties to the sub-contract under item 7 Interim Payments – Percentages of value of the sub-contract particulars.

If no rate is inserted the default percentages are as follows:

(1) Pre-practical completion: 5% (i.e. 95% of the applicable value).
(2) Post-practical completion: 2.5% (i.e. 97.5% of the applicable value).

These percentages relate to the practical completion of the sub-contract works, not the main contract works.

ShortSub

Retention is again simple and straightforward.

Under the sub-contract, clause 12.2 provides that the sub-contractor is paid 95% of sums calculated as due in interim payment, reducing to 97.5% seventeen (17) days post-practical completion of the sub-contract works (clause 12.3).

SubSub

No provision for retention or a retention bond is allowed for in this sub-subcontract. No doubt, the JCT drafting committee deemed these unnecessary for the relatively small package sizes that are likely to occur.

Are there any items which retention is not to be deducted against?

DBSub

Yes. Clause 4.15.1 of DBSub/C does not permit retention to be deducted against the gross valuation (appropriate at that time) of items covered by clauses 4.13.4 to 4.13.7.

ICSub, ICSub/D and ICSub/NAM

Yes. Clause 4.10.1 records the gross valuation items that are subject to 'the relevant percentage' (i.e. retention – see clauses 4.10.1.1 and 4.10.1.2) as well as those that are not (see clause 4.10.2).

MWSub/D and ShortSub

These sub-contracts are silent on this matter.

SubSub

No provisions for retention exist under the SubSub.

When is the final balance of retention released?

DBSub

This is dealt with at clause 4.15.2 of DBSub/C.

Provided there are no defects in the sub-contract works (or such works in a Section) on the date of the expiry of the rectification period of the main contract for the main contract works (or relevant Section) the balance of any retention deducted and retained by the contractor shall be included in the next interim payment following the expiry of the applicable rectification period.

It is to be noted that the sub-contractor's final release of retention is not dependent upon there being no defects in the overall main contractor's works, nor is it tied to the release of the retention under the main contract works. It is dealt with entirely independently, i.e. no defects in the sub-contractor's works at the expiry of the main contract rectification period.

What is the main contract rectification period referred to in DBSub Clause 4.15.2?

The rectification period for the main contract works is the period inserted into the main contract particulars (against the reference to clause 2.38 of the main contract) and relates to the period after practical completion of the main contract works (or a Section thereof) where the contractor is liable for any defects, shrinkages or other faults in the works.

The rectification period entered into the main contract particulars is whatever is agreed between the parties to the main contract. This period is frequently entered as 12 months, but the default position in the main contract (i.e. the period that is assumed to apply if no entry is made against clause 2.38 in the main contract particulars) is six months.

Under DBSub how is retention release dealt with if, at the end of the applicable rectification period, the sub-contractor has defects outstanding in respect of his works?

This is dealt with at clause 4.15.3 of DBSub/D/C.

First, the sub-contractor's applicable defects are to be stated in a list issued by the contractor to the sub-contractor for making good by the sub-contractor. Two possible scenarios are then allowed for in the sub-contract:

(1) The sub-contractor attends to the defects and the contractor and the sub-contractor subsequently agree that all of the defects on the list of defects have been cleared. The contractor is then to write to the sub-contractor to confirm the date when the defects were all cleared.

(2) The sub-contractor attends to the defects (or does not attend to the defects as the case may be) but the contractor and the sub-contractor are unable to agree that the defects have all been cleared (which would often be the case where the sub-contractor considers that the defects have been cleared and the contractor considers that they have not been cleared).

In the latter case, the dispute regarding this matter may be referred to adjudication for a decision. However, given that a party cannot (normally) recover its own costs in an adjudication action, the effectiveness of this particular provision, where relatively small amounts of retention monies have been deducted and retained by a contractor against a particular sub-contractor, is open to question.

Once the contractor and the sub-contractor have agreed that all of the defects have been cleared, or once the clearance of all defects has been determined by adjudication, the balance of any retention monies deducted and retained by the contractor shall be included in the next interim payment following the date (either as agreed or as determined by adjudication) that the defects have all been cleared.

Under DBSub, what is the position if the contractor and the sub-contractor are unable to agree that the defects have all been cleared but neither party wish to refer the dispute to adjudication?
In these circumstances the sub-contract provides that the retention monies deducted and retained by the contractor are to be included in the next interim payment following the issue of the certificate of making good (defects) under the main contract for the main contract works or relevant Section thereof.

ICSub, ICSub/D and ICSub/NAM

The final balance of any retention held by the contractor and due to the sub-contractor is released by way of the issue of the final certificate.

MWSub/D and ShortSub

The final release of retention to the sub-contractor is dealt with at clause 12.5 which says that the retention is due for release seven days after the date of the rectification period (previously known as the 'defects liability period') under the main contract. If the sub-contractor has not, by that time, made good defects in his works, the final release of retention is seven days after the completion of such defects.

The sub-contract is silent on what happens in situations where the sub-contractor disputes defects being asserted by the contractor in the sub-contract works. Whilst adjudication is one possible option, the cost of the proceedings may be prohibitively expensive in relation to the defects complained of and the monies withheld.

SubSub

As previously noted, the SubSub has no provision for retention.

Retention bond

As an alternative to retention monies being held, some of the JCT sub-contracts make provision for a retention bond.

DBSub

DBSub does include provisions for a retention bond in lieu of retention.

However, a retention bond only applies where item 9 of the sub-contract particulars is completed, recording that a retention bond is to apply.

The form of the bond (which has been agreed between the JCT and the British Bankers' Association) is set out under schedule 3 part 2 of the DBSub/C.

MPSub

The MPSub does not contain any retention provisions. The sub-contract guide published by the JCT states that if additional security is required by the contractor a bond or parent company guarantee should be obtained prior to entering into the sub-contract.

ICSub, ICSub/D, ICSub/NAM, MWSub/D and ShortSub

No provisions for a retention bond are included.

SubSub

No provisions for retention exist under the SubSub.

What is the information that needs to be inserted in the JCT standard retention bond?

When completing the bond the information that needs to be inserted is:

- The date of the bond.
- The name and address of the surety (i.e. the party that guarantees payment of the bond amount).
- The name and address of the contractor.
- The name and address of the sub-contractor.
- The surety's maximum aggregate liability amount under the bond, stated as a sum in pounds. This will be the same sum as stated in the sub-contract under item 9 of the sub-contract particulars.
- The surety's address where a demand under the bond should be sent.
- The surety's address where a copy of the written notice to the sub-contractor of his liability for the amount demanded should be sent.
- The (default) expiry date of the bond. This will be the date as stated in the sub-contract under item 9 of the sub-contract particulars.

The bond then needs to be signed as a deed by or on behalf of the surety.

How does the retention bond work?

The bond operates in conjunction with clause 4.16 of DBSub/C.

When a retention bond is in place, the contractor (at the date of each interim payment) is to prepare a statement which specifies the deduction in respect of retention that would have been made had the retention bond not been in place (refer to clause 4.16.1 of DBSub/C).

When does the bond have to be in place?

The retention bond needs to be in place and needs to be provided by the sub-contractor to the contractor on or before the date of commencement of the sub-contract works. In addition and in line with clause 4.16.2 of DBSub/C the proposed surety must be approved by the contractor.

How long must the retention bond be in place for?

The retention bond must be maintained by the sub-contractor until the expiry date of the bond.

If the sub-contractor does not provide or maintain the retention bond as required, then the contractor is entitled to deduct and retain retention monies in line with the provisions of clauses 4.10.1.1 and 4.15 in the next applicable interim payment to the sub-contractor. If the sub-contractor subsequently provides and thereafter maintains the required retention bond the contractor shall (*rather than may*), in the next interim payment after such compliance, release to the sub-contractor the retention deducted during the period of the breach (refer to clause 4.16.3 of DBSub/C).

What is the position if the value of the sub-contract works subsequently exceeds the sub-contract sum (i.e. variations)?

The surety's maximum aggregate liability amount under the bond is normally based upon the retention percentage applied to the original scope of the sub-contract works. Clearly, if variations are issued, the value of the sub-contract works to which the retention percentage would apply may increase, and in such a case the maximum retention monies may exceed the maximum aggregate liability amount under the bond.

If such circumstances apply, clause 4.16.4 of DBSub/C makes it clear that the sub-contractor can either arrange with the surety for the aggregate liability amount to be increased so that it equates to such an increased retention sum, or alternatively, the contractor may deduct and retain the retention monies due that is in excess of the maximum aggregate liability amount on the retention bond.

Item 2 of the Notes section to the retention bond (which will not appear on the bond issued by the surety) states that it is understood that a surety will, at additional cost to the sub-contractor and possibly subject to other terms and conditions of the surety, provide for a greater sum than that originally inserted as the maximum aggregate liability amount inserted in the bond.

What benefit does the contractor receive from the bond?

The purpose of the retention bond is to allow the contractor to recover from the surety:

(1) The costs actually incurred by the contractor by reason of the failure of the sub-contractor to comply with the directions of the contractor under the sub-contract (as clause 4.3.1 of the retention bond).
(2) Any expenses or any direct loss and/or damage caused to the contractor as a result of the termination of the sub-contractor's employment by the contractor (as clause 4.3.2 of the retention bond).
(3) Any costs, other than the amounts referred to in items 1 and 2 above, which the contractor has actually incurred and which, under the sub-contract, he is entitled to deduct from monies otherwise due or to become due to the sub-contractor (as clause 4.3.3 of the retention bond).

The second paragraph to item 3 of the Notes to the retention bond states that any demand under clause 4 of the retention bond must not exceed the costs actually incurred by the contractor, therefore (and accordingly) the contractor cannot add profit to any costs incurred when making a claim under the retention bond.

How does the contractor make a claim under the bond?

Before making a claim under the retention bond, the contractor is to give written notice to the sub-contractor of his liability for the amount demanded, and is to request the sub-contractor to discharge his liability. At the time when such a notice is sent to the sub-contractor, a copy is to be sent to the surety at the address noted under clause 4.4 of the retention bond. A claim under the retention bond cannot be made until 14 days after the said written notice has been provided to the sub-contractor, and can only be made if the sub-contractor has not complied with the requirements of the written notice.

When making a claim against the bond, the contractor must:

- provide details of the retention bond;
- provide the date of the demand;
- state the amount of the retention that would have been held by the contractor at the date of the demand had retention been deductible (refer to clause 4.2 of the retention bond);
- state the amount demanded, and identify which one or more of clauses 4.3.1, 4.3.2 or 4.3.3 of the retention bond is being relied upon for the demand being made;
- incorporate a certification that the sub-contractor has been given 14 days' written notice of his liability for the amount demanded by the contractor and that the sub-contractor has not discharged that liability;
- provide evidence that a copy of the said written notice to the sub-contractor was simultaneously sent to the surety at the address noted under clause 4.4 of the retention bond.

The demand must be signed by persons who are authorised by the contractor to act for and on his behalf, and those signatures must be authenticated by the contractor's bankers; and the demand must be sent to the surety at the address noted under clause 4.1 of the retention bond.

The last paragraph of clause 4 of the retention bond states:

'Such demand as above shall, for the purposes of this Bond but not further or otherwise, be conclusive evidence (and admissible as such) that the amount demanded is properly due and payable to the Contractor by the Sub-contractor.'

Item 3 of the Notes section to the retention bond (which will not appear on the bond issued by the surety) states that the inclusion in the above quoted paragraph of the words 'but not further or otherwise' is to make clear that the sub-contractor would not be prevented by the terms of clause 4 (of the retention bond) from alleging, under the sub-contract, that the sub-contractor was not in breach of any of the matters stated in clauses 4.3.1 to 4.3.3 of the bond.

In other words, a demand made under clauses 4.3.1 to 4.3.3 of the retention bond will be taken as being conclusive evidence to the surety that the amount demanded is properly due and payable (therefore, the retention bond is, in effect, an 'on-demand' bond) but any such demand will not prevent the sub-contractor from pursuing his claim under the terms of the sub-contract.

What is the position if a performance bond is also in place?

Finally, clause 4.16.5 of DBSub/C notes that where the contractor has required the sub-contractor to provide a performance bond, then, in respect of any default to which that performance bond refers which is also a matter for which the contractor could make under the terms of the retention bond, the contractor shall first have recourse to the retention bond.

Counterclaim, set-off and abatement

What is the difference between a counterclaim, set-off and abatement?

It is settled law that an effective withholding notice is necessary in order to be able to set-off sums against monies otherwise payable. However, in terms of abatement, case law (whilst not entirely conclusive) suggests that the common law defence of abatement survives the absence of such a notice.

It is, therefore, important to be aware of the difference between counterclaim, set-off and abatement as at law there is a legal difference between set-off and abatement of price.

What is a counterclaim?

A counterclaim is a common feature in litigation proceedings. In response to the proceedings brought by the claimant the defendant is required to serve a defence.

Additionally, the defendant may also serve, at the same time, a counterclaim against the claimant. By way of example, in a litigated claim for payment commenced by a sub-contractor, the contractor will serve a defence and may additionally counterclaim for damages relating to defective work and/or delays caused by the contractor.

In respect of counterclaims, Lord Denning said[1]:

> 'The word "counterclaim" is not defined in section 28, but, . . . [e] I think it is: any claim that could be the subject of an independent action. It is not confined to money claims . . . [e] and . . . [e] it need not relate to or be connected with the original subject of the cause or matter.'

Therefore, a counterclaim must be capable of an existence independent of the other party's claim being made, although it can be used to set-off if it meets the necessary requirements.

What is a set-off?

A set-off acts as a mechanism to absolve a party from actually paying all or part of another party's claim (i.e. cancelling out their entitlement to payment). There are three types of set-off:

- A set-off at common law is available in respect of opposing claims for readily ascertainable liquidated debts or monetary demands (but not damages), which do not need to be related to the same transaction.
- Equitable set-off applies in the case of a cross-claim that is so closely associated with the claim that justice requires it to be taken into account. Unlike set-off at common law, it can include a claim for damages but not a separate or independent counterclaim.
- Contractual set-off is made in accordance with the express machinery of the contract.

For a set-off to be valid an effective withholding notice must be issued.

What is abatement?

The principle of abatement of price was described by Lord Morris in *Gilbert Ash* v. *Modern Engineering*[2] as follows:

> 'It has long been an established principle of law that if one man does work for another, the latter, when sued, may defend himself by showing that the work was badly done and that the claim made in respect of it should be dismissed.'

---

[1] *Henriksens A/S* v. *Rolimpex THZ* [1974] 1 QB 233, CA.
[2] *Gilbert Ash Ltd* v. *Modern Engineering (Bristol) Ltd* [1974] AC 689.

How does abatement affect the value of work?

Most contracts provide that payment is to be only for work properly carried out and, even if adequate payment terms are not incorporated into the contract, the fallback provision under the Scheme[3] is for payment in respect of work 'performed in accordance with [the] contract'.

Therefore, it can readily be argued that the 'value' of work is only arrived at after discounting that part of the work that is not in accordance with the contract (i.e. after abatement) by reason of breach.

If a sub-contractor applies for more money than he is entitled to because he has not yet carried out part of the work then there is no set-off or abatement, he is simply not entitled to all he has claimed (i.e. the whole sum is not due).

Assuming that there is no contractual provision under which the sum claimed is the amount due, no withholding notice is required to reduce the sub-contractor's entitlement.

What is the correct quantification of abatement?

The measure of any abatement is the amount by which the worth of the subject matter is less by reason of the breach of contract. However, this creates some difficulties. In particular, should the cost of remedial works be taken into account?

In practice it may be that a realistic measure of abatement is dependent upon the circumstances, in particular whether the contractor has a right to remedy defects and whether he has failed to exercise or has been denied that right.

The recent case of *Multiplex* v. *Cleveland Bridge*[4] is of some assistance here. In this case Mr Justice Jackson undertook a thorough review of the relevant law relating to abatement, and at paragraph 652 (set our below) he provided seven legal principles concerning abatement:

'652. Although there is not a complete harmony of approach to be discerned from this line of cases, I derive seven legal principles from the authorities cited:

(i)   In a contract for the provision of labour and materials, where performance has been defective, the employer is entitled at common law to maintain a defence of abatement.

(ii) The measure of abatement is the amount by which the product of the contractor's endeavours has been diminished in value as a result of that defective performance.

(iii) The method of assessing diminution in value will depend upon the facts and circumstances of each case.

(iv) In some cases, diminution in value may be determined by comparing the current market value of that which has been constructed with the market value which it ought to have had. In other cases, diminution in value may be determined by reference to the cost of remedial works. In the latter situation, however, the cost of remedial works does not become the measure of abatement. It is merely a factor

---

[3] The Scheme for Construction Contracts (England and Wales) Regulations 1998.
[4] *Multiplex* v. *Cleveland Bridge* [2006] EWHC 1341 (TCC).

which may be used either in isolation or in conjunction with other factors for determining diminution in value.

(v) The measure of abatement can never exceed the sum which would otherwise be due to the contractor as payment.

(vi) Abatement is not available as a defence to a claim for payment in respect of professional services.

(vii) Claims for delay, disruption or damage caused to anything other than that which the contractor has constructed cannot feature in a defence of abatement.'

Interest

What is interest?

Interest means a financial charge that is made (in this case by a sub-contractor) against another party (in this case the contractor) in respect of outstanding monies owed (which normally relates to overdue or late payments in part or in full).

Until fairly recently, a party had no automatic right or implied right to interest, unless there was an express term in the contract which dealt with the question of interest.

The position has, however, now changed by way of the introduction of the Late Payment of Commercial Debts (Interest) Act 1998 and the Late Payment of Commercial Debts Regulations 2002.

There is now an implication in construction sub-contracts (amongst many other contracts) that debts carry a right to simple interest and a right to compensation.

Is there an express right to interest under the JCT sub-contracts?

DBSub

The DBSub/C deals with the matter of interest under clauses 4.10.5 (for interim payments) and 4.12.4 (for the final payment).

In respect of both clause 4.10.5 and clause 4.12.4, the sub-contract states that if the contractor fails properly to pay the amount due, or any part of it, to the sub-contractor by the final date of payment, the contractor shall pay to the sub-contractor, in addition to the amount not properly paid, simple interest thereon at the interest rate for the period until such payment is made.

Both clauses also add that the non-payment of such interest shall be treated as a debt due to the sub-contractor by the contractor.

The key points to note are:

- Interest is only payable on amounts that are 'due' to the sub-contractor. Interest is not payable on the basis of a sub-contractor's application for payment (unless, of course, this is actually the amount that is due to the sub-contractor).
- Interest is payable from the date of the final date for payment.
- Interest stops accruing when the amount 'due' to a sub-contractor has been paid.
- Interest can only be claimed on the basis of simple interest, not compound interest.

- The non-payment of any interest due shall be treated as a debt due to the sub-contractor by the contractor, and may be pursued by the sub-contractor in the same way as any other debt due under the sub-contract.
- The sub-contractor does not need to claim interest to be entitled to be paid interest (although it would be sensible for a sub-contractor to make an application for interest payments, since it is highly unlikely that a contractor will make a payment for interest to a sub-contractor unless he has first been prompted by the sub-contractor).
- The acceptance of any payment of interest by the sub-contractor shall not in any circumstances be construed as a waiver by the sub-contractor of his right to proper payment of the principal amounts due from the contractor, to suspend performance of his obligations, or to terminate his employment under the sub-contract.

MPSub

The MPSub deals with the matter of interest under clause 33. Under that clause the sub-contract states that if either party fails to make payment in accordance with the sub-contract the other party shall be entitled to simple interest on the amount outstanding until payment is made.

The key points to note are:

- Interest is payable to either party.
- Interest is only payable on amounts outstanding in accordance with the sub-contract.
- Interest stops accruing when the amount outstanding has been paid.
- Interest can only be claimed on the basis of simple interest, not compound interest.
- Neither party needs to claim interest to be entitled to be paid interest.

ICSub

The ICSub deals with the matter of interest in exactly the same way as the DBSub. The only difference being that clause 4.12.4 deals with interest on interim payments (rather than clause 4.10.5 in DBSub) and clause 4.14.4 deals with interest on the final payment (rather than clause 4.12.4 in DBSub).

ICSub/D

All as ICSub above.

ICSub/NAM

All as ICSub above.

MWSub/D and ShortSub

These sub-contracts deal with the matter of interest under clause 12.9.

The sub-contract states that if the contractor fails to pay any sum due to the sub-contractor by the final date of payment, the contractor shall pay to the sub-contractor interest on such an overdue sum.

It should be noted that the sub-contractor does not need to claim interest to be entitled to be paid interest (although it would be sensible for a sub-contractor to make an application for interest payments, since it is highly unlikely that a contractor will make a payment for interest to a sub-contractor unless he has first been prompted by the sub-contractor).

SubSub

The SubSub deals with the matter of interest under clause 12.6.

The sub-contract states that if the sub-contractor fails to pay any sum due to the sub-subcontractor by the final date of payment, the sub-contractor shall pay to the sub-subcontractor interest on such overdue sum.

It should be noted that the sub-subcontractor does not need to claim interest to be entitled to be paid interest (although it would be sensible for a sub-subcontractor to make an application for interest payments, since it is highly unlikely that a sub-contractor will make a payment for interest to a sub-subcontractor unless he has first been prompted by the sub-subcontractor).

What is the rate of interest applicable?

DBSub

The interest rate that applies (as defined under clause 1.1) is 'a rate 5% per annum above the official dealing rate of the Bank of England correct at the date that a payment due under this contract becomes overdue'.

From the above, it appears that the interest rate applicable to a particular overdue payment is fixed, (i.e. at 5% per annum above the official dealing rate of the Bank of England) at the date that the payment became due, and does not, as one might expect, change during the period of time that the debt remains outstanding to suit any fluctuations (whether these be up or down) to the official dealing rate of the Bank of England.

MPSub

The interest rate that applies (as stated under clause 33.1) is a rate of 5% per annum in excess of the Bank of England's base rate for the period until payment is made. Unlike DBSub, it would appear that the interest rate applicable to a particular overdue payment varies in line with the change of the Bank of England base rate rather than being fixed for the period that the debt is outstanding.

ICSub

All as DBSub above.

ICSub/D

All as DBSub above.

ICSub/NAM

All as DBSub above.

MWSub/D and ShortSub

The interest rate that applies (as defined under clause 12.9) is 'at the rate of 5% per annum above the official dealing rate of the Bank of England correct at the final date for payment.'

From the above, it appears that the interest rate applicable to a particular overdue payment is fixed (i.e. at 5% per annum above the official dealing rate of the Bank of England) at the final date for payment, and does not, as one might expect, change during the period of time that the debt remains outstanding to suit any fluctuations (whether these be up or down) to the official dealing rate of the Bank of England.

SubSub

All as MWSub/D and ShortSub above.

Is the interest rate in the JCT sub-contracts compliant with the Late Payment of Commercial Debts (Interest) Act 1998?

This particular question has not been tested in the courts. The Late Payment of Commercial Debts (Interest) Act 1998 makes it clear that parties may only agree interest rates as express terms of their contract if such rates are 'substantial'. The rate of interest under the Act is 8% over the Bank of England base rate.

It is envisaged by the Act that some parties may agree a specific interest rate, or compensation, in their contract. If so the Act will not apply and the agreed express terms will apply.

However, to avoid larger companies abusing their bargaining power any agreed interest rate or compensation for late payment must be 'substantial'; if not the Act will apply instead.

Given that the interest rate in the DBSub, MPSub, ICSub, ICSub/D, ICSub/NAM, MWSub/D, ShortSub and SubSub is 3% lower than the interest rate in the Act, and given also that those sub-contracts do not allow for any 'compensation' payments as detailed under section 5A of the Act, there must be some question as to whether or not the interest rate in those sub-contracts would be considered to be 'substantial' if this were tested in court.

If this matter were tested in court, and if the interest rates in the above noted sub-contracts were found not to be 'substantial', then the terms of the Act (with a higher interest rate) would be imported into the sub-contracts.

It should be noted that in respect of the MPSub, in particular, clause 33.2 says that it is agreed that the provisions of clause 33.1 constitute a substantial remedy for the purposes of section 9(1) of the Late Payment of Commercial Debts (Interest) Act 1998. This provision is intended to defeat the argument (as outlined above) that a rate 5% above the Bank of England base rate is not substantial, and that a rate of 8% above the Bank of England base rate should apply.

Sub-contractor's right of suspension

Can a sub-contractor suspend performance on site where the sum due under the contract has not been paid in full?

It is important to note that there is no common law right to suspend performance where the sum due under the contract has not been paid in full.

However, because of the perceived payment abuse in the construction industry, section 112 of the Housing Grants, Construction and Regeneration Act 1996 (HGCRA 1996) now confers an express statutory right of suspension to parties on a qualifying construction contract. This introduced a potentially powerful sanction against non-payment.

Is there an express right of suspension in the sub-contracts?

DBSub

The DBSub deals with this matter under clause 4.11.

MPSub

The right to suspend is not an express term of the sub-contract. However, as noted above, the HGRA creates a statutory right to suspend in the prescribed circumstances, which is implied as a sub-contract term.

ICSub, ICSub/D and ICSub/NAM

In the above sub-contracts this matter is dealt with at clause 4.13. Except for reference therein to clause 4.12.2 in lieu of clause 4.10.2 the clause is identically drafted to clause 4.11 of DBSub.

MWSub/D, ShortSub and SubSub

In the above sub-contracts this matter is dealt with at clause 13.

Must a warning notice be given first before the sub-contractor suspends?

Yes. See clause 4.11 of DBSub; clause 4.13 of ICSub, ICSub/D and ICSub/NAM; see clause 13 of MWSub/D, ShortSub and SubSub; and see section 12(2) of HGCRA 1996 for MPSub.

The sub-contractor is required to give a seven-day written notice to the contractor of his intention to suspend the performance of his obligations under the sub-contract, and the ground or grounds on which it is intended to suspend such performance. The sub-contractor cannot suspend his performance before the expiry of that seven-day notice period.

If the contractor does not make payment in full (of the amount due) within the seven-day period, then the sub-contractor can suspend performance until payment in full occurs.

Note that the trigger for validly suspending performance runs from the notice, not the date when payment should have been made. The notice of suspension may often be issued some time after the final date for payment, not least due to promises of 'the cheque is in the post'.

Are there any traps for the unwary?

Yes. There are a few important points that the sub-contractor needs to be aware of:

(1) It cannot be over-emphasised that a sub-contractor cannot suspend performance until after he has given a seven-day written notice, and until after that seven-day notice period has expired.
(2) The sub-contractor can only suspend in respect of sums identified and calculated as due for payment under the sub-contract provisions. Any disputes as to the quantum or calculation of this sum are not valid grounds to suspend.
(3) As a sub-contractor can only suspend performance if the amount that is due has not been made at the correct time, he must be certain that the amount is actually due before taking this step. The amount actually due may be affected by things such as:

    (a) abatement (e.g. a reduction in the value of the works because of defective or incomplete works); and
    (b) set-off.

    A sub-contractor may be at risk in operating this section where the contractor relies in any future action upon an abatement or a set-off.
(4) A correctly notified payment and/or withholding notice could validly reduce the amount due for payment.

Does suspension of performance affect any of the sub-contractor's other rights and remedies?

No. The sub-contractor's right to suspend performance does not affect any other rights and remedies the sub-contractor holds under the sub-contract. By way of example, the sub-contractor would be entitled to give notice of termination under the contract as well as suspend performance. The sub-contractor would also be entitled to claim interest on late payment.

What happens if the sub-contractor wrongfully suspends?

A wrongful suspension by the sub-contractor may amount to a repudiation, which, if accepted by the contractor, may lead to the sub-contractor suffering all of the contractor's damages arising from that repudiation.

It may also constitute a ground for termination of the sub-contract (e.g. see clause 7.4.1.1 of DBSub, i.e. 'without reasonable cause wholly or substantially suspends the carrying out of the Sub-contract Works').

Is the sub-contractor entitled to an extension of time for the suspended period?

If the sub-contractor does (correctly) suspend performance, he is entitled to an extension of time to the period for completion. This will either be an express term of the sub-contract (see, for example, clause 2.19.5 of DBSub/C; clause 22.1.7 of MPSub, etc.), or if the sub-contract is silent it will be an implied contract term (as section 112(4) of HGCRA 1996).

How is the extension of time calculated?

The extension of time period will be the period when the suspension starts (which cannot occur until after the expiry of the seven-day notice period) to the time when payment in full occurs.

When does payment in full occur?

There is no guidance as to when 'payment in full' occurs; however, in most cases this should be obvious.

In terms of cheques, however, it is a well-settled principle of law that a cheque is to be treated as cash, and thus is not susceptible to set-off[5]. It is, therefore, considered that the courts are likely to decide that 'payment in full' is made when the sub-contractor receives a cheque rather than (the later date) of when the cheque had been cleared through the sub-contractor's account. It should be noted, however, that in rare cases the courts have accepted that an employer could cancel an issued cheque to a contractor[6].

Is there a period of grace to mobilise the sub-contractor's resources following the cessation of suspension?

There is no guidance here; however, it is submitted that no lead in time would be permitted for the sub-contractor to return to site after 'payment in full' has been made. If, for example, payment in full was made on Thursday, the sub-contractor would be obliged to return to site on the day following that (i.e. the Friday).

The sub-contractor would not (it is submitted) be entitled to an extension of the period for completion for any delay in the return to work after payment in full has been made.

---

[5] *Nova (Jersey) Knit Ltd* v. *Kammgarn Spinnerei* [1977] 1 WLR 713, HL; *Isovel Contracts Ltd* v. *ABB Building Technologies Ltd* [2002] 1 BCLC 390.
[6] *Willment Brothers Ltd* v. *North West Thames Regional Health Authority* (1984) 26 BLR 51.

If the sub-contractor does (correctly) suspend performance, is he entitled to recover the loss and expense associated with the period of such suspension?

DBSub

Provided the suspension was not 'frivolous or vexatious' then the suspension is a relevant matter entitling the sub-contractor to claim loss and/or expense. The qualification of the sub-contractor's entitlement to recovery of loss and/or expense by these two terms may obviously be a source for dispute.

MPSub

Clause 30.2.3 provides that the valid exercise of the sub-contractor of his rights under section 112 of HGRA 1996 is a matter for which the contractor will be liable to the sub-contractor in respect of loss and/or expense.

ICSub, ICSub/D and ICSub/NAM

Clause 4.17.3 of the above sub-contracts is identically drafted to clause 4.20.3 of DBSub.

MWSub/D, ShortSub and SubSub

These sub-contracts make no reference to recovery of loss and expense associated with this matter. Neither does the HGCRA 1996.

However, given that the grounds for suspension are due to a breach of contract by the sub-contractor, the sub-contractor may be able to seek any loss caused as contractual damages in the ordinary way.

Fluctuations

What fluctuations are recoverable?

DBSub

Under clause 4.17 of DBSub/C, three fluctuation provisions are referred to:

- Fluctuations option A: contribution, levy and tax fluctuations.
- Fluctuations option B: labour and materials cost and tax fluctuations.
- Fluctuations option C: formula adjustment.

The fluctuations option which is applicable to any particular sub-contract is indicated under item 10 of the sub-contract particulars in DBSub/A.

Footnote 20 to item 10 of the sub-contract particulars states that all but one fluctuations option is to be deleted. Therefore, the intention is that one fluctuations option (and only one fluctuations option) will apply to every sub-contract.

Footnote 19 to item 10 of the sub-contract particulars states that where fluctuations option A or B applies in the main contract it is a requirement of that contract that the same option applies to any sub-contract.

Clause 4.18 of DBSub/C notes that, irrespective of which fluctuations option applies to the sub-contract works generally, none of the fluctuations options will apply in respect of work for which a schedule 2 quotation has been accepted by the contractor or in respect of a variation to an accepted schedule 2 quotation.

Details of the three fluctuations options are set out in schedule 4 to the DBSub/C.

MPSub

The MPSub does not make any provision for the recovery of any fluctuations.

ICSub and ICSub/D

Under clause 4.15, fluctuations (on a contribution, levy and tax fluctuations basis only) as dealt with by the application of schedule 2 to the conditions, applies, unless shown as deleted in the sub-contract particulars under item 9.

Footnote 13 to item 9 of the sub-contract particulars states that where the contribution, levy and tax fluctuations option applies in the main contract it is a requirement of that contract that that same fluctuations option applies to any sub-contract.

ICSub/NAM

Under clause 4.15, fluctuations on a contribution, levy and tax fluctuations basis (option A) or on a formula adjustment basis (option C), as identified at item IT14 of the ICSub/NAM/IT, applies in accordance with the provisions of schedule 2 of ICSub/NAM/C, unless both of these options are deleted.

MWSub/D, ShortSub and SubSub

None of these sub-contracts make any provision for the recovery of any fluctuations.

Chapter 9
Loss and Expense

What is loss and expense?

In the context of this book these are generally financial claims submitted by the sub-contractor (or sub-subcontractor under SubSub) associated with the prolongation and/or disruption of the works, where the recovery of such additional costs is not possible under any other provision of the sub-contract (or sub-subcontract in the case of the SubSub).

Are claims restricted to the sub-contract terms?

The answer is generally no. When considering loss and expense, the first thing that must be noted is that claims in the construction industry generally fall under two differing legal heads. That is:

(1) claims for breach of contract or for other events for which specific provision is made within the conditions of the contract (normally referred to as loss and expense claims); and
(2) claims for breach of contract for which no specific provision has been included in the conditions of contract (normally referred to as common law damages claims).

Why include loss and expense provisions in sub-contracts?

Loss and expense claims have certain advantages over common law damages claims, and some of these advantages are:

- There is a clear definition of responsibilities and procedures.
- There is a clarification of the events and the circumstances for claims to be made.
- There is a right to interim payments and/or decisions.
- There is the avoidance of the uncertainties involved in claiming at common law.
- Loss and expense claims create a right under the contract to a debt, rather than a claim for damages.

Normally, the inability to recover monies under the express terms of the contract for reasons such as not conforming with the procedures required, does not preclude a claim at common law on the same matters, provided there is the entitlement at common law to claim damages, and provided that the sub-contractor's common law rights have not been specifically excluded by the sub-contract.

If a matter stated in the sub-contract as giving rise to an entitlement to loss and expense is a breach of contract, then the sub-contractor will normally have a right to common law damages. An example of this would be the late issue of information by the contractor.

However, if the matter relied on does not amount to a breach of contract (i.e. issuing variations, etc.) then common law damages may not be applicable.

On the other hand, the preservation of the sub-contractor's common law rights may entitle the sub-contractor to make common law damages claims in respect of matters for which there are no express provisions within the loss and expense clauses (for example the breach of any applicable implied terms by the contractor).

The authority on the application of damages arising from the contract both in addition to and, where appropriate, instead of the express provisions of the contract for loss and expense, is clearly stated in the case of *Stanley Hugh Leach* v. *Merton*[1].

It is important to note that should the sub-contractor need to pursue his claim at common law because he has not followed the procedures under the sub-contract, then it is possible that he may be penalised by the court if his non-conformity with the sub-contract procedure did not allow the contractor to mitigate his loss.

What this means in practice is that if the contractor had been made aware of the pending claim at an earlier stage (in line with the express provisions of the sub-contract) he may have been able to have taken steps to reduce the effects of the breach. If the contractor was not given the opportunity to mitigate his loss in this way, the sub-contractor may have any future award in damages reduced by a court because of this failing.

Whereas under the main contract, the architect/contract administrator cannot, without the express authority of the employer, decide on claims that are not expressly provided for under the contract and therefore cannot decide on common law damages claims, no such restriction exists in the contractor/sub-contractor relationship.

Can common law claims be excluded by the contract?

In some circumstances (albeit relatively uncommon) the parties may seek to exclude (via a contract term) the parties' common law rights, i.e. so that only the terms of the contract itself governs the relationship between the parties.

However, it must be remembered that the courts view this course of action with some caution and, consequently, there must be a very clear worded contract term in the contract excluding the parties' common law rights for it to be effective.

---

[1] *Stanley Hugh Leach* v. *London Borough of Merton* (1985) 32 BLR 51.

DBSub

For the avoidance of any doubt, in respect of the above, clause 4.22 of DBSub/C preserves the common law damages rights and the remedies of the contractor and the sub-contractor.

MPSub

Unlike clause 4.22 of DBSub above, MPSub does not include an express term preserving the sub-contractor's common law rights to claim damages for breach of contract as an alternative remedy to the sub-contract's express loss and expense provisions.

However, it is submitted that the sub-contract's silence on this matter does not alter the sub-contractor's right to bring a claim based on his common law rights. As noted above, common law rights can only be excluded by clear and unambiguous wording in the sub-contract to this effect.

ICSub

All as DBSub above. Clause 4.19 of ICSub/C contains identical wording to clause 4.22 of DBSub/C.

ICSub/D

All as ICSub above.

ICSub/NAM

All as ICSub above.

MWSub/D and ShortSub

These sub-contracts do not contain an express term preserving the sub-contractor's common law rights to claim damages for breach of contract; however, it is considered that their silence on this point does not affect the sub-contractor's right to bring a claim based on his common law rights.

SubSub

All as MWSub/D and ShortSub above. A sub-subcontractors right to bring a claim based upon common law damages is unaffected by the absence of express provision preserving common law rights.

Are claims for extensions of time and loss and/or expense linked?

People in the construction industry often make the mistake (contractually) of linking time and money, in that they assume that financial claim recovery can *only* be secured if an extension of time has been granted and, conversely, that it will *certainly* be due if an extension has been given.

In reality, neither of these propositions is entirely accurate.

Extension of time clauses and those for additional financial recovery (i.e. for loss and expense) are separate matters in the JCT sub-contracts considered in this book, and indeed in all other JCT standard forms.

Additional monies may be recovered even if prolongation is not present, or alternatively, the sub-contractor may receive a comprehensive extension of time and still not be entitled to additional financial recovery.

Therefore, it is advisable that delay and money notices are kept separate.

The express contractual provisions permitting the recovery of direct loss and/or expense are considered below.

What are the grounds/requirements for loss and expense?

DBSub

DBSub provides for the payment of direct loss and/or expense at clauses 4.19 to 4.21.

Clause 4.19 provides that if the regular progress of the sub-contract works is materially affected or is likely to be materially affected by any of the Relevant Sub-contract Matters set out in clause 4.20 (as detailed below), the sub-contractor may make written application to the contractor for the recovery of the resultant loss and expense incurred.

A sub-contractor is not permitted to recover his additional costs twice. Accordingly, clause 4.19 states that any such loss and expense must be that which the sub-contractor would not be reimbursed by a payment under any other provision in the DBSub (e.g. loss and/or expense would not be recoverable under these provisions where it is being paid elsewhere as part of an accepted schedule 2 quotation).

Assuming that the relevant criteria for claiming loss and/or expense applies, then the contractual provisions are triggered by a written application being made by the sub-contractor. Accordingly, if the sub-contractor makes a written application for loss and expense, then the amount of the payable loss and expense agreed between the contractor and the sub-contractor shall be taken into account in the calculation of the final sub-contract sum or shall be recoverable by the sub-contractor from the contractor as a debt, provided that:

- the sub-contractor's application is made as soon as it has become, or should reasonably have become apparent to him that the regular progress has been or is likely to be affected;
- the sub-contractor submits (upon request from the contractor) such information as is reasonably necessary to show that the regular progress has been or is likely to be affected; and
- the sub-contractor submits details of the loss and expense as reasonably requested by the contractor.

MPSub

MPSub provides limited rights for the sub-contractor to be reimbursed for loss and/or expense due to the regular progress of the sub-contract works being materially affected by the three qualifying categories of 'matters' listed at clause 30.2 (detailed below) for which the contractor might be liable to pay loss and/or expense.

However, unlike the other JCT sub-contracts MPSub expressly provides for loss and/or expense resulting from any change to be included separately under clause 29. This point is emphasised at clause 30.8. Both the sub-contract and its accompanying guide make clear that the intention of the limited rights under clause 30 is that most matters, which would otherwise give rise to claims for loss and/or expense, should be addressed as part of the valuation of changes.

Because of this, clause 30.8 states that the contractor's ascertainment of loss and expense must not include any element contributed to by a cause which does not constitute a change, or one of the matters listed at clause 30.2.

Clause 30.3, in line with other JCT standard sub-contracts, is triggered by a timely notification given by the sub-contractor to the contractor. The sub-contractor is further obliged to take all practicable steps to reduce the loss and/or expense to be incurred – what this means in practice will depend on the particular facts, and it would be sensible to seek directions and/or instructions from the contractor.

The contractor is liable to pay the sub-contractor any loss and/or expense ascertained in accordance with clause 30 provided that:

- The loss and expense is not attributable to a change (clause 30.3 of MPSub). Loss and/or expense related to change is valued separately (clause 29).
- The sub-contractor's notice is issued as soon as he becomes aware that the regular progress of the sub-contract works is being or is likely to be materially affected as a result of any of the three matters listed at clause 30.3.
- The sub-contractor has also provided his assessment of the loss and expense incurred, and that likely to be incurred. Information submitted in respect of loss and/or expense actually incurred must be that reasonably necessary for the contractor's ascertainment purposes (clause 30.4).
- The sub-contractor's assessments and information are to be continually updated at monthly intervals until the contractor has been provided with all of the information reasonably necessary to permit 'the whole of the loss and/or expense incurred to be ascertained' by the contractor (clause 30.4).
- Post-practical completion – within 28 days of practical completion, the sub-contractor is to provide documentation in support of any further ascertainment the sub-contractor considers should be made by the contractor in respect of any notified loss and/or expense matter (clause 30.6).

Again, as noted above, recovery of loss and expense incurred is dependent upon the regular progress of the sub-contract works being materially affected.

Upon receipt of the sub-contractor's information regarding any loss and/or expense incurred (clause 30.4) the contractor must, within a period of 28 days, notify the sub-contractor of his ascertainment of the loss and expense incurred.

It should be noted that:

- ascertainment is made on the basis of information provided by the sub-contractor; and

- the contractor's notification must be sufficiently detailed to enable the sub-contractor to identify any differences between the contractor's and the sub-contractor's assessments (clause 30.5).

The contractor is liable to pay the sub-contractor any loss and/or expense ascertained by the contractor in accordance with clause 30 (clause 30.7).

Post-practical completion (clause 30.6) states that upon receipt any further documentation issued by the sub-contractor under clause 30.6 in respect of any notified matter, the contractor then has 56 days in which to review the documentation, his previous ascertainment, and to notify the sub-contractor of any further ascertainment the contractor considers appropriate.

ICSub

All as DBSub above. Clause 4.16 of ICSub/C contains identical wording to clause 4.19 of DBSub/C.

ICSub/D

All as ICSub above.

ICSub/NAM

All as ICSub above.

MWSub/D and ShortSub

Clause 10.3 restricts the sub-contractor's right to payment of any direct loss and/or expense to circumstances due to the regular progress of the sub-contract works being affected by the compliance with any written variation. Such recovery is also subject to the sub-contractor notifying the contractor of the same as soon as it is 'reasonably practicable'. Under this provision the contractor is obliged to determine a fair and reasonable amount of direct loss and/or expense.

The limited right to payment of loss and/or expense under the sub-contract may cause some problems in practice, e.g. not all instructions may constitute variations. However, if additional costs due to matters other than variations are incurred by the sub-contractor, then these may be pursued as a claim for common law damages in appropriate circumstances.

SubSub

All as MWSub/D and ShortSub above. Clause 10.3 contains identical provisions albeit in respect of the sub-subcontractor.

What are Relevant Sub-contract Matters?

As noted above, the majority of the JCT sub-contract provisions regarding loss and expense only provide an entitlement where it can be shown that a Relevant Sub-contract Matter has affected the regular progress of the sub-contract works.

DBSub

DBSub lists Relevant Sub-contract Matters at clauses 4.20. The Relevant Sub-contract Matters, by and large, replicate the Relevant Sub-contract Events detailed under clause 2.19 of DBSub.

These Relevant Sub-contract Matters are:

Clause 4.20.1 'Variations (excluding any for which a Schedule 2 Quotation has been accepted by the Contractor but including any other matters or directions which under these Conditions are to be treated as, or as requiring, a Variation).'

Clause 4.20.2 'Directions of the Contractor, including those which pass on instructions of the Architect/Contract Administrator:
  .1 for the expenditure of Provisional Sums included in the Contractor's Requirements or any Bills of Quantities, excluding an instruction for expenditure of a Provisional Sum for defined work;
  .2 for the opening up for inspection or testing of any work, materials or goods under clause 3.17 of the Main Contract Conditions (including making good), unless the inspection or test shows that the work, materials or goods were not in accordance with the Sub-contract;
  .3 for the opening up for inspection or testing of any work, materials or goods under clause 3.10 (including making good), unless the inspection or test shows that the work, materials or goods were not in accordance with this Sub-contract;
  .4 in relation to any discrepancy in or divergence between any of the Numbered Documents or any discrepancy in or divergence between any of those documents and the Contract Documents under the Main Contract;
  .5 for the postponement of any work to be executed under this Sub-contract (whether in connection with a postponement under the Main Contract or otherwise);
  .6 with respect to any find of antiquities.'

Clause 4.20.3 'Suspension by the Sub-contractor under clause 4.11 of the performance of his obligations under this Sub-contract, *provided the suspension was not frivolous or vexatious.*'

Clause 4.20.4 'Suspension by the Contractor under clause 4.11 of the Main Contract Conditions of the performance of his obligations under the Main Contract.'

Clause 4.20.5 'Where there are Bills of Quantities, the execution of work for which an Approximate Quantity is included in those bills which is not a reasonably accurate forecast of the quantity of work required.'

Clause 4.20.6 'Delay receipt of any permission or approval for the purposes of Development Control Requirements necessary for the sub-contract works to be carried out or proceed, which delay the sub-contractor has taken all practical steps to avoid or reduce.'

Clause 4.20.7 'Any impediment, prevention or default, whether by act or by omission, by the Employer or any of the Employer's Persons except to the extent caused or contributed to by any default, whether by act or omission, of the Sub-contractor or of any of the Sub-contractor's Persons.'

Clause 4.20.8 'Any impediment, prevention or default, whether by act or by omission, by the Contractor or any of the Contractor's Persons (including, where the

Contractor is the Principal Contractor, any default, whether by act or omission, in that capacity) except to the extent caused or contributed to by any default, whether by act or omission, of the Sub-contractor or of any of the Sub-contractor's Persons.'

In respect of the words: 'except to the extent caused or contributed to by any default, whether by act or omission, of the Sub-contractor or of any of the Sub-contractor's Persons,' within clauses 4.20.7 and 4.20.8 of DBSub/C, it should be particularly noted that clause 2.7.3 and 2.7.4 of DBSub/C requires that:

- such further drawings, details, information and direction referred to in clauses 2.7.1 and 2.7.2 of DBSub/C shall be provided or given at the time it is reasonably necessary for the recipient party to receive them, having regard to the progress of the sub-contract works and the main contract works; and
- where the recipient party has reason to believe that the other party is not aware of the time by which the recipient needs to receive such further drawings, details, information or directions, he shall, so far as is reasonably practicable, advise the other party sufficiently in advance to enable him to comply with the requirements of clause 2.7 of DBSub/C.

MPSub

MPSub at clause 30.2 lists three limited matters entitling the sub-contractor to claim reimbursement of any loss and expense incurred. Obviously, to apply, one of these matters must be the cause of the loss and/or expense incurred by the sub-contractor. As noted earlier, loss and expense caused by a change is dealt with separately under clause 29 (not clause 30).

The three loss and expense matters, by and large, replicate three of the events listed within clause 22.1 of MPSub, i.e. events which can give rise to an adjustment to the completion date.

The loss and expense matters are:

Clause 30.2.1 'A breach or act of prevention on the part of the Contractor, or by any of his sub-contractors and suppliers or any other person under the control and direction of the Contractor, other than the Sub-contractor *and other than any matters or actions that are expressly permitted by this Sub-contract and that are stated not to give rise to a Change.*'

Clause 30.2.2 'Interference with the Sub-contractor's regular progress of the Sub-contract Works by Others on the Site.'

Clause 30.2.3 'The valid exercise by the Sub-contractor of his rights under section 112 of the HGCRA 1996'.

ICSub

The Relevant Sub-contract Matters are listed under clause 4.17 of ICSub/C. These are considered below.

The Relevant Sub-contract Matters, by and large, replicate the Relevant Sub-contract Events detailed under clause 2.13 of ICSub/C.

Clause 4.17.1 'Variations (including any other matters or directions which under these Conditions are to be treated as, or as requiring, a Variation).'

Clause 4.17.2 'Directions of the Contractor, including those which pass on instructions of the Architect/Contract Administrator:

  .1 for the expenditure of Provisional Sums included in any Bills of Quantities, excluding an instruction for expenditure of a Provisional Sum for defined work;

  .2 for the opening up for inspection or testing of any work, materials or goods under clause 3.14 of the Main Contract Conditions (including making good), unless the inspection or test shows that the work, materials or goods were not in accordance with the Sub-contract;

  .3 for the opening up for inspection or testing of any work, materials or goods under clause 3.9 (including making good), unless the inspection or test shows that the work, materials or goods were not in accordance with this Sub-contract.

  .4 in relation to any discrepancy in or divergence between any of the Numbered Documents or any discrepancy in or divergence between any of those documents and the Contract Documents under the Main Contract;

  .5 for the postponement of any work to be executed under this Sub-contract (whether in connection with a postponement under the Main Contract or otherwise).'

Clause 4.17.3 'Suspension by the Sub-contractor under clause 4.13 of the performance of his obligations under this Sub-contract, *provided the suspension was not frivolous or vexatious.*'

Clause 4.17.4 'Suspension by the Contractor under clause 4.11 of the Main Contract Conditions of the performance of his obligations under the Main Contract.'

Clause 4.17.5 'The execution of work for which an Approximate Quantity included in the Contract Bills which is not a reasonably accurate forecast of the quantity of work required.'

Clause 4.17.6 'Where there are Bills of Quantities, the execution of work for which an Approximate Quantity included in those bills is not a reasonably accurate forecast of the quantity of work required.'

Clause 4.17.7 'Any impediment, prevention or default, whether by act or by omission, by the Employer, the Architect/Contract Administrator, the Quantity Surveyor or any of the Employer's Persons except to the extent caused or contributed to by any default, whether by act or omission, of the Sub-contractor or of any of the Sub-contractor's Persons.'

Clause 4.17.8 'Any impediment, prevention or default, whether by act or by omission, by the Contractor or any of the Contractor's Persons (including, where the Contractor is the Principal Contractor, any default, whether by act or omission, in that capacity) except to the extent caused or contributed to by any default, whether by act or omission, of the Sub-contractor or of any of the Sub-contractor's Persons.'

In respect of the words: 'except to the extent caused or contributed to by any default, whether by act or omission, of the Sub-contractor or of any of the Sub-contractor's Persons' within clauses 4.17.7 and 4.17.8 of ICSub/C, it should be particularly noted that clause 2.5.1 and 2.5.2 of ICSub/C requires that:

- Clause 2.5.1 obliges the contractor, 'from time to time', to provide two copies of such further drawings, details, information and directions as are reasonably necessary to explain and amplify the numbered documents furnished by the contractor and to issue any directions (including those for or in regard to the expenditure of provisional sums) as are necessary to enable the sub-contractor to carry out and complete the sub-contract works in accordance with the contract.
- Clause 2.5.2 obliges the sub-contractor to provide the contractor with working/setting out drawings, and any other information necessary, to facilitate the contractor in undertaking any appropriate preparations required to enable the sub-contractor to carry out and complete the sub-contract works in accordance with the sub-contract.

ICSub/D

All as ICSub above. Clause 4.17 of ICSub/D/C contains identical wording to clause 4.17 of ICSub/C, except for the inclusion of an additional reference to Contractor's Requirements within the wording of clause 4.17.2.1 (see below):

> Clause 4.17.2 'Directions of the Contractor, including those which pass on instructions of the Architect/Contract Administrator:
> .1 for the expenditure of Provisional Sums included in the Contractor's Requirements or any Bills of Quantities, excluding an instruction for expenditure of a Provisional Sum for defined work.'

ICSub/NAM

All as ICSub above. Clause 4.17 of ICSub/NAM/C contains identical wording to clause 4.17 of ICSub/C, except for the inclusion of an additional reference to NAM design requirements within in the wording of clause 4.17.2.1 (see below):

> Clause 4.17.2 'Directions of the Contractor, including those which pass on instructions of the Architect/Contract Administrator:
> .1 for the expenditure of Provisional Sums included in the Contractor's Requirements or any Bills of Quantities, excluding an instruction for expenditure of a Provisional Sum for defined work.

MWSub/D and ShortSub

Unlike the other JCT sub-contract forms noted above, the MWSub/D and the ShortSub do not list any relevant matters. However, as noted earlier in this chapter, clause 10.3 of these sub-contracts provides for the sub-contractor to be paid direct loss and/or expense due to the regular progress of the sub-contract works being affected by the compliance with any variation instructed in writing.

SubSub

As MWSub/D and ShortSub above.

Can the contractor recover direct loss and/or expense from the sub-contractor?

DBSub

Clause 4.21 of DBSub/C gives the contractor a right to recover from the sub-contractor the agreed amount of any direct loss and expense caused to the contractor where the 'regular progress of the main contract works' is materially affected by any act, omission or default of the sub-contractor, or any of the sub-contractor's persons.

Clearly the contractor has the burden of proof in demonstrating that the above qualifying criteria exists in order for this provision to apply. This may be problematic in practice, especially if there are other potentially competing causes of delay and/or disruption caused by the contractor and/or his other sub-contractors.

Clause 2.21 of DBSub/C also deals with the recovery by the contractor of any direct loss and/or expense suffered or incurred by the contractor caused by the culpable failure of the sub-contractor to complete the sub-contract works on time.

In respect of clause 4.21, the contractor is obliged to notify the sub-contractor in writing with reasonable particulars of the effects or likely effects on the regular progress of the main contract works resulting from any act, omission or default of the sub-contractor, or any of the sub-contractor's persons, and shall also give details of the resultant loss and expense as the sub-contractor reasonably requests.

Any amount of loss and expense agreed between the contractor and the sub-contractor relating to clause 4.21.1 may be deducted from any monies due to or to become due to the sub-contractor or shall be recoverable by the contractor from the sub-contractor as a debt.

If the contractor follows the option of deducting the loss and expense from monies due to the sub-contractor, it is important that a withholding notice required by clause 4.10.3 and/or 4.12.3 of DBSub/C is issued timeously.

MPSub

Clause 21 of MPSub/C gives the contractor a right to recover from the sub-contractor direct loss and expense incurred by the contractor as a consequence of the sub-contractor's failure to proceed with and complete all of the sub-contract works in accordance with the sub-contract, provided that the contractor has, as soon as is reasonably practicable:

- notified the sub-contractor that loss and/or expense has been or is likely to be incurred by the contractor (clause 21.1); and
- notified the sub-contractor of the loss and/or expense incurred by the contractor, and provided supporting details (clause 21.2).

As noted in clause 21, the recovery of direct loss and/or expense from the sub-contractor can include any liquidated damages in circumstances where:

- these have been paid or allowed to the employer by the contractor; or
- the contractor is liable to pay these to the employer as a consequence of their inclusion in a payment advice issued under the main contract.

ICSub

As DBSub above. Clause 4.18 of ICSub/C is an identical provision to clause 4.21 of DBSub/C.

ICSub/D

As ICSub above.

ICSub/NAM

As ICSub above.

MWSub/D and ShortSub

MWSub/D and ShortSub do not include an express provision for the contractor to recover loss and/or expense from a sub-contractor.

SubSub

SubSub does not include an express provision for the sub-contractor to recover loss and/or expense from sub-subcontractors.

In pursuing a loss and expense claim, what does a sub-contractor need to prove?

Most claims, whether for time or money, involve establishing what is often called the nexus of cause and effect (i.e. the link between cause and effect). This means that what needs to be proved is that, because of the occurrence of a particular event certain things happened and as a direct result of those events happening this in turn led to one of the parties incurring delays or costs which were not previously contemplated by the parties and which it would not have been reasonable for the parties to have contemplated.

In order to establish this nexus (or link), which is never an easy task in a construction situation, records need to be available which show that the circumstances that existed before an event occurred changed after that event occurred, and that that change in circumstances could only have been as a result of the event in question.

Burden of proof

Ordinarily, the burden of proof lies with the party who makes an assertion, hence the legal maxim 'he who asserts must prove'.

Accordingly, it is not for a party defending a claim to disprove the claim, it is for the party pursuing a claim to prove the claim.

However, if the party pursuing a claim adduces sufficient evidence to raise a presumption that what is claimed is true, the burden of proof will pass to the other party. It is then for that party to adduce sufficient evidence to rebut the presumption.

Therefore, any claim document should be prepared with the above basic principle in mind.

Standard of proof – what needs to be provided?

The standard of proof required in civil proceedings is referred to as the 'balance of probabilities' principle, whereby the court/tribunal makes its decision on the basis that something is more likely to have occurred than not.

Therefore, if a set of scales is imagined as being the balance of probabilities, there only needs to be 51% of probability on one side of the scales to enable that side of the scales to succeed over the other side of the scales. Naturally, it is necessary for the party with the 'burden of proof' at any particular time to push the scales down to (at least) the 51% level.

This is to be contrasted with the 'standard of proof' in criminal proceedings, which is set at a much higher level of 'beyond reasonable doubt'.

When preparing a claim (in addition to the burden of proof principle outlined above) the required standard of proof needs to be recognized, particularly as, in certain situations, the party receiving the claim (i.e. before proceedings are commenced) may appear to expect a standard of proof more akin to 'beyond reasonable doubt' rather than on the balance of probabilities.

Part of this reason may be because of the apparent dichotomy between the standard of proof required for civil proceedings (i.e. the balance of probabilities) and the requirement under most of the JCT sub-contracts (e.g. clause 4.19.3 of DBSub/C) which assumes that the contractor will *ascertain* the loss and/or expense value.

The word 'ascertain' is usually understood to mean 'to find out for certain'. However, this is much closer to the standard of proof for criminal proceedings (i.e. beyond reasonable doubt) than it is to the standard of proof for civil proceedings (i.e. the balance of probabilities).

There have been conflicting court cases regarding this particular matter. For example, in the case of *Alfred McAlpine* v. *Property and Land*[2], it was found that when an architect is required to ascertain he is obliged to find out for certain and not merely to make a general assessment. However, in the *How* v. *Lindnor*[3] case, it was held that assessment of loss and expense was akin to assessment of damages, requiring no special standard of proof, and that the exercise of judgment was not only permissible but required where loss had not been proved with absolute certainty.

Further, in the case of *Norwest Holst* v. *Co-op*[4], Judge Thornton found that under a sub-contract the parties were normally required to agree loss and expense and,

---

[2] *Alfred McAlpine Homes North Ltd* v. *Property and Land Contractors Ltd* (1995) 76 BLR 59.
[3] *How Engineering Services Ltd* v. *Lindnor Ceilings Ltd* [1999] CILL 1521.
[4] *Norwest Holst Construction* v. *Co-op Wholesale Society* (1997/1998) (unreported).

in the event that they could not agree, it was in order for an arbitrator to determine such reasonable loss and expense (on a balance of probabilities basis).

Another point to note is that when assessing probabilities, the court or tribunal will have in mind, to whatever extent is appropriate in the particular case, that the more serious the allegation the less likely it is that the event occurred and, hence, the stronger should be the evidence before the court concludes that the allegation is established on the balance of probabilities.

This does not mean that different standards of proof are required where different assertions are made, but it does mean that the inherent probability or improbability of an event is itself a matter to be taken into account when weighing the probabilities and deciding whether, on balance, the event occurred.

The more improbable the event, the stronger must be the evidence that it did occur before, on the balance of probability, its occurrence will be established[5].

It should be noted that 'more likely than not' is not necessarily a hard test to pass and facts can often be proved to this standard merely by circumstantial evidence, or by one person's word being preferred to another's, although contemporaneous written records are always more influential.

In all circumstances, some admissible evidence must be produced to support everything claimed.

What needs to be proved in a loss and expense claim document?

Following on from the above, the following initial points need to be proved in a loss and expense claim document:

- that an event actually occurred;
- that the event was one expressly catered for within the sub-contract;
- that the notices required under the sub-contract had been given; and
- what the effect was, in financial terms, of the specified event.

In respect of the above list, what then needs to be shown is the connection between the *cause* and the *effect*, i.e. showing how the latter item on the above list (the financial effect) resulted directly from the first item on the above list (the event that occurred).

This is the part of the process that normally causes most difficulty.

The above connection is a matter of evidence and it is the part of a claim for which records are absolutely crucial.

Indeed, this is when the quality and extent of the records and documentation come into their own.

With some of the heads of claim (outlined later in this chapter) such a connection is relatively easy to identify, whilst with others it is rather more problematic.

It is because of the difficulty of establishing the required nexus of cause and effect that it is not uncommon for global claims to be produced.

Before considering the individual heads of claim in detail it is worth looking at the background of global claims generally.

---

[5] Ref Lord Nicholls in Re H (Minors) [1996] AC 563.

What is a global claim?

A global claim is a claim for financial loss which arises from various different events; however, the requirement to link cause and effect is absent (i.e. individual sums of money are not claimed for each individual event). Instead a single global sum is claimed in respect of the alleged cumulative effect of all of the events.

Quite often claimants will simply base their claims in terms of quantum on the difference between estimated cost and actual cost, normally with an adjustment for any recovery made within the variation account. A simple example is as follows:

Total costs incurred by the sub-contractor	£1,000,000
Less: sub-contract order value	£600,000
Less: recovery from variations	£200,000
Claim for loss and/or expense (i.e. the shortfall)	**£200,000**

The basis of the global claim being advanced is that the total extra cost incurred, as calculated above, has resulted from numerous events whose consequences had such a complex interaction, that it is impossible or impractical to disentangle them to show cause and effect.

Unsurprisingly, global claims have been the subject of much controversy and have been considered in many court cases, remaining a difficult and developing area of law.

The recent case of *John Doyle Construction Ltd* v. *Laing Management (Scotland) Ltd*[6] reviewed and provided useful guidance in respect of the submission of global claims. Whilst a Scottish case, it is likely to be followed by the English Courts.

In that case, Laing's contentions that the global claim should be struck out were not accepted by the Inner House of the Court of Session, and the judgment included important guidance regarding the current law on global claims, as set out in the following six points:

(1) That, in the normal case, individual causal links must be demonstrated between each of the events for which the employer is responsible and particular items of loss and expense.

(2) That, in circumstances where it is impossible to separate the specific loss and expense caused by a number of different events which are the responsibility of the employer, then these can be pleaded as producing a cumulative effect. In these circumstances, it is not necessary to break down each event and isolate the loss caused by each.

(3) That, where, however, a significant cause of the loss and expense is a matter for which the employer is not liable, the global claim must fail.

(4) That, where it is shown that some of these events (albeit not a 'significant' amount in causal terms) are not actually the responsibility of the employer, the global claim should not necessarily fail, since it may be possible for the judge, or arbitrator, to apportion the loss as between the causes for which the employer is responsible and other causes.

(5) That it was acknowledged that this could lead to 'rough and ready' results.

(6) That, when pleading the claim, the particular events and heads of loss should be set out in reasonable detail. There will, however, usually be no need in

---

[6] *John Doyle Construction Ltd* v. *Laing Management (Scotland) Ltd* [2004] CILL 2135.

commercial cases to do more than simply to plead the proposition that the particular events caused the relevant heads of loss. Causation is largely 'a matter of inference' and is frequently based upon experts' reports. The consideration of the global claim should wait until this evidence is before the tribunal.

Submitting a claim/global claim

Generally, the best practice to follow in preparing a claim involving multiple variations, delay, disruption and extra cost is to follow the guidelines laid down by the courts strictly, so far as it is possible, and to present a cause and effect claim where possible, and a global claim only in respect of any element of the claim where it is simply not possible or practical to do otherwise.

One way of presenting this type of claim is to have a points of claim document which deals with the legal issues, setting out the terms of the contract which are applicable, in particular the variations provisions, any terms which give rise to a claim for breach of contract and setting out in summary form in what respects the breaches occurred. This document would be relatively short and the detailed factual matters supporting the claim would be set out in a separate document.

The separate document could be in the form of a 'cause and effect' schedule which, for a claim based mainly on variations, might be set out as follows:

Column 1 – Event
Column 2 – Clause No.
Column 3 – Other Term
Column 4 – Effect (Disruption)
Column 5 – Delay (From and To)
Column 6 – Claim/Loss (£)

Where:

Column 1 would set out the event relied on, e.g. a variation order or an event which is said to be a breach of the sub-contract.

Column 2 would simply refer to an appropriate and relevant sub-contract clause number or, if the clause has sub-clauses, the appropriate and relevant sub-contract sub-clause number.

Column 3 would be used if some other express or implied terms were relied upon for events other than variations.

Column 4 would be used to demonstrate the effect. This would most likely be a narrative if the claim related to disruption, as opposed to a prolongation claim, explaining what activity was affected by the event complained of and in what way.

Column 5 would deal with prolongation, giving the precise start date of delay and the precise finish date of delay. It would also be appropriate to indicate in that column whether delay was continuous or intermittent, since it is a common failing in this type of claim to adopt a notional rather than a historical analysis in which it is assumed that where something goes wrong, a continuous delay to the contract as a whole occurs until that matter is put right, which is often not the case. It is also necessary in assessing prolongation to include an analysis of concurrent delay, so that critical delay is identified (another common failing in this type of claim being to assume that all delay is cumulative or failing to identify what is concurrent and what is cumulative).

Column 6 would include an amount which is claimable under the terms of the contract, for each individual item noted in column 1.

In attempting to produce this type of schedule it will be found in some cases that it is not possible with multiple variations, and possibly other claims, to give a precise delay period or claim figure for each one, because of the complex inter-action of events.

In that event, all one can do is to break it down *as far as possible* and to present the remainder as a global claim.

Approaching the matter in this way is not only a good discipline but is also more likely to result in success, as it closely replicates the latest requirements of the courts.

If faced with a global claim, it is fairly standard to request a 'cause and effect' analysis as this will highlight the weaknesses of the other party's case.

What is the purpose of a claim for loss and/or expense?

A claim for loss and/or expense is the means of putting a sub-contractor back into the position in which he would have been but for the delay or disruption, it is not a means of turning a loss into a profit (unless, of course, the loss is because of the non-payment of loss and/or expense).

The settlement in effect amounts to common law damages which is dealt with later in this chapter; therefore an exact establishment of the sub-contractor's addi-tional costs must be made. At first sight it might be thought that establishment of those *additional* costs is relatively easy but in fact it can be very difficult.

What are common heads of a loss and expense claimed?

The common heads of a loss and expense claim are described below.

On-site establishment costs (site overheads)

These are also commonly referred to as 'preliminary items'. Additional expense under this heading is usually easily ascertainable from the sub-contractor's cost records.

Such costs may include the following examples:

- Project manager (site based)
- Site foreman
- Site QS
- Site engineer and assistants
- Site offices
- Site administration costs
- Site stores
- Welfare facilities (if provided by the sub-contractor)
- Cranage/forklifts
- General and small plant
- Scaffolding/towers

- Telephone
- Electricity
- Insurance

What the sub-contractor is seeking to establish is that, because of events that have occurred and for which the sub-contract provides an entitlement, actual resources of the type listed above (amongst others), additional to those which otherwise would have been necessary, have had quite reasonably to be engaged on the project.

What is the basis of the claim under this heading?
The sub-contractor is essentially seeking to recover site overhead costs, i.e. those costs which are not related to one specific item of work but which are necessary for the site establishment generally.

The site overheads part of the claim is commonly broken down into delay costs (due to prolongation) and 'thickening' (or extra resource) costs (due to disruption).

The claim is usually for time-related costs, i.e. the additional cost associated with the prolonged use of a particular resource, for example a five-week delay will clearly affect the sub-contractor's time-related costs, such as site supervision, cabins, hired equipment, etc.

However, this may not always be the case. It may be that the sub-contractor incurs *additional* supervision costs without the overall completion date for the sub-contract works being affected. For example, *additional* supervision costs may be incurred by the sub-contractor during the original sub-contract period, or a sub-contractor may be delayed or disrupted by an instruction ordering additional work to non-critical items (e.g. scaffold to an area of work which was not on the critical path). In such circumstances, it may be entitled to loss and expense incurred (i.e. thickening as noted above) regardless of whether the additional work delayed the sub-contractor in the completion of the works.

Are the delay costs (prolongations) taken as those in the over-run period?
The prolongation part of the claim must relate to those periods when delay occurred; this will not necessarily be the over-run period.

For example, if a sub-contractor had a sub-contract period of six weeks, but was delayed in week two for one week, the site overhead costs would need to be claimed for week two (i.e. the week when the delay occurred) and not for week seven (i.e. the week when the delay effect became apparent).

What costs are claimed?
The actual costs incurred by the sub-contractor are to be claimed. A pro-rata of bill/tendered allowances does not represent actual cost and may therefore not be successful.

Is a sub-contractor entitled to additional site overhead costs where he programmes to complete a project before the contract completion date but is prevented from doing so?
There is much general debate surrounding early finish programmes.

The sub-contract on-site period may be, for example, eight weeks, but the sub-contractor may decide that he wishes to complete the sub-contract works in six weeks. However, the contractor issues instructions that prevent the sub-contractor from completing the sub-contract works until the end of week eight. The sub-contractor then wishes to recover the additional site overhead costs for weeks seven and eight.

In the *Glenlion Construction* v. *The Guinness Trust*[7] case, the position in respect of JCT contracts was made reasonably clear. In that case (which related to a contractor/employer relationship), it was found that the contractor was not entitled to an extension of time in such a situation. Under JCT contracts extensions of time are only due if completion of the works is delayed beyond the completion date.

Despite the above, if a sub-contractor can show that he has been caused loss and expense by reason of the delay, he may be entitled to claim for loss and expense.

What points need to be considered in assessing such a claim?
In assessing such a claim several points should be kept in mind:

- What has to be established in the first instance is that additional resources were required.

- The sub-contractor needs to establish with sufficient particularity:
  - — what was included in his tender; and
  - — that the tender allowance was sufficient.

- Just because resources are engaged during a period of overrun does not mean they are additional. For example, if a sub-contractor originally expected to have two foremen on site for ten weeks but actually only had one foreman on site for 20 weeks there would not (on the face of it) be any additional costs.

- The sub-contractor needs to establish that the additional resources were required due to the reasons that he is relying on.
  This may be done in several ways:

  - — If the claim is for prolongation then it is necessary to show that resources were retained on site during the periods of delay for reasons outside of his control. This is usually one of the easier points of the claim.
    However, the sub-contractor does not need to have received an extension of time before he can pursue such prolongation costs during a period in excess of the original period.
    Conversely, not all extensions of time circumstances give rise to financial entitlement, and, even if they do, not all the additional resources may be recoverable.

  - — A claim for site overhead costs may be pursued in the absence of prolongation.
    If an appropriate resource is required for longer during the original sub-contract period, or more of them are required for the same period, then they should be recovered if they can be associated with a specific event.
    This may apply, for example, in the situation where a variation is issued which results in certain work activities, which were programmed to be executed consecutively now needing to be carried out concurrently, thus requiring two or more work faces to be operating and needing two (or more) sets of supervisors, etc., where previously one would have been adequate.

---

[7] *Glenlion Construction Ltd* v. *The Guinness Trust* (1987) 39 BLR 89.

- The sub-contractor then has to establish the cost of the resources.
  With supervisory staff this would be ascertained using salary levels, pensions, national insurance, car costs, petrol, bonuses, etc.

  If agency staff have been used then their actual costs will be recoverable providing that it was reasonable to use them.

Inefficient or increased use of labour and plant (commonly referred to as a disruption claim)

Delay and disruption can lead to increased expenditure on plant and labour in two ways. One is that it may be necessary to employ extra resources and the other is that existing resources may be used inefficiently.

This head of claim in particular may be difficult to ascertain due to lack of records and the difficulty of establishing the nexus of cause and effect.

However, just because it is difficult to establish the above, it does not mean that the sub-contractor should not submit a claim. Equally it does not mean that the contractor should not make a reasonable assessment (see comments earlier within this chapter regarding 'global' claims).

As noted earlier in this chapter, the 'global' claim approach should be used very much as the exception rather than the rule. It is applicable and permitted only where the events are so numerous/complex/interrelated that no amount of analysis or record keeping would establish or evidence the connection of cause and effect on individual matters. However, it is not good enough that such analysis cannot be done simply because the sub-contractor has not kept reasonable records.

How can disruption (loss of productivity) be assessed?

There are a number of methods of assessing this head of claim and indeed few people agree on the approach to use. However, in all cases, good records are essential.

Ideally, what should be established is that, *before* a particular event occurred productive resources were achieving a certain level of productivity/financial income for the sub-contractor. However, *after* that particular event occurred and indeed, because of it, the same level of productivity/financial income could not be achieved.

Take for example the case of a bricklaying sub-contractor.

In the sub-contract the sub-contractor is told that he will be required to build Wall A, followed by Wall B, and followed by Wall C. In his tender, the sub-contractor allowed for an output of 60 bricks per hour per bricklayer.

- Week one: things progress well. The bricklayers worked on Wall A only, as planned; output achieved 65 bricks per hour. The sub-contractor was exceeding his tender expectations and this illustrated the sufficiency of his tender.
- Week 2: things start to go wrong. The sequence of work changed, and the sub-contractor was instructed by the contractor to switch his bricklayers from Wall A to Wall B to Wall C, often all in the same day. As a consequence, the output of the bricklayers at the end of the second week reduced to 50 bricks per hour.

In such a case, the disruption claim would be based on the loss of production caused by the disruption event. In other words, the sub-contractor expected to achieve 60 bricks per hour (or the sub-contractor might argue that he actually achieved 65 bricks per hour) before the disruption event occurred, but because of the disruption event the sub-contractor's bricklayers only achieved 50 bricks per hour.

Although the above is a very simple example, the principle remains the same for much more complicated claims. Clearly, for this head of claim, more than any other, good, complete and accurate records are absolutely crucial.

As noted above, there are several types of approach in respect of a disruption claim which may be applicable, depending on the nature of the work, the circumstances and the records available. The main approaches used being:

- To show that resources were standing idle when they should otherwise have been working – thus taking longer to do the same work.
- To illustrate how the work of a particular trade has been prolonged and to apply a disruption factor in the same proportions to the cost of the trade.
- To estimate a disruption factor.
- To do an overall global claim based on costs of all trades assessed in the tender and to claim the difference between that and actual costs.
- A mini global claim using the approach as last but targeting specific trades.
- To illustrate the resources employed in producing specific and evidenced amount of work prior to an event and to claim the extra resources for doing equivalent work after the event.
- To show that more expensive resources were required to do the same value of work because of an event, even though the level of resources was the same, i.e. agency resources, emergency hire of plant, expensive gangs, etc.

Winter working

This heading amounts to a different aspect of loss of productivity resulting from the need to work in less favourable climatic conditions. This is a recoverable head of claim[8].

What must be shown?

What the sub-contractor has to establish in this case is that for reasons providing an express contractual entitlement he has had to undertake productive work in climatic conditions less favourable than would otherwise have been the case.

It is not enough to show only that work was undertaken during the winter (for example), it must also be shown that this fact did actually affect the work.

Of course, this type of claim could be problematic for sub-contractors, particularly if their actual period on site could have commenced at any time, within wide parameters.

---

[8] *Bush* v. *Whitehaven Port & Town* (1888) 52 JP 392.

Head office overheads and profit

What are head office overheads?

Head office overheads are sometimes called 'off-site overheads' or 'establishment charges'.

Head office overheads are those administrative and management costs of running the head office of a sub-contractor's business over and above the site costs.

Generally, head office overheads will include for:

- Purchase and/or rent of office and yards
- Maintenance and running costs
- Transport and mobilisation costs
- Company cars
- Depreciation
- Director's emoluments and expenses
- Salary and other cost of staff
- Administration costs
- Legal and audit fees, etc.

What is the basis of the claim of additional head office overheads costs?

Most sub-contractor's head office resources exist to support operations undertaken on site, and such head office overheads are normally taken as being recovered out of the income from his business as a whole.

Where completion of one project has been delayed a sub-contractor may claim to have suffered a loss arising from the diminution of his income from the project reducing the turnover of his business. Despite this, the sub-contractor continues to incur expenditure on head office overheads which he cannot materially reduce or, in respect of the project, can only reduce, if at all, to a limited extent.

Therefore, but for the delay and disruption, the sub-contractor's workforce would have had the opportunity of being employed on another project which would have had the effect of contributing to the head office overheads and profit during the overrun period.

There is some authority that a claim on this basis is sustainable.

In *JF Finnegan* v. *Sheffield City Council*[9], Judge Sir William Stabb (talking about a contractor in that case) said:

'It is generally accepted that, on principle, a contractor who is delayed in completing a contract due to the default of his employer, may properly have a claim for head office or off-site overheads during the period of delay, on the basis that the workforce, but for the delay, might have had the opportunity of being employed on another contract which would have had the effect of funding the overheads during the overrun period.'

It should be noted that the entitlement to such general overheads is in itself an arguable point. However, the consensus of opinion is in favour, providing, of course, it can be established that such overheads would have been recovered were it not for the delay.

---

[9] *JF Finnegan Ltd* v. *Sheffield City Council* (1988) 43 BLR 124.

However, substantial claims of this kind are rarely made because most sub-contractors are able to cope with delay on a particular contract using their existing resources whose cost is reasonably constant.

What must the sub-contractor seek to demonstrate?

In terms of first establishing liability, the sub-contractor must seek to prove that head office overhead costs have been increased as a result of a delay and disruption suffered by the sub-contractor, for which he is entitled to recompense under the sub-contract. Examples may range from the cost of extra staff recruited because the particular project was in difficulties, to the cost of extra telephone calls and postage in the period of delay.

It is suggested that, in order to succeed, a sub-contractor must provide evidence to show:

(1) that the profit or overhead contribution was capable of being earned elsewhere at the time of delay;
(2) that the profit and overhead percentage is a reasonable one; and
(3) that work of the same level of profitability and/or overhead recovery was available during the period of delay.

In terms of evidencing that there was other work available which, but for the delay, he would have secured, the sub-contractor might seek to evidence this by producing declined invitations to tender, with evidence that the reason for declining was that the delay in question left him insufficient capacity to undertake other work.

He might alternatively show from his accounts a drop in turnover and establish that this resulted from the particular delay rather than from extraneous causes. If loss of turnover resulting from delay is not established, the effect of the delay is only that receipt of the money is delayed. It is not lost.

If liability is established, then the sub-contractor must establish quantum. When pursuing a claim for head office overheads and profit, a sub-contractor will frequently rely on the use of a formula to evidence such quantum.

What formulae are used?

There are three formulae that are in common usage (and, in addition, there are many other derivatives from these formulae).

These formulae are:

- the Hudson formula;
- the Emden's formula; and
- the Eichleay formula.

Whilst the use of formulae has been heavily criticised by many commentators, there is judicial authority for their use in certain instances. However, whatever formula may be used, it must be noted that the use of a formula merely provides the means for assessing quantum. The fact that there was an actual loss, and that this loss flows directly from the relevant matter relied upon must first be established.

In addition, when submitting claims for head office overheads and profit, care needs to be taken to avoid any double recovery or overlap with other claims or payments obtained by the sub-contractor, such as variations which have been computed by using the contract prices as a basis. Normally, in such a situation the valuation of the variation will include for an element of additional head office overheads and profit recovery.

Is loss of profit recoverable?
Although there has been some debate about whether loss of profit is recoverable as part of a loss and expense claim, if the words 'direct loss and expense' are to be interpreted as equivalent to the measure of damages at common law for breach of contract (which is generally accepted to be the case), then there can really be no doubt that a sub-contractor can claim for the loss of profit that he would have earned on other sub-contracts had there been no delay and disruption to the project in question, provided that such loss of profit was foreseeable.

Clearly, for a claim of loss of profit to be made, a sub-contractor would need to show, as with head office overheads, that at the time of the delay he could have used the lost turnover profitably. A claim for loss of profit does not, it is submitted, fail merely because the contract in question was unprofitable. The question is what the contractor would have done with the money if he had received it at the proper time. Even if, at that time, the contractor's business was making a loss, a sum analogous to loss of profit is, it is submitted, recoverable if the loss of turnover increased the loss of business.

Increased costs

What are increased costs?
This head of claim usually arises on a fixed price sub-contract (although it can equally apply where a fluctuation recovery has been agreed) where expenditure on resources due to increased costs during an extended period is sought.

How are increased costs claimed?
It is submitted that the correct measure is the difference between what the sub-contractor would have spent on resources and what he has actually spent due to delays suffered. There are, however, a number of different methods of calculating this head of claim.

The best way is for the claim to be based on substantiated details of the level of costs that would have been expended on resources and the actual costs paid for those same resources backed up by invoices, etc. However, it is rare in practical terms that such an analysis could be economically undertaken.

Therefore an alternative method would be to use some type of formula approach.

Using a formula approach, the calculation would broadly be as follows.

Ascertain the value of resources in the original tender but using actual work values.

Find out the level of inflation for such resources over the original sub-contract period, normally by using some standard published indices.

Calculate how much the cost of resources would have increased during the original sub-contract period based on the indices obtained. This will then be taken as the amount deemed to be allowed in the rates.

Undertake the same calculation but using the indexed inflation figure over the extended sub-contract period.

The extra over cost between the above two calculations will constitute the increased cost claim.

Cost of claim

Where the preparation of a claim is undertaken by a sub-contractor's in-house staff, this cost will often be partly (or completely) reflected in the sub-contractor's loss and expense claim for site overheads or head office overheads.

Often, a sub-contractor may employ a consultant to assist in the preparation of a loss and expense claim, and the general rule is that the costs incurred in taking this action are generally not recoverable, except and until the matter proceeds to arbitration[10] or litigation.

The reasoning behind disallowing this head of claim is that the sub-contractor is obliged, under the sub-contract, to provide information, etc., to resolve financial issues. Therefore the submission of a claim and the work involved must have been included in his price.

It may be the case, however, that such monies might be recovered if the contractor requested further information from the sub-contractor which is judged to be beyond the sub-contract requirements.

Interest and finance charges

When a sub-contractor incurs loss and expense, this has to be financed by him either from his own capital resources or alternatively by increased borrowing. In either case, it is clear that the use of money costs money.

Depending on the level of inflation (and the interest rates applicable at any point in time), and the amount outstanding in respect of the loss and expense claim, this matter will either be a relatively small matter or will have major implications. In either event, it is a matter of great importance to the sub-contractor how such interest or finance charges should be recovered.

Interest in respect of monies owed is dealt with as an express term of the sub-contracts. Therefore, if it is proven that monies were outstanding in respect of the non-payment or under payment of a loss and expense claim simple interest would be applied in line with the sub-contract in question.

In the alternative case, finance charges, or the cost of being stood out of one's money, are a recoverable head of a loss and expense claim in any event. This was established beyond doubt in the Court of Appeal case of *FG Minter Ltd* v. *WHTSO*[11].

In respect of common law damages, to recover finance charges, it is necessary for the claimant to show that the finance charges fell within the second limb of *Hadley* v. *Baxendale* (dealt with later in this chapter), i.e. that the finance charges are special damages that were in the contemplation of the parties at the time that the contract was formed[12].

---

[10] *James Longley & Co. Ltd* v. *South West Thames Regional Health Authority* (1984) 25 BLR 56.
[11] *FG Minter Ltd* v. *WHTSO* (1980) 13 BLR 1, CA.
[12] Refer to *President of India* v. *Lips Maritime* [1988] AC 395, HL; *Holbeach Plant Hire* v. *Anglian Water Authority* (1988) 14 Con LR 101.

Acceleration

What is acceleration?

This may best be explained by considering a typical example. A delay is caused by a failing of the contractor. Mindful of the financial consequences on the contractor of finishing late (e.g. liquidated damages and prolongation costs), the contractor may be more interested in the sub-contractor completing his works in line with the original dates rather than in extending the sub-contractor's period of time on site.

In such a situation, it is not uncommon for a contractor to attempt to reach some form of acceleration agreement with the sub-contractor, whereby the sub-contractor will take measures (for example weekend working, double shifts, increased resources, etc.) to complete 'on time' (or, at least, earlier than would have been achieved under the normal course of events) and, if the earlier required date is achieved, the contractor will make an agreed acceleration payment to the sub-contractor.

Are there provisions for acceleration in the sub-contract?

DBSub
In the case of delay, the only express action available to the contractor under DBSub is to extend the sub-contractor's period of time on site.

It is debatable whether a schedule 2 quotation provided by a sub-contractor incorporates the possibility of agreeing acceleration to the sub-contract.

On the face of it, this would be possible because clause 2.2 of the said schedule 2 quotation states that the quotation shall comprise 'any adjustment required to the time period'. This 'adjustment' could naturally include for both an increase and a decrease to the previously agreed time period. A decrease to the previously agreed time period could be as a result of acceleration measures, and those acceleration measures could be priced within a schedule 2 quotation.

The reason why the word 'debatable' is used above, is that clause 2.2 of the schedule 2 quotation for the main contract specifically states 'including, where relevant, stating an earlier Completion Date than the Date for Completion given in the Contract Particulars'; but no such similar wording is included in clause 2.2 of the schedule 2 quotation document within the DBSub/C.

Nevertheless, this is a matter for the parties to agree, and it is therefore questionable whether the omission of the words in question has any real practical effect.

Apart from the above possibility, the DBSub/C does not specifically allow for acceleration agreements, and great care needs to be taken by both the contractor and the sub-contractor when entering into any such agreement.

MPSub
Clause 23 facilitates acceleration but requeres both the agreement of the contractor and the sub-contractor, i.e. acceleration of the sub-contract works cannot be unilaterally instructed under clause 23.

Should the contractor want to investigate the possibility of achieving practical completion before the expiry of the period for completion, clause 23.1 permits him to invite proposals from the sub-contractor.

Following the invitation, the sub-contractor must:

- provide proposals to accelerate the sub-contract works, detailing time saved and the associated additional costs to be paid (clause 23.1.1); or alternatively
- explain why it is impractical to achieve practical completion at an earlier date (clause 23.1.2). There appears to be no action the contractor can take if this occurs.

Under clause 23.2, the proposals can either be accepted or further revised proposals sought from the sub-contractor. If accepted, the contractor must issue an instruction recording therein the agreed adjustment to the period for completion and, as applicable, the relevant additional cost to be paid to the contractor. The instruction is to be treated as a change.

It should be noted that if any sum is agreed for undertaking acceleration measures, then, when agreeing upon that sum, allowance should be made for any bonus that may be payable for 'early' completion pursuant to clause 24.

ICSub, ICSub/D, ICSub/NAM, MWSub/D, ShortSub and SubSub

The above sub-contracts do not contain acceleration provisions.

In the case of delay, the only express action available to the contractor under these sub-contracts is to extend the sub-contractor's period of time on site.

Accordingly, great care needs to be taken by both the contractor and the sub-contractor if they enter into any such agreement.

What if the sub-contractor accelerates without an agreement?

A sub-contractor should not undertake acceleration measures (unless those measures are in mitigation of delays that the sub-contractor has caused) without reaching some form of agreement with the contractor, otherwise the probability is that he will not receive any payment for the acceleration measures undertaken.

Common law damages claims (in respect of a breach of contract)

As noted earlier within this chapter, clause 4.22 of DBSub, clause 4.19 of ICSub/C, etc., preserve the common law rights of the contractor and the sub-contractor.

In respect of these common law rights, a breach of contract which has not been excused gives the injured party the right to bring an action for damages.

What are damages?

There are many classic definitions but essentially, damages are awarded 'so far as money can do it'[13] to put the claimant as nearly as possible 'in the same position as he would have been in if he had not sustained the wrong for which he is now getting compensation or reparation'[14].

However, the courts set a limit to the loss for which damages are recoverable, and loss beyond such limit is said to be too remote.

In this regard the rule of remoteness is set out in the case of *Hadley* v. *Baxendale*, and this is:

---

[13] *Robinson* v. *Harman* (1848) 1 Ex 850.
[14] Lord Blackburn in *Livingstone* v. *Rawyards Coal Company* (1880) 5 App Cas 25, HL.

'Where two parties have made a contract which one of them has broken, the damages which the other party ought to receive in respect of such breach of contract should be such as may fairly and reasonably be considered either:

(1) arising naturally, i.e. according to the usual course of things from such breach of contract itself; or
(2) such as may reasonably be supposed to have been in the contemplation of both parties at the time they made the contract, as the probable result of the breach of it.'

The rule as stated above is recognised as having two limbs, and these limbs have been indicated in the above quoted text by the insertion of the numbers (1) and (2).

The first limb of Hadley v. Baxendale[15]

There have been two leading cases where the *Hadley* v. *Baxendale* rule has been considered. These are the *Victoria Laundry* v. *Newman Industries*[16] case and the *Koufos* v. *Czarnikow*[17] case.

In the *Victoria Laundry* case, Lord Justice Asquith stated several propositions that emerged, he said, from the authorities as a whole. Taking these into account together with the opinions of the House of Lords in the *Koufos* v. *Czarnikow* case, and other following cases, the first limb of the *Hadley* v. *Baxendale* rule may be elaborated into the following propositions:

(1) The aggrieved party is only entitled to recover such part of the loss actually resulting as may fairly and reasonably be considered as arising naturally, that is according to the usual course of things, from the breach of contract.
(2) The question is to be judged as at the time of the contract.
(3) In order to make the contract breaker liable it is not necessary that he should actually have asked himself what loss was liable to result from a breach of the kind which subsequently occurred. It suffices that, if he had considered the question, he would as a reasonable man have concluded that the loss of the type in question, not necessarily the specific loss, was 'liable to result'.
(4) The words 'liable to result' should be read in the sense conveyed by the expressions 'a serious possibility' and 'a real danger' and 'not unlikely to occur'.

For the first limb, therefore, knowledge of certain basic facts according to the usual course of things is imputed, but not special knowledge.

The second limb of Hadley v. Baxendale

The second limb of *Hadley* v. *Baxendale* depends on additional special knowledge by the defendant.

The passage from *Hadley* v. *Baxendale* quoted above is followed by:

---

[15] *Hadley* v. *Baxendale* (1854) 9 Ex 341.
[16] *Victoria Laundry (Windsor) Ltd* v. *Newman Industries Ltd* [1949] 2 KB 528.
[17] *Koufos* v. *Czarnikow* [1969] 1 AC 350.

'If the special circumstances were communicated by the plaintiffs to the defendants, and thus known to both parties, the damages resulting from the breach of such a contract, which they would reasonably contemplate, would be the amount of injury which would ordinarily follow from a breach of contract under these special circumstances so known and communicated.'

As with the first limb, the question is to be judged at the time of the contract so that damages claimed under the second limb will not be awarded unless the claimant has particular evidence to show that the defendant then knew the special circumstances relied on.

Date of assessment

The general rule, both in contract and in tort, is that damages should be assessed as at the date when the cause of action arises[18]. But there are many exceptions and it has been said that 'this so-called general rule . . . [e] has been so far eroded in recent times . . . [e] that little of practical reality remains of it'[19].

Thus, 'where it is necessary in order adequately to compensate the plaintiff for the damage suffered by reason of the defendant's wrong a different date of assessment can be selected.'[20]

Measure of damages

As noted earlier, the purpose of damages is to put the innocent party (i.e. the claimant), 'so far as money can do it'[21] back to the same position as if the contract had been properly performed.

However, sometimes the proper measure of damages is not the cost of reinstatement but the difference in value between the work actually produced and the work that should have been produced[22]. This will particularly be so where the claimant has no prospect or intention of rebuilding, or where it would be unreasonable to award the cost of reinstatement.

The frequently quoted *Ruxley* v. *Forsyth*[23] House of Lords case was regarding a swimming pool that was not constructed to the required depth.

The swimming pool should have been constructed to a depth of seven feet six inches; however, when the pool was completed its depth was only six feet nine inches.

Forsyth sought damages for the reinstatement costs (i.e. to re-build the swimming pool so that it was seven feet six inches deep). However, the House of Lords found that where it would be unreasonable for the claimant to insist on reinstatement because the cost of the work involved would be out of all proportion to the benefit obtained, the claimant's measure of damages would simply be the difference in value (sometimes referred to as the loss of amenity value).

---

[18] *Miliangos* v. *George Frank (Textiles) Ltd* [1976] AC 443, HL; *Dodd Properties* v. *Canterbury City Council* [1980] 1 WLR 433, CA.

[19] Lord Justice Ormrod in *Cory & Son* v. *Wingate Investments* (1980) 17 BLR 104, CA.

[20] Lord Browne-Wilkinson in *Smith New Court Ltd* v. *Scrimgeour Vickers* [1997] AC 254, HL.

[21] *Robinson* v. *Harman* (1848) 1 Ex 850.

[22] *Dodd Properties* v. *Canterbury City Council* [1980] 1 WLR 433, CA.

[23] *Ruxley Electronics and Construction Ltd* v. *Forsyth* [1995] 3 WLR 118.

Mitigation of loss

The award of damages as compensation is qualified by a principle, 'which imposes on a plaintiff the duty of taking all reasonable steps to mitigate the loss consequent on the breach, and debars him from claiming any part of the damage which is due to his neglect to take such steps'[24]. But this 'does not impose on the plaintiff an obligation to take any step which a reasonable and prudent man would not ordinarily take in the course of his business'.[25]

Liquidated damages

Whilst not normally specifically relevant to a contractor/sub-contractor relationship, and whilst not catered for within the JCT sub-contracts under consideration within this book, a brief note should be made of liquidated damages.

The parties to a contract often agree that a liquidated (i.e. fixed and agreed) sum shall be paid as damages for some breach of a contract.

A typical clause provides that if a contractor fails to complete by a date stipulated in the contract, or by any extended date, he shall pay or allow the employer to deduct liquidated damages at the rate of a certain amount per week for the period during which the works are uncompleted.

The basic rules in respect of liquidated damages are that they cannot be a 'penalty' and they must be a genuine pre-estimate of the level of damages that would be incurred.

It should be noted that a contractor may be able to recover from a sub-contractor the liquidated damages that he has incurred as part of a common law damages claim (or, indeed, as part of a loss and expense claim) against the sub-contractor.

---

[24] Lord Haldane in *British Westinghouse* v. *Underground Railways Co.* [1912] AC 673; *Andros Springs (Owners)* v. *World Beauty (Owners)* [1969] 3 All ER 158, CA; *Sotiros Shipping* v. *Sameiet Solholt* [1983] 1 Lloyds Rep 605, CA; *Kaines* v. *Osterreichische* [1993] 2 Lloyds Rep 1, CA.
[25] *British Westinghouse* v. *Underground Railways Co.* [1912] AC 673 referring to Lord Justice James in *Dunkirk Colliery Co.* v. *Lever* (1878) 9 Ch D 20, CA.

Chapter 10
Variations

What are variations under the sub-contract?

Variations under the sub-contract are changes to the scope of the work or conditions under which the work is carried out, only to the extent that terms in the sub-contract provide for such changes.

This is not the same as variations to the sub-contract. Variations to the sub-contract are changes to the sub-contract itself (i.e. changes to the agreement, including the terms and conditions), and such changes can only be made by express agreement between the parties.

Why are variation clauses under the sub-contract included in sub-contracts?

The nature of construction works makes the possibility of changes to the scope of the works or the conditions under which the work is carried out susceptible to change.

It is therefore necessary that the sub-contract conditions incorporate provisions for variations to the original scope of the work or the conditions under which the work is to be carried out.

If such a provision were not included within the sub-contract then every time a change occurred this would constitute a breach of the sub-contract or would require separate sub-contracts to be raised for all additional works, etc.

What is a variation?

DBSub

The definitions section (clause 1.1) defines variation as 'see clause 5.1'.

Under clause 5.1 a variation is defined as being:

'5.1.1 The alteration or modification of the design, the quality or (except where the Remeasurement basis applies) the quantity of the Sub-contract Works including:

.1   the addition, omission or substitution of any work;
.2   the alteration of the kind or standard of any of the materials or goods to be used in the Sub-contract Works;
.3   the removal from the site of any work executed or materials or goods brought thereon by the Sub-contractor for the purposes of the Sub-contract Works other

than work, materials or goods which are not in accordance with this Sub-contract.

5.1.2 The imposition in an instruction of the Architect/Contract Administrator issued under the Main Contract (or a direction of the Contractor passing on that instruction) of any obligation or restrictions in regard to the matters set out in this clause 5.1.2 or the addition to or alteration or omission of any such obligations or restrictions so imposed or imposed by the Numbered Documents and the Schedule of Information and its annexures in regard to:

.1 access to the site or use of any specific parts of the site;
.2 limitations of working space;
.3 limitations of working hours;
.4 the execution or completion of the work in any specific order.'

MPSub

In the MPSub, a variation is called a 'Change'.
A change is defined under clause 1 as being:

- any alteration in the design requirements and/or proposals that give rise to an alteration in the design, quality or quantity that is required to be executed;
- any alteration to or addition by the contractor of any restriction or obligation set out in the design requirements and/or proposals as to the manner in which the sub-contractor is to execute the sub-contract works; or
- any matter that the sub-contract requires to be treated as giving rise to a change.

However, a change does not arise where the alteration or matter in question results from any negligence or default on the part of the sub-contractor. Clearly, the use of the word 'any' before the phrase 'negligence or default on the part of the sub-contractor' may have a significant effect on what is considered as being a change under the sub-contract terms.

ICSub

All as DBSub above.

ICSub/D

All as DBSub above.

ICSub/NAM

All as DBSub above.

The MWSub/D and ShortSub

These sub-contracts do not define variations; however, it is considered that the matters included within the definition of a variation under the DBSub form above would also be considered to be variation items under these sub-contract forms.

SubSub

All as MWSub/D and ShortSub.

How is a variation instructed?

DBSub

All variation instructions are to be issued in writing in line with clause 3.4.

If the contractor purports to give any direction to the sub-contractor, or his authorised representative, otherwise than in writing (e.g. orally), then clause 3.7 makes it clear that that direction shall have no immediate effect.

However, a procedure is set out under clause 3.7 that can be followed for the confirmation of instructions issued otherwise than in writing.

This procedure is:

(1) The sub-contractor is to confirm the direction to the contractor in writing, within seven days of the direction being issued.
(2) If the contractor does not dissent in writing to the sub-contractor that the direction was issued, (within seven days from receipt of the sub-contractor's confirmation), then the direction shall take effect as from the expiry of that said seven-day period.

If, within seven days of giving a direction other than in writing, the contractor confirms the direction in writing, then the sub-contractor shall not be obliged to confirm the direction and the direction shall take effect as from the date of the contractor's written confirmation (as clause 3.7.1).

It should be noted that if action is taken on the basis of an oral instruction (more commonly referred to as a verbal instruction), without following the above procedure, then such action is taken entirely at the sub-contractor's risk.

However, clause 3.7.2 confirms that if a sub-contractor complies with an oral direction without following the procedure set out under clause 3.7 the contractor may at any time prior to the final payment under the sub-contract confirm the direction in writing with retrospective effect (refer to clause 3.7.2).

The operative word in the foregoing phrase is 'may', and there is no obligation for a contractor to retrospectively confirm a purported direction at all, and, in reality, it is extremely unlikely that a contractor would confirm a purported direction where it disagrees that such a direction was issued in the first place.

It should be noted that any written instruction of the architect/contract administrator issued under the main contract that affects the sub-contract works which is then issued by the contractor to the sub-contractor shall be deemed to be a direction of the contractor (refer to clause 3.4).

How this latter requirement will be implemented in practice, in terms of disclosure by the contractor, remains to be seen.

It should be noted that certain directions issued by the contractor under clauses 6.7.3, 6.8.3.1 and 6.14.1 (in respect of particular insurance related matters) are to be treated as variations in line with clauses 6.7.4, 6.8.3.2 and 6.14.2.

MPSub

All variations ('changes') are to be notified in writing – as clauses 23, 28 and 29.1. Although clause 29.1 does not specifically refer to the notice being in writing, clause 5.1 states that all communications required to be given by one party to the other under the sub-contract are to be given in writing.

ICSub

All variation instructions are to be issued in writing in line with clause 3.4. There is no provision for oral/verbal instructions.

It should be noted that certain directions issued by the contractor under clauses 6.7.3, 6.8.3.1 and 6.12.1 (in respect of particular insurance related matters) are to be treated as variations in line with clauses 6.7.4, 6.8.3.2 and 6.12.2.

ICSub/D

All as ICSub above.

ICSub/NAM

All as ICSub above.

MWSub/D and ShortSub

All variation instructions are to be issued in writing in line with clause 9.1.

If the contractor gives oral instructions, those instructions are to be confirmed by the contractor within two working days of being given (as clause 9.2).

SubSub

All variation instructions are to be issued in writing in line with clause 9.1.

If the sub-contractor gives oral instructions, those instructions are to be confirmed by the sub-subcontractor within two working days of being given (as clause 9.2).

Can a variation vitiate a contract?

To vitiate means to make invalid or ineffectual.

It is quite common for contracts to contain a clause which expressly states that no variation shall vitiate the contract. Although the above phrase is common in most standard forms, it is submitted that there must be some limit to the nature and extent of a variation which can be ordered, particularly where the variation requires the sub-contractor to carry out design works or works on site for which it professes no skill or experience whatsoever.

Must a sub-contractor comply with all variation directions issued?

DBSub

Clause 3.4 expressly provides that no variation directed by the contractor or subsequently sanctioned by him shall vitiate the sub-contract, and clause 3.5 notes that the sub-contractor is to comply with all directions issued to him under clause 3.4.

Clause 3.5.1 notes that where a contractor issues a direction which requires a variation of the type referred to in clause 5.1.2 (i.e. the imposition, alteration or omission of any obligation or restriction in regard to access to the site, limitation of working space, limitation of working hour, etc.), the sub-contractor need not comply to the extent that he makes reasonable objection to it in writing to the contractor.

Of course, there could be many disputes regarding what 'reasonable objections' are, and any such dispute could have a major impact on the actions that a sub-contractor can or should sensibly attempt to take.

In the case where a direction is given which pursuant to clause 5.3.1 requires the sub-contractor to provide a schedule 2 quotation (see further below regarding schedule 2 quotations), the variation shall not be carried out until the contractor has in relation to the schedule 2 quotation either issued written acceptance, or has issued a further direction under clause 5.3.2 (i.e. that the work is to be valued by a valuation by the contractor in accordance with the valuation rules pursuant to clause 5.2).

Under clause 3.5.3 if, in the sub-contractor's opinion, compliance with any direction of the contractor may injuriously affect the efficacy of the design of the sub-contractor's design, the sub-contractor has the right to object (provided that such objection is made within five days from receipt of the contractor's direction).

The contractor's direction will then not have any effect unless confirmed by the contractor.

Naturally, given the strict liability placed upon the sub-contractor in respect of design, the sub-contractor would need to make clear to the contractor that, by confirming the earlier direction that was (in the opinion of the sub-contractor) injurious to the sub-contractor's design, the contractor would be removing the liability for such design from the sub-contractor.

MPSub

Clause 9.1 notes that the sub-contractor shall comply with all written instructions issued by the contractor in connection with the design, execution and completion of the sub-contract works, except to the extent that the terms of the sub-contract restrict the contractor's right to issue any particular instruction.

ICSub

Clause 3.4 expressly provides that no variation directed by the contractor or subsequently sanctioned by him shall vitiate the sub-contract, and clause 3.5 notes that the sub-contractor is to comply with all directions issued to him under clause 3.4.

However, where a contractor issues a direction which requires a variation of the type referred to in clause 5.1.2 (i.e. the imposition, alteration or omission of any obligation or restriction in regard to access to the site, limitation of working space, limitation of working hours, etc.), clause 3.5 notes that the sub-contractor need not comply to the extent that he makes reasonable objection to it in writing to the contractor.

Of course, there could be many disputes regarding what reasonable objections are, and any such dispute could have a major impact on the actions that a sub-contractor can or should sensibly attempt to take.

ICSub/D

Clause 3.4 expressly provides that no variation directed by the contractor or subsequently sanctioned by him shall vitiate the sub-contract, and clause 3.5 notes that the sub-contractor is to comply with all directions issued to him under clause 3.4.

However, where a contractor issues a direction which requires a variation of the type referred to in clause 5.1.2 (i.e. the imposition, alteration or omission of any obligation or restriction in regard to access to the site, limitation of working space, limitation of working hours, etc.), clause 3.5.1 notes that the sub-contractor need not comply to the extent that he makes reasonable objection to it in writing to the contractor.

Of course, there could be many disputes regarding what reasonable objections are, and any such dispute could have a major impact on the actions that a sub-contractor can or should sensibly attempt to take.

In addition, under clause 3.5.2 if, in the sub-contractor's opinion, compliance with any direction of the contractor may injuriously affect the efficacy of the design of the sub-contractor's design, the sub-contractor has the right to object (provided that such objection is made within five days from receipt of the contractor's direction).

The contractor's direction will then not have any effect unless confirmed by the contractor.

Naturally, given the strict liability placed upon the sub-contractor in respect of design, the sub-contractor would need to make clear to the contractor that, by confirming the earlier direction that was (in the opinion of the sub-contractor) injurious to the sub-contractor's design, the contractor would be removing the liability for such design from the sub-contractor.

ICSub/NAM

All as ICSub/D above.

MWSub/D

Clause 10.1 requires the sub-contractor to carry out any reasonable variation of the sub-contract works that is instructed in writing by the contractor. Of course, there could be many disputes regarding what a 'reasonable variation' is, and the wording of clause 10.1 is therefore not entirely satisfactory. Clause 9.1 notes that the contractor is not to issue any instructions affecting the design of the sub-contractor's design portion works without first obtaining the consent of the sub-contractor (consent that the sub-contractor shall not unreasonably withhold).

ShortSub

Clause 10.1 requires the sub-contractor to carry out any reasonable variation of the sub-contract works that is instructed in writing by the contractor. Of course, there could be many disputes regarding what a reasonable variation is, and the wording of clause 10.1 is therefore not entirely satisfactory.

SubSub

Clause 10.1 requires the sub-subcontractor to carry out any reasonable variation of the sub-subcontract works that is instructed in writing by the sub-contractor. Of course, there could be many disputes regarding what a reasonable variation is, and the wording of clause 10.1 is therefore not entirely satisfactory.

What happens if a sub-contractor does not comply with a direction issued?

DBSub

Clause 3.6 makes it clear that if, within seven days after receipt of a written notice from the contractor which requires a sub-contractor to comply with a direction, the sub-contractor does not comply with that direction, then the contractor may employ and pay other persons to execute any work whatsoever which may be necessary to give effect to that direction.

In such a situation, the sub-contractor would be liable for all additional costs incurred by the contractor in connection with such employment and an appropriate deduction shall either be taken into account in the calculation of the final sub-contract sum or shall be recoverable by the contractor from the sub-contractor as a debt.

MPSub

Clause 9.3 states that where a sub-contractor fails to comply with an instruction, the contractor may engage others to give effect to the instruction provided that he has first given seven days' notice in writing to the sub-contractor of his intention to do so. In such a situation, the sub-contractor shall be liable to pay the contractor's costs of engaging others, but only after taking into account the amount that would have been payable to the sub-contractor under the terms of the sub-contract if it had complied with the instruction.

ICSub

All as DBSub, above.

ICSub/D

All as DBSub, above.

ICSub/NAM

All as DBSub, above.

MWSub/D

Clause 9.4 states that if a sub-contractor unreasonably delays or withholds his consent to an instruction referred to in clause 9.1, or if, within seven days after receipt of a written notice from the contractor which requires a sub-contractor to comply with a direction, the sub-contractor does not comply with that direction, then the contractor may employ and pay other persons to carry out the work in question and may recover all additional costs incurred from the sub-contractor.

ShortSub

Clause 9.4 states that if, within seven days after receipt of a written notice from the contractor which requires a sub-contractor to comply with a direction, the sub-contractor does not comply with that direction, then the contractor may employ and pay other persons to carry out the work and adds that all additional costs incurred shall be due to the contractor.

SubSub

Clause 9.4 states that if, within five days after receipt of a written notice from the sub-contractor which requires a sub-subcontractor to comply with a direction, the sub-subcontractor does not comply with that direction, then the sub-contractor may employ and pay other persons to carry out the work and adds that all additional costs incurred shall be due to the sub-contractor.

How should variations be valued?

DBSub

Unless otherwise agreed by the contractor and the sub-contractor, clause 5.2 states that variations are valued either by way of an accepted schedule 2 quotation, or in accordance with clauses 5.6 to 5.12 (known as the Valuation Rules).

If the sub-contract is based on the adjustment basis then, in line with clause 5.2.1 of the conditions, the valuation rules will apply to:

- all variations including any sanctioned in writing by the contractor but excluding any variation to an accepted schedule 2 quotation;
- all work which under the conditions is to be treated as a variation;
- all work executed by the sub-contractor in accordance with the directions of the contractor as to the expenditure of Provisional Sums which are included in the sub-contract documents; and
- all work executed by the sub-contractor for which an approximate quantity has been included in any bills of quantities or in the contract requirements.

If the sub-contract is based on the remeasurement basis then the valuation rules will apply to all work executed by the sub-contractor in accordance with the sub-contract documents and the directions of the contractor, including any direction requiring a variation or in regard to the expenditure of a Provisional Sum included in the sub-contract documents (as clause 5.2.2 of the conditions).

Such valuation, in so far as it relates to the sub-contractor's designed works, shall be in accordance with clause 5.10 and references in clauses 5.6 and 5.7 shall (thus) exclude the valuation of variations in respect of the sub-contractor's designed works analysis.

Effect to the valuation of variations shall be given in the calculation of the final sub-contract sum (clause 5.5).

MPSub

Clause 29.2 states that under MPSub variations are valued either by way of an accepted quotation (as clause 29.5), or by way of a fair valuation (as clause 29.6).

ICSub

Unless otherwise agreed by the contractor and the sub-contractor, clause 5.2 states that under ICSub variations are to be valued in accordance with clauses 5.3 to 5.8 (known as the Valuation Rules).

If the sub-contract is based on the adjustment basis then, in line with clause 5.2.1 of the conditions, the valuation rules will apply to:

- all variations, including any sanctioned in writing by the contractor;
- all work which under the conditions is to be treated as a variation;
- all work executed by the sub-contractor in accordance with the directions of the contractor as to the expenditure of Provisional Sums which are included in the sub-contract documents; and
- all work executed by the sub-contractor for which an approximate quantity has been included in any bills of quantities.

If the sub-contract is based on the remeasurement basis then the valuation rules will apply to all work executed by the sub-contractor in accordance with the sub-contract documents and the directions of the contractor, including any direction requiring a variation or in regard to the expenditure of a Provisional Sum included in the sub-contract documents (as clause 5.2.2 of the conditions).

ICSub/D

Unless otherwise agreed by the contractor and the sub-contractor, clause 5.2 states that under ICSub/D variations are to be valued in accordance with clauses 5.3 to 5.9 (known as the Valuation Rules).

If the sub-contract is based on the adjustment basis then, in line with clause 5.2.1 of the conditions, the valuation rules will apply to:

- all variations including any sanctioned in writing by the contractor;
- all work which under the conditions is to be treated as a variation;
- all work executed by the sub-contractor in accordance with the directions of the contractor as to the expenditure of Provisional Sums which are included in the sub-contract documents; and
- all work executed by the sub-contractor for which an approximate quantity has been included in any bills of quantities or in the contract requirements.

If the sub-contract is based on the remeasurement basis then the valuation rules will apply to all work executed by the sub-contractor in accordance with the sub-contract documents and the directions of the contractor, including any direction requiring a variation or in regard to the expenditure of a Provisional Sum included in the sub-contract documents (as clause 5.2.2 of the conditions).

Such valuation, in so far as it relates to the sub-contractor's designed works, shall be in accordance with clause 5.10 and references in clauses 5.3 and 5.4 shall (thus) exclude the valuation of variations in respect of the sub-contractor's designed portion analysis.

ICSub/NAM

All as ICSub above.

MWSub/D and ShortSub

Clause 10.2 notes that variations shall be valued by the contractor on a fair and reasonable basis, with reference, where available and relevant, to rates and prices in the pricing document (being any document identified in the second recital of the contract which shows rates and prices).

SubSub

Clause 10.2 notes that variations shall be valued by the sub-contractor on a fair and reasonable basis, with reference, where available and relevant, to rates and prices in the pricing document (being any document identified in the second recital of the contract which shows rates and prices).

What is a schedule 2 quotation?

A schedule 2 quotation is a quotation that is provided in line with the provisions of schedule 2 of the sub-contract. In respect of the sub-contracts under consideration this only applies to the DBSub.

What is the procedure to be followed in respect of a schedule 2 quotation?

Clause 5.3.1 of the DBSub/C notes that if the contractor in his direction states that the sub-contractor is to provide a schedule 2 quotation, the sub-contractor shall, subject to receipt of sufficient information, provide a quotation in accordance with the procedure outlined below, unless, within four days of his receipt of that direction (or such longer period as is either stated in the direction or is agreed between the contractor and the sub-contractor), he notifies the contractor that he disagrees with the application of that procedure to that direction.

If the sub-contractor does issue such a notice of disagreement, then the sub-contractor shall not be obliged to provide a schedule 2 quotation, and the work shall not be carried out unless and until the contractor gives a further direction that the work is to be carried out and is to be valued using one of the valuation rules.

Submission of quotation

In summary, and assuming that the contractor and sub-contractor do not agree upon extended time periods, then where schedule 2 quotations are required:

- The contractor in his direction is to state that the sub-contractor is to provide a schedule 2 quotation.
- If the sub-contractor disagrees with the application of that procedure to that direction, he is to notify the contractor of that fact within four days of his receipt of the direction.
- If the sub-contractor does issue such a notice of disagreement, then the sub-contractor shall not be obliged to provide a schedule 2 quotation, and the work shall not be carried out unless and until the contractor gives a further direction that the work is to be carried out and is to be valued using one of the valuation rules.
- If the sub-contractor does not issue such a notice of disagreement, but considers that the information provided is not sufficient, then, not later than four days from the date of the instruction, he shall notify the contractor who shall supply the necessary information.
- The sub-contractor shall submit his schedule 2 quotation to the contractor not later than 14 days from the later of:

    — the date of receipt of the instruction; or
    — the date of receipt by the sub-contractor of sufficient information as referred to above.

- The schedule 2 quotation shall remain open for acceptance by the contractor for 14 days from its receipt by the contractor.

Content of the quotation

It is important to note that the schedule 2 quotation is to separately comprise:

(1) the amount of the adjustment to the final sub-contract sum (excluding any amount to be paid in lieu of any ascertainment for direct loss and/or expense, but including allowance, where appropriate, for preliminary items) supported by all necessary calculations, which shall be made by reference, where relevant, to the rates and prices in the sub-contract sum or the sub-contract tender sum;
(2) any adjustment to the time required for completion of the sub-contract works and/or of any works in any Section by reference to the period or periods stated in the sub-contract particulars (item 5) to the extent that such adjustment is not included in any revision to the completion date previously issued by the contractor or in the contractor's confirmed acceptance of any other schedule 2 quotation;
(3) the amount to be paid in lieu of any ascertainment, under clause 4.19 of DBSub/C, of direct loss and expense not included in any other accepted schedule 2 quotation or in any previous ascertainment under clause 4.19;
(4) a fair and reasonable amount in respect of the cost of preparing the schedule 2 quotation; and
(5) where specifically required by the instruction, the sub-contractor shall provide indicative information in statements on:

    (a) the additional resources (if any) required to carry out the variation; and
    (b) the method of carrying out the variation.

Each part of the schedule 2 quotation shall contain reasonably sufficient supporting information to enable that part to be evaluated by or on behalf of the contractor.

Acceptance of the quotation

If the contractor wishes to accept a schedule 2 quotation the contractor is to notify the sub-contractor in writing not later than the last day of the period of acceptance.

If the contractor accepts a schedule 2 quotation, the contractor shall, immediately upon that acceptance, confirm such acceptance in writing to the sub-contractor:

(1) that the sub-contractor is to carry out the variation;
(2) the adjustment to the final sub-contract sum;
(3) any adjustment to the time required by the sub-contractor for completion of the sub-contract works or (where applicable) such works in any relevant Section.

Quotation not accepted

If the contractor does not accept the schedule 2 quotation by the expiry of the period of acceptance, the contractor shall, on the expiry of that period either:

(1) instruct that the variation is to be carried out and is to be valued under the valuation rules; or
(2) instruct that the variation is not to be carried out.

Cost of quotation

If a schedule 2 quotation is not accepted, a fair and reasonable amount shall be added to the final sub-contract sum in respect of the preparation of the schedule 2 quotation provided that the schedule 2 quotation has been prepared on a fair and reasonable basis.

Although there is no definition of when a schedule 2 quotation has been prepared on a fair and reasonable basis, it is clear that the non-acceptance by the contactor shall not, of itself, be evidence that the quotation was not prepared on a fair and reasonable basis.

Restriction on use of quotation

Unless the contractor accepts a schedule 2 quotation, neither the contractor nor the sub-contractor may use that quotation for any purpose whatsoever.

What happens if work covered by a schedule 2 quotation that has been accepted by the contractor is itself varied?

Where a schedule 2 quotation has been accepted by the contractor, then, if the contractor subsequently issues a direction varying the work that was the subject matter of that schedule 2 quotation, the contractor shall make a valuation of that variation on a fair and reasonable basis having regard to the content of the schedule 2 quotation and shall include in that valuation the direct loss and/or expense, if any, incurred by the sub-contractor because the regular progress of the sub-contract works or any Section thereof is materially affected by compliance with the direction (refer to clause 5.3.3 of DBSub/C).

Can other quotations be used in respect of the valuation of variations?

Except for the MPSub, none of the other sub-contracts under consideration expressly allow for quotations to be used in respect of the valuation of variations.

What is the procedure to be followed in respect of a variation quotation under the MPSub?

Clause 29.3 of the MPSub notes that the contractor may, prior to instructing any change, provide details of the proposed change and request the sub-contractor to submit a quotation in respect of the change. The sub-contractor shall provide the quotation within ten days of the request, or such longer period as the contractor states in his request.

Content of the quotation

It is important to note that, in line with clause 29.4, the schedule 2 quotation is to provide:

(1) a fair valuation of the change made in line with clause 29.6, including stating separately any loss and expense applicable;
(2) any adjustment to the period for completion that will be required as a consequence of the change;
(3) sufficient detail for the contractor to assess the amounts and periods required.

For how long is the quotation open for acceptance?

Clause 29.4.4 notes that the quotation is to identify the period that the quotation is to remain open for acceptance, and adds that that period is not to be less than 21 days.

Acceptance/non-acceptance of quotation

Clause 29.5 allows the contractor to accept a quotation by issuing an instruction identifying the quotation that is being accepted.

The contractor can, alternatively, ask for a revised quotation to be submitted, which will then, itself, be subject to acceptance or non-acceptance by the contractor.

Where agreement is not reached in respect of a submitted quotation, then clause 29.6 notes that a fair valuation shall be made of the change (in line with the provisions of clause 29.6).

What are the valuation rules?

DBSub

The valuation rules are applicable to the extent that a valuation relates to the execution of additional or substituted work which can properly be valued by measurement.

It should be noted that in respect of all the valuation rules, the contractor is to give the sub-contractor an opportunity to be present and to allow the sub-contractor to take such notes and measurements as the sub-contractor may require at the time when it is necessary to measure work for the purpose of valuation (refer to clause 5.4).

There are certain general rules that apply (refer to clause 5.8), namely:

(1) Where there are bills of quantities, the measurement of variations shall be in accordance with the same principles as those governing the preparation of those bills of quantities.

   The default position in respect of the preparation of bills of quantities is that the *Standard Method of Measurement of Building Works*, 7th edition, produced by the Royal Institution of Chartered Surveyors and the Construction Confederation (SMM7) will have been used.

   Clause 2.9.1 notes that if there is any unstated departure from SMM7 in the preparation of the bills of quantities, or if there is any error in description or in quantity or any omission of items, the departure, error or omission shall be corrected, and any such correction shall be treated as a variation.

(2) When valuing variations, allowances shall be made for any percentage or lump sum adjustments in any bills of quantities and/or other sub-contract documents.

(3) Where the adjustment basis applies, an allowance (where appropriate) shall be made for any addition to or reduction from any preliminary items of the type referred to in SMM7. However, the above principle does not apply where a variation relates to the contractor's direction for the expenditure of a Provisional Sum for defined work, for the reasons outlined below.

   The SMM7 General Rule 10.3 defines a Provisional Sum for defined work as being a sum for work which is not completely designed but for which the following information is provided:

   (a) The nature of the construction of the work.
   (b) A statement of how and where the work is fixed to the building and what other work is to be fixed thereto.
   (c) A quantity or quantities which indicate the scope and extent of the works.
   (d) Any specific limitations on the method or the sequence or the timing of the works.

   If the information specified in rule 10.3 is not available, the contract bills should describe the Provisional Sum as undefined.

   General Rule 10.4 of SMM7 states that: 'where Provisional Sums are given for defined work the Contractor (and the sub-contractor) will be deemed to have made due allowance in programming, planning and pricing preliminaries', whilst general rule 10.6 of SMM7 states that the contractor is deemed to have made *no such allowance* for undefined work.

Therefore the categorisation of defined or undefined Provisional Sums has significant implications.

If a Provisional Sum is incorrectly described as defined, then clause 2.9.1 makes it clear that if there is any error in or omission of information in any item which is the subject of a Provisional Sum for defined work, the description shall be corrected so that it does provide that information. Clause 2.9.3 then adds that any such correction, alteration or modification to the Provisional Sum shall be treated as a variation.

(4) Where the remeasurement basis applies, any amounts priced in the preliminaries section of any bills of quantities shall be adjusted (where appropriate) to take into account any variations of any contractor's directions for the expenditure of Provisional Sums for undefined work included in the sub-contract documents.

The valuation rule dealt with under clauses 5.6.1.1 and 5.7.1 – variation works of similar character, similar conditions and similar quantity

Under this valuation rule, where variation work is carried out that is of similar character to, is executed under similar conditions as, and does not significantly change the quantity of work set out in any bills of quantities and/or other sub-contract documents, the rates and prices for the work in those documents shall determine the valuation.

This is subject only to clause 5.10 in the case of sub-contractor design works as described below.

Similar conditions are those conditions which are to be derived from the express provisions of the sub-contract. Extrinsic evidence of, for instance, the parties' subjective expectations is not admissible[1].

The valuation rule dealt with under clauses 5.6.1.2 and 5.7.2 – variation works of similar character, but dissimilar conditions and/or dissimilar quantity

Under this valuation rule, where variation work is carried out that is of similar character to work set out in any bills of quantities and/or other sub-contract documents, but is not executed under similar conditions thereto and/or significantly changes its quantity, the rates and prices for the work set out in the documents above shall be the basis for determining the valuation and the valuation value shall include a fair allowance for such difference in conditions and/or quantity.

It must be noted that where a sub-contractor has simply submitted a rate in error, it has been found that such a mistake would not prevent the use of those rates to value a subsequent variation[2].

The valuation rule dealt with under clauses 5.6.1.3 and 5.7.3 – variation works not of a similar character

Under this valuation rule, where variation work is carried out that is not of similar character to work set out in any bills of quantities and/or other sub-contract documents, the work shall be valued at fair rates and prices.

Fair rates and prices was considered in the *Crittall Windows* v. *TJ Evers*[3] case. In that case, Judge Humphrey Lloyd QC stated that 'a fair valuation generally means

[1] *Wates Construction* v. *Bredero Fleet* (1993) 63 BLR 128.
[2] *Henry Boot Construction Ltd* v. *Alstom* [1999] BLR 123.
[3] *Crittall Windows* v. *TJ Evers Ltd* (1996) 54 Con LR 66.

a valuation which will not give the contractor more than his actual costs reasonably and necessarily incurred plus similar allowance for overheads and profit'. Judge Humphrey Lloyd QC has since repeated this approach in two other cases[4].

However, a more common view is that fair rates and prices must have regard to the contractor's general pricing level, and therefore a valuation below actual cost would be fair where the contract price is below actual costs or market pricing.

The valuation rule dealt with under clause 5.6.1.4 – reasonably accurate approximate quantity
This valuation rule is only applicable when the sub-contract is let on the adjustment basis (i.e. where a sub-contract sum is used as the basis of the sub-contract rather than where a sub-contract tender sum is used). This rule is that where an approximate quantity is included in the bills of quantities and/or other sub-contract documents, and that approximate quantity is a reasonably accurate forecast of the quantity of work actually required, then the rate or price for that approximate quantity shall determine the valuation of the varied works.

This rule only applies where the work required has not been altered or modified in any way from that specified other than in terms of quantity.

The valuation rule dealt with under clause 5.6.1.5 – not reasonably accurate approximate quantity
This valuation rule is again also only applicable when the sub-contract is let on the adjustment basis. This rule is that where an approximate quantity is included in the bills of quantities and/or other sub-contract documents, and that approximate quantity is not a reasonably accurate forecast of the quantity of work actually required, then the rate or price for that approximate quantity shall be used as the basis of determining the valuation of the varied works, with a fair allowance being made for such difference in quantity.

Again, this rule only applies where the work required has not been altered or modified in any way from that specified other than in terms of quantity.

The valuation rule dealt with under clause 5.6.2 – omission of work
This valuation rule is only applicable when the sub-contract is let on the adjustment basis. This rule states that to the extent that a valuation relates to the omission of work set out in any bills of quantities and/or other sub-contract documents, the rates and the prices for such work set out therein (and no adjustment to those rates) shall determine the valuation of the work omitted. This is subject only to clause 5.10, in the case of sub-contractor design works.

Change of conditions for other work
This valuation rule (under clause 5.11) notes that if there is a substantial change in the conditions under which any other work (including sub-contractor design works) is executed, as a result of:

(1) compliance with any direction requiring a variation (except a variation for which a schedule 2 quotation has been accepted or where the variation is to

---

[4] *Floods Queensferry Ltd* v. *Shand Construction Ltd* [1999] BLR 315; *Weldon Plant Ltd* v. *The Commissioner for the New Towns* [2000] BLR 496.

works covered by a schedule 2 quotation; and except for any direction as a result of which work due to be executed in the sub-contract on the remeasurement basis is not executed);

(2) compliance with any direction as to the expenditure of a Provisional Sum for undefined work;

(3) compliance with any direction as to the expenditure of a Provisional Sum for defined work, to the extent that the direction for that work differs from the description given for such work in any bills of quantities; or

(4) where the adjustment basis applies, the execution of the work for which an approximate quantity is included in any bills of quantities, to the extent that the quantity is more or less than the quantity ascribed to that work in those bills;

then such other work shall be treated as if it had been the subject of a direction of the contractor requiring a variation and shall be valued on the basis of the rates and prices for the work set out in any bills of quantities and/or other sub-contract documents together with a fair allowance for the difference in conditions.

Sub-contractor's designed works – valuation

Valuations relating to the sub-contractor's designed works are to be made under clause 5.10.

In such valuation:

- Allowance shall be made in such valuation for work involved in the preparation of the relevant design work (clause 5.10.1).
- The valuation shall be consistent with the values of work of a similar character set out in the sub-contractor's designed works analysis making due allowance for any change in the conditions under which work is carried out and/or any significant change in the quantity of work so set out.

  Where there is no work of a similar character set out in the sub-contractor's designed works analysis a fair valuation shall be made (clause 5.10.2).
- The valuation of the omission of the work set out in the sub-contractor's designed works analysis in accordance with the values therein for such work (clause 5.10.3).
- Clauses 5.8.2, 5.8.3, 5.8.4, 5.9 and 5.11 (as dealt with elsewhere in this chapter) shall apply so far as is relevant (as clause 5.10.4).

Daywork

'Daywork' is a means of valuing variations which is based upon recorded time, and recorded material and plant usage. Clause 5.9 makes it clear that Daywork should only be used when a variation cannot properly be valued by measurement.

The time and resources spent on a variation that it is proposed is valued on a Daywork basis should be recorded by the sub-contractor on a Daywork voucher (or sheet), and that Daywork voucher should have sections to record the operatives' names and trades, the time spent by the operatives on the variation, and also the plant and materials used.

The completed Daywork voucher must be delivered for verification to the contractor not later than the Wednesday following the week in which the work was executed (clause 5.9). The reason for this relatively short timetable is clearly to enable the contractor to contemporaneously check the details on the Daywork voucher.

The various elements of the Daywork voucher are valued on the basis of the schedule of Daywork rates included in the numbered documents. If no such schedule is included in the numbered documents then the various elements of the Daywork voucher are valued on the following basis:

- Labour

  Either at the all-in rate stated for each grade of operative as noted under item 11 of the sub-contract particulars (refer to clause 5.9.1 of SBCSub/C and SBCSub/D/C), or if an all-in rate is not stated then the prime cost for each grade of operative calculated in accordance with the Definition of Prime Cost of Daywork carried out under a Building Contract together with percentage additions to the prime cost at the rates set out in the sub-contract particulars item 11 (as clause 5.9.1)

  The Definition of Prime Cost of Daywork carried out under a Building Contract being a document issued by the Royal Institution of Chartered Surveyors and the Construction Confederation, with the edition to be used being that current at the sub-contract base date (as stated in the sub-contract particulars).

  If the operatives in question are within the province of any specialist trade where the Royal Institution of Chartered Surveyors and the appropriate body representing the employers in that trade have agreed and issued a definition of prime cost of Daywork, then the prime cost for each grade of operative calculated in accordance with that definition is to be used together with the percentage addition to the prime cost at the rates set out in the sub-contract particulars (as clause 5.9.2).

  There are currently three definitions to which clause 5.9.2 refers, being those agreed between the Royal Institution of Chartered Surveyors and the Electrical Contractors Association, the Electrical Contractors Association of Scotland, and the Heating and Ventilating Association, respectively.

- Plant

  Plant is valued in accordance with one or more of the three definitions noted above to obtain the plant Daywork base rate at the time that the Daywork was carried out, and then the appropriate percentage addition is applied as inserted against the appropriate definition (as clause 5.9.1).

  When tendering for work, it is important to note that the prime cost for plant on a Daywork basis may, on occasion, be based on a basic plant schedule which may pre-date the date when the Dayworks are being carried out by several years, and the percentage addition allowed by the sub-contractor needs to reflect this fact.

- Materials

  Materials are priced at the prime cost for materials, together with the percentage addition to the prime cost at the rates set out in the sub-contract particulars item 11 (as clause 5.9.1).

Fair valuation

If a valuation does not relate to the execution of work or the omission of work, or if the valuation of any work or liabilities directly associated with a variation cannot reasonably be effected in the valuation by any of the valuation rules noted above, then clause 5.12.1 notes that a fair valuation shall be made.

It must be noted that this additional provision does not relate to the execution of additional work or the omission of work, and only relates to any work or liabilities directly associated with a variation.

Non-recovery of loss and/or expense

Clause 5.12.2 notes that no allowance for any effect upon the regular progress of the works (or any part of them) or for any other direct loss and/or expense for which the sub-contractor would be reimbursed by any other provision in the sub-contract is to be made under the valuation rules.

MPSub

When a valuation has not been agreed following the quotation procedure under clauses 29.3 to 29.5 inclusive, clause 29.6 requires a fair valuation to be made which has regard to the following:

(1) The nature and timing of the change.
(2) The effect of the change on other parts of the sub-contract works.
(3) The prices and principles set out in the pricing document (as set out under schedule 2 to the sub-contract), so far as applicable.
(4) Any loss and/or expense that will be incurred as a consequence of the change, provided that no loss and/or expense contributed to by any cause other than a change or a matter set out in clause 30.2 is included.

Unless an instruction has been issued under clause 29.5, the sub-contractor, within ten days of the change being identified, is to provide details of his proposed valuation of the change together with such information as is reasonably necessary to permit a fair valuation to be made by the contractor (clause 29.7).

The contractor is to respond with his valuation of the change (together with sufficient detail) within 21 days of receipt of the sub-contractor's valuation and supporting information (clause 29.8).

There is also a review of the valuation of changes that is undertaken after practical completion.

Under this procedure, the sub-contractor is to provide particulars of any further valuation of any change no later than 28 days after practical completion; and the contractor is to review and notify the sub-contractor of the outcome of his review within 56 days of receipt of the particulars from the sub-contractor.

ICSub

All as ICSub/D below, except for any valuation rules relating to the sub-contractor's designed works, design portion, or design portion analysis, as these are not applicable to the ICSub form of sub-contract.

ICSub/D

The valuation rules are applicable to the extent that a valuation relates to the execution of additional or substituted work which can properly be valued by measurement.

There are certain general rules that apply (refer to clause 5.5), namely:

(1) Where there are bills of quantities, the measurement of variations shall be in accordance with the same principles as those governing the preparation of those bills of quantities.

The default position in respect of the preparation of bills of quantities is that the *Standard Method of Measurement of Building Works*, 7th edition, produced by the Royal Institution of Chartered Surveyors and the Construction Confederation (SMM7) will have been used.

Clauses 2.7.2 and 2.8.1 note that if there is any unstated departure from SMM7 in the preparation of the bills of quantities, or if there is any error in description or in quantity or any omission of items, the departure, error or omission shall be corrected, and any such correction shall be treated as a variation.

(2) When valuing variations, allowances shall be made for any percentage or lump sum adjustments in any bills of quantities and/or other sub-contract documents.

(3) Where the adjustment basis applies, an allowance (where appropriate) shall be made for any addition to or reduction from any preliminary items of the type referred to in SMM7. However, the above principle does not apply where a variation relates to the contractor's direction for the expenditure of a Provisional Sum for defined work, because general rule 10.4 of SMM7 states that 'where Provisional Sums are given for defined work the Contractor (and the sub-contractor) will be deemed to have made due allowance in programming, planning and pricing preliminaries', whilst general rule 10.6 of SMM7 states that the contractor is deemed to have made *no such allowance* for undefined work.

(4) Where the remeasurement basis applies, any amounts priced in the preliminaries section of any bills of quantities shall be adjusted (where appropriate) to take into account any variations of any contractor's directions for the expenditure of Provisional Sums for undefined work included in the sub-contract documents.

The valuation rule dealt with under clauses 5.3.1.1 and 5.4.1 – variation works of similar character, similar conditions and similar quantity

Under this valuation rule, where variation work is carried out that is of similar character to, is executed under similar conditions as, and does not significantly change the quantity of work set out in any bills of quantities and/or other sub-contract documents, the rates and prices for the work in those documents shall determine the valuation.

This is subject only to clause 5.9 in the case of sub-contractor design portion works described (see below).

Similar conditions are those conditions which are to be derived from the express provisions of the sub-contract. Extrinsic evidence of, for instance, the parties' subjective expectations is not admissible[5].

---

[5] *Wates Construction v. Bredero Fleet* (1993) 63 BLR 128.

The valuation rule dealt with under clauses 5.3.1.2 and 5.4.2 – variation works of similar character, but dissimilar conditions and/or dissimilar quantity

Under this valuation rule, where variation work is carried out that is of similar character to work set out in any bills of quantities and/or other sub-contract documents, but is not executed under similar conditions thereto and/or significantly changes its quantity, the rates and prices for the work set out in the documents above shall be the basis for determining the valuation and the valuation value shall include a fair allowance for such difference in conditions and/or quantity.

It must be noted that where a sub-contractor has simply submitted a rate in error, it has been found that such a mistake would not prevent the use of those rates to value a subsequent variation[6].

The valuation rule dealt with under clauses 5.3.1.3 and 5.4.3 – variation works not of a similar character

Under this valuation rule, where variation work is carried out that is not of similar character to work set out in any bills of quantities and/or other sub-contract documents, the work shall be valued at fair rates and prices.

Fair rates and prices was considered in the *Crittall Windows* v. *TJ Evers*[7] case.

In that case, Judge Humphrey Lloyd QC stated that: 'a fair valuation generally means a valuation which will not give the contractor more than his actual costs reasonably and necessarily incurred plus similar allowance for overheads and profit'.

Judge Humphrey Lloyd QC has since repeated this approach in two other cases[8].

However, a more common view is that fair rates and prices must have regard to the contractor's general pricing level, and therefore a valuation below actual cost would be fair where the contract price is below actual costs or market pricing.

The valuation rule dealt with under clause 5.3.1.4 – reasonably accurate approximate quantity

This valuation rule is only applicable when the sub-contract is let on the adjustment basis (i.e. where a sub-contract sum is used as the basis of the sub-contract rather than where a sub-contract tender sum is used). This rule is that where an approximate quantity is included in the bills of quantities and/or other sub-contract documents, and that approximate quantity is a reasonably accurate forecast of the quantity of work actually required, then the rate or price for that approximate quantity shall determine the valuation of the varied works.

This rule only applies where the work required has not been altered or modified in any way from that specified other than in terms of quantity.

The valuation rule dealt with under clause 5.3.1.5 – not reasonably accurate approximate quantity

This valuation rule is again also only applicable when the sub-contract is let on the adjustment basis. This rule is that where an approximate quantity is included in

---

[6] *Henry Boot Construction Ltd* v. *Alstom* [1999] BLR 123.
[7] *Crittall Windows* v. *TJ Evers Ltd* (1996) 54 Con LR 66.
[8] *Floods Queensferry Ltd* v. *Shand Construction Ltd* [1999] BLR 315; *Weldon Plant Ltd* v. *The Commissioner for the New Towns* [2000] BLR 496.

the bills of quantities and/or other sub-contract documents, and that approximate quantity is not a reasonably accurate forecast of the quantity of work actually required, then the rate or price for that approximate quantity shall be used as the basis of determining the valuation of the varied works, with a fair allowance being made for such difference in quantity.

Again, this rule only applies where the work required has not been altered or modified in any way from that specified other than in terms of quantity.

The valuation rule dealt with under clause 5.3.2 – omission of work

This valuation rule is only applicable when the sub-contract is let on the adjustment basis.

This rule states that to the extent that a valuation relates to the omission of work set out in any bills of quantities and/or other sub-contract documents, the rates and the prices for such work set out therein (and no adjustment to those rates) shall determine the valuation of the work omitted. This is subject only to clause 5.10, in the case of sub-contractor design works.

Change of conditions for other work

This valuation rule (under clause 5.7) notes that if there is a substantial change in the conditions under which any other work (including sub-contractor design portion works) is executed, as a result of:

(1) compliance with any direction requiring a variation, or any direction as a result of which work due to be executed in the sub-contract on the remeasurement basis is not executed;
(2) compliance with any direction as to the expenditure of a Provisional Sum for undefined work;
(3) compliance with any direction as to the expenditure of a Provisional Sum for defined work, to the extent that the direction for that work differs from the description given for such work in any bills of quantities; or
(4) where the adjustment basis applies, the execution of the work for which an approximate quantity is included in any bills of quantities, to the extent that the quantity is more or less than the quantity ascribed to that work in those bills;

then such other work shall be treated as if it had been the subject of a direction of the contractor requiring a variation and shall be valued on the basis of the rates and prices for the work set out in any bills of quantities and/or other sub-contract documents together with a fair allowance for the difference in conditions.

Sub-contractor's designed works – valuation

Valuations relating to the sub-contractor's designed portion are to be made under clause 5.9.

In such valuation:

- Allowance shall be made in such valuation for work involved in the preparation of the relevant design work (clause 5.9.1).
- The valuation shall be consistent with the values of work of a similar character set out in the sub-contractor's designed portion analysis making due allowance

for any change in the conditions under which work is carried out and/or any significant change in the quantity of work so set out.

Where there is no work of a similar character set out in the sub-contractor's designed portion analysis a fair valuation shall be made (clause 5.9.2).

• The valuation of the omission of the work set out in the sub-contractor's designed works analysis in accordance with the values therein for such work (clause 5.9.3).

• Clauses 5.5.2, 5.5.3, 5.5.4, 5.6 and 5.7 (as dealt with elsewhere in this chapter) shall apply so far as is relevant (as clause 5.9.4).

Daywork

Daywork is a means of valuing variations which is based upon recorded time, and recorded material and plant usage. Clause 5.6 makes it clear that daywork should only be used when a variation cannot properly be valued by measurement.

The time and resources spent on a variation that it is proposed is valued on a Daywork basis should be recorded by the sub-contractor on a Daywork voucher (or sheet), and that Daywork voucher should have sections to record the operatives' names and trades, the time spent by the operatives on the variation, and also the plant and materials used.

The completed Daywork voucher must be delivered for verification to the contractor not later than the Wednesday following the week in which the work was executed (clause 5.6). The reason for this relatively short timetable is clearly to enable the contractor to contemporaneously check the details on the Daywork voucher.

The various elements of the Daywork voucher are valued on the basis of the schedule of daywork rates included in the numbered documents. If no such schedule is included in the numbered documents then the various elements of the Daywork voucher are valued on the following basis:

• Labour
Either at the all-in rate stated for each grade of operative as noted under item 10 of the sub-contract particulars (refer to clause 5.6.1 of SBCSub/C and SBCSub/D/C), or if an all-in rate is not stated then the prime cost for each grade of operative calculated in accordance with the Definition of Prime Cost of Daywork carried out under a Building Contract, together with percentage additions to the prime cost at the rates set out in the sub-contract particulars item 10 (as clause 5.6.1).

The Definition of Prime Cost of Daywork carried out under a Building Contract being a document issued by the Royal Institution of Chartered Surveyors and the Construction Confederation, with the edition to be used being that current at the sub-contract base date (as stated in the sub-contract particulars).

If the operatives in question are within the province of any specialist trade where the Royal Institution of Chartered Surveyors and the appropriate body representing the employers in that trade have agreed and issued a definition of prime cost of Daywork, then the prime cost for each grade of operative calculated in accordance with that definition is to be used, together with the percentage addition to the prime cost at the rates set out in the sub-contract particulars (as clause 5.6.2).

There are currently three definitions to which clause 5.6.2 refers, being those agreed between the Royal Institution of Chartered Surveyors and the Electrical Contractors Association, the Electrical Contractors Association of Scotland, and the Heating and Ventilating Association, respectively.

- Plant

Plant is valued in accordance with one or more of the three definitions noted above to obtain the plant daywork base rate at the time that the Daywork was carried out, and then the appropriate percentage addition is applied as inserted against the appropriate definition (as clause 5.6.1).

When tendering for work, it is important to note that the prime cost for plant on a Daywork basis may, on occasion, be based on a basic plant schedule which may pre-date the date when the Dayworks are being carried out by several years, and the percentage addition allowed by the sub-contractor needs to reflect this fact.

- Materials

Materials are priced at the prime cost for materials together with the percentage addition to the prime cost at the rates set out in the sub-contract particulars item 10 (as clause 5.6.1).

Fair valuation

If a valuation does not relate to the execution of work or the omission of work, or if the valuation of any work or liabilities directly associated with a variation cannot reasonably be effected in the valuation by any of the valuation rules noted above, then clause 5.8.1 notes that a fair valuation shall be made.

It must be noted that this additional provision does not relate to the execution of additional work or the omission of work, and only relates to any work or liabilities directly associated with a variation.

Non-recovery of loss and/or expense

Clause 5.8.2 notes that no allowance for any effect upon the regular progress of the works (or any part of them) or for any other direct loss and/or expense for which the sub-contractor would be reimbursed by any other provision in the sub-contract is to be made under the valuation rules.

ICSub/NAM

All as ICSub above.

MWSub/D and ShortSub

Clause 10.2 notes that variations shall be valued by the contractor on a fair and reasonable basis, with reference, where available and relevant, to rates and prices in the pricing document (being any document identified in the second recital of the contract which shows rates and prices).

SubSub

Clause 10.2 notes that variations shall be valued by the sub-contractor on a fair and reasonable basis, with reference, where available and relevant, to rates and prices in the pricing document (being any document identified in the second recital of the contract which shows rates and prices).

Chapter 11
Injury, Damage and Insurance

What is the sub-contractor's liability for personal injury or death?

DBSub

The sub-contractor is liable for, and shall indemnify the contractor against, any expense, liability, loss, claim or proceedings whatsoever in respect of personal injury to or the death of any person arising out of, or in the course of, or caused by the carrying out of the sub-contract works (as clause 6.2 of the sub-contract conditions). Clause 6.5.2 notes that this indemnity does not include for personal injury to or the death of any person due to the effect of an Excepted Risk as defined under clause 6.1 of the sub-contract conditions.

Under clause 6.2, the sub-contractor is not liable where the personal injury to or the death of any person arising out of, or in the course of, or caused by the carrying out of the sub-contract works is due to any act of neglect, breach of statutory duty, omission or default of:

(1) the contractor;
(2) any of the contractor's persons;
(3) the employer;
(4) any of the employers persons; or
(5) any statutory undertaker.

In addition to the sub-contractor's obligation to indemnify the contractor as detailed above, the sub-contractor is also to take out and maintain insurance in respect of claims arising under this liability (as clause 6.5.1 of the sub-contract conditions). This is an entirely sensible requirement because the indemnity of a sub-contractor that is insubstantial may be effectively worthless. The fact that the sub-contractor is required to take out insurance does not affect or lessen the sub-contractor's obligation to indemnify the contractor under clause 6.2.

The level of the insurance cover required under clause 6.5 is to be entered in the sub-contract particulars in the sub-contract agreement. The insurance cover to be provided is to be not less than that entered in the sub-contract particulars, but may be for a greater sum. The insurance cover is to be for any one occurrence or series of occurrences arising out of one event.

Under clause 6.5.3 of the sub-contract conditions, the contractor has the right to require the sub-contractor to produce evidence that he has taken out the insurances required by clause 6.5.1 and that they are being maintained. Clause 6.5.4 adds that if the sub-contractor does not take out or maintain the insurance as required by clause 6.5.1, then the contractor may take out an insurance to cover the sub-contractor's default and the cost of doing so shall either be deducted from any monies due or to become due to the sub-contractor under the sub-contract, or shall be recoverable from the sub-contractor as a debt.

MPSub

Clause 35.1 notes that the sub-contractor is liable for, and shall indemnify the contractor against, any expense, liability, loss, claim or proceedings arising under statute or at common law in respect of personal injury to or the death of any person to the extent that such personal injury or death arises out of or in the course of carrying out of the sub-contract works, provided that the personal injury or death is not as a consequence of some act of neglect of the contractor or the employer (or persons that the contractor or the employer are responsible for). The sub-contractor's liability excludes any amount recoverable by the contractor or the employer under an applicable insurance policy under the main contract.

Clause 35.2 states that the contractor is liable for, and shall indemnify the sub-contractor against, any expense, liability, loss, claim or proceedings arising under statute or at common law in respect of personal injury to or the death of any person to the extent that such personal injury or death arises out of or in the course of carrying out the project works as a consequence of some act of neglect of the contractor or the employer (or persons that the contractor or the employer are responsible for). The sub-contractor's liability excludes any amount recoverable by the sub-contractor under an applicable insurance policy under the main contract.

Where an insurance policy under the main contract includes the sub-contractor as an insured party, the sub-contractor can ask the contractor to provide documentary evidence that the policy has been taken out and remains in place. If the contractor does not provide the required documentary evidence within 14 days of being asked to, the sub-contractor may take out the insurance and the contractor will be liable to pay the sub-contractor the costs incurred in taking out and maintaining that insurance.

The sub-contractor is to advise the contractor if an event arises that may give rise to a claim under the insurance policy. Any such claim is to be disregarded in the computation of the amount due to the sub-contractor in accordance with the sub-contract. Any payment received by the sub-contractor or the contractor will be by way of the insurance policy. If a sub-contractor makes a claim against the insurance policy, then he is to pay the policy excess as stated in the sub-contract particulars to the contractor.

Where an insurance policy includes terrorism cover, but that terrorism cover is no longer available, then the risk of loss that would otherwise have been covered by that terrorism cover rests with the contractor. Any additional works necessary to complete the project as a consequence of a loss due to terrorism that would otherwise have been covered by the policy, shall be treated as a change (i.e. a variation).

ICSub

All as DBSub.

ICSub/D

All as DBSub.

ICSub/NAM

All as DBSub, except that the level of the insurance cover required under clause 6.5 is to be entered in the tender documents at item IT15.1.

MWSub and ShortSub

This is not specifically dealt with under these sub-contracts, but it should be noted that clause 7.3 requires the sub-contractor to assume the liabilities of the main contractor to the extent that such obligations and liabilities relate to the sub-contract works.

SubSub

This is not specifically dealt with under the SubSub, but it should be noted that clause 7.3 requires the sub-subcontractor to assume the liabilities of the sub-contractor to the extent that such obligations and liabilities relate to the sub-subcontract works.

What is the sub-contractor's liability for injury or damage to property?

DBSub

The sub-contractor is liable for, and shall indemnify the contractor against, any expense, liability, loss, claim or proceedings in respect of any loss, injury or damage whatsoever to any property real or personal in so far as such loss, injury or damage arises out of or in the course of or by reason of the carrying out of the sub-contract works (as clause 6.3).

Clause 6.5.2 notes that this indemnity does not include for loss, injury or damage due to the effect of an Excepted Risk as defined under clause 6.1 of the sub-contract conditions.

In line with clause 6.3 of the sub-contract conditions, the sub-contractor is only liable where the said loss, injury or damage is due to any negligence, breach of statutory duty, omission or default of the sub-contractor or any of the sub-contractor's persons.

Clause 6.4 of the sub-contract conditions makes it clear that the above liability and indemnity shall not include any liability or indemnity in respect of loss or damage to the main contract works (including the sub-contract works) and/or to any site materials (including the sub-contract site materials) by any of the Specified Perils (as defined under clause 6.1 of the sub-contract conditions) even if this loss or damage is caused by the negligence, breach of statutory duty, omission or default of the sub-contractor or any of the sub-contractor's persons for the period up to and including the relevant terminal date (as defined under clause 6.1 of the sub-contract conditions).

In addition to the sub-contractor's obligation to indemnify the contractor as detailed above, the sub-contractor is also to take out and maintain insurance in respect of claims arising under this liability (as clause 6.5.1 of the sub-contract conditions).

Although this is an entirely sensible requirement, because the indemnity of a sub-contractor that is insubstantial may be effectively worthless, clause

6.5.1 makes it clear that such insurance is without prejudice to the sub-contractor's obligations to indemnify the contractor under clause 6.3 of the sub-contract conditions.

In respect of the loss or damage to the sub-contract works and/or to any of the sub-contract site materials by any of the Specified Perils, clause 6.5.1 of the sub-contract conditions makes it clear that the sub-contractor is not obliged to take out and maintain insurance for this eventuality. The reason for this is because the sub-contract works (when they form part of the main contract works) and the sub-contract site materials (when they form part of the main contract site materials) are covered by the main contract joint names policy for loss or damage by the Specified Perils up to and including the terminal date (as defined under clause 6.1 of the sub-contract conditions).

However, the sub-contractor needs to particularly note that the main contract joint names policy referred to above is only in respect of the Specified Perils. Other risks, such as subsidence, impact, theft or vandalism (but not including Excepted Risks) are not covered. Also, the main contract joint names policy referred to above does not cover the sub-contract works or the sub-contract site materials when they are still in the custody and control of the sub-contractor.

It should be noted that the parties are to list under the sub-contract particulars those elements of the sub-contract works that the contractor is prepared to regard as fully, finally and properly incorporated into the main contract works prior to practical completion of the sub-contract works or section as applicable; and also to detail the extent to which each of the listed elements needs to be carried out to achieve the status of being fully, finally and properly incorporated into the main contract works. The provision of this information reduces the probability of a dispute arising as to whether the sub-contract works and/or the sub-contract site materials are still in the custody and control of the sub-contractor or not.

It should be noted that the joint names policy referred to in paragraph C.1 of insurance option C in schedule 3 of the main contract conditions does not provide any cover to sub-contractors. This particular joint names policy relates to the insurance by the employer of existing structures and works in extensions to them. The sub-contractor will be liable for any loss or damage caused to existing structures and/or their contents by the negligence, omission or default of the sub-contractor or any person for whom he is responsible.

Such liability is a third party liability for which the sub-contractor is expressly liable under clause 6.3 of the sub-contract conditions and against which he is required to insure under clause 6.5.1. The level of the insurance cover required under clause 6.5 is to be entered in the sub-contract particulars in the sub-contract agreement, but may be for a greater sum. The insurance cover is to be for any one occurrence or series of occurrences arising out of one event.

Under clause 6.5.3 of the sub-contract conditions, the contractor has the right to require the sub-contractor to produce evidence that he has taken out the insurances required by clause 6.5.1 and that they are being maintained. Clause 6.5.4 adds that if the sub-contractor does not take out or maintain the insurance as required by clause 6.5.1, then the contractor may take out an insurance to cover the sub-contractor's default and the cost of doing so shall either be deducted from any monies due or to become due to the sub-contractor under the sub-contract, or shall be recoverable from the sub-contractor as a debt.

MPSub

Clause 35.1 notes that the sub-contractor is liable for, and shall indemnify the contractor against, any expense, liability, loss, claim or proceedings arising under statute or at common law in respect of the loss, injury or damage to any property real or personal to the extent that such loss, etc., arises out of or in the course of carrying out of the sub-contract works, provided that the loss, etc., is not as a consequence of some act of neglect of the contractor or the employer (or persons that the contractor or the employer are responsible for). The sub-contractor's liability excludes any amount recoverable by the contractor or the employer under an applicable insurance policy under the main contract.

Clause 35.2 states that the contractor is liable for, and shall indemnify the sub-contractor against, any expense, liability, loss, claim or proceedings arising under statute or at common law in respect of the loss, injury or damage to any property real or personal to the extent that such loss, etc., arises out of or in the course of carrying out the project works as a consequence of some act of neglect of the contractor or the employer (or persons that the contractor or the employer are responsible for). The sub-contractor's liability excludes any amount recoverable by the sub-contractor under an applicable insurance policy under the main contract.

Where an insurance policy under the main contract includes the sub-contractor as an insured party, the sub-contractor can ask the contractor to provide documentary evidence that the policy has been taken out and remains in place. If the contractor does not provide the required documentary evidence within 14 days of being asked to, the sub-contractor may take out the insurance and the contractor will be liable to pay the sub-contractor the costs incurred in taking out and maintaining that insurance.

The sub-contractor is to advise the contractor if an event arises that may give rise to a claim under the insurance policy. Any such claim is to be disregarded in the computation of the amount due to the sub-contractor in accordance with the sub-contract. Any payment received by the sub-contractor or the contractor will be by way of the insurance policy. If a sub-contractor makes a claim against the insurance policy, then he is to pay the policy excess as stated in the sub-contract particulars to the contractor.

Where an insurance policy includes terrorism cover, but that terrorism cover is no longer available, then the risk of loss that would otherwise have been covered by that terrorism cover rests with the contractor. Any additional works necessary to complete the project as a consequence of a loss due to terrorism that would otherwise have been covered by the policy, shall be treated as a change (i.e. a variation).

ICSub

All as DBSub.

ICSub/D

All as DBSub.

ICSub/NAM

All as DBSub, except that the level of the insurance cover required under clause 6.5 is to be entered in the tender documents at item IT15.1.

MWSub/D and ShortSub

This is not specifically dealt with under these sub-contracts, but it should be noted that clause 7.3 requires the sub-contractor to assume the liabilities of the main contractor to the extent that such obligations and liabilities relate to the sub-contract works.

SubSub

This is not specifically dealt with under the SubSub, but it should be noted that clause 7.3 requires the sub-subcontractor to assume the liabilities of the sub-contractor to the extent that such obligations and liabilities relate to the sub-subcontract works.

What Specified Perils cover, for loss or damage to works and site materials, does the sub-contractor obtain under the joint names all risks policies?

DBSub

The cover that the sub-contractor obtains from the all risks policy taken out under the main contract is set out under clause 6.6.1 of the sub-contract conditions. This cover is limited to loss or damage to the sub-contract works and the sub-contract site materials by the Specified Perils (as defined under clause 6.1 of the sub-contract conditions).

The cover which the sub-contractor obtains from the all risks policy is provided for by clause 6.9 of the main contract conditions. The sub-contractor has the right (under clause 6.6.2 of the sub-contract conditions) to require the contractor to produce evidence that he has taken out the insurances required by clause 6.6.1 and that they are being maintained.

In the case where the employer is a local authority and insurance options B or C apply (under the main contract) the contractor's obligation is limited to the production of a certificate certifying that terrorism cover is being provided. The reason for this exclusion is because under the main contract conditions both clause B.2.1 and C.3.1 make it clear that if the employer is a local authority, then evidence of insurance does not need to be provided to the contractor.

Clause 6.6.3 of the sub-contract conditions adds that if the contractor does not take out or maintain the insurance as required by clause 6.6.1, then the sub-contractor may take out an insurance to cover the contractor's default and the cost of doing so shall either be taken into account in the calculation of the final sub-contract sum, or shall be recoverable from the contractor as a debt.

MPSub

Not applicable.

ICSub

All as DBSub.

ICSub/D

All as DBSub.

ICSub/NAM

All as DBSub.

MWSub/D, ShortSub and SubSub

Not applicable.

What is the sub-contractor's liability for damage to the sub-contract works?

DBSub

The sub-contractor's responsibility in the event that there is damage to the sub-contract works is set out under clause 6.7 of the sub-contract conditions.

The sub-contractor is responsible for the cost of restoration of the sub-contract works and replacement or repair of any sub-contract site materials that are lost or damaged before the terminal date and of the removal and disposal of any resultant debris except to the extent that the loss or damage to the sub-contract works or sub-contract site materials is due to any of the Specified Perils, or any Excepted Risk (as defined under clause 6.1 of the sub-contract conditions), or is due to any negligence, breach of statutory duty, omission or default of the contractor, or any of the contractor's persons, or of the employer, or of any of the employer's persons, or of any statutory undertaker executing work solely in pursuance of his statutory rights or obligations.

Clause 6.7.2 of the sub-contract conditions makes it clear that if, during the progress of the sub-contract works, sub-contract materials or goods have been fully and finally incorporated into the main contract works before the terminal date, then the sub-contractor is only responsible for the cost of restoration of such work lost or damaged and the removal and disposal of any debris to the extent that such loss or damage is caused by the negligence, breach of statutory duty, omission or default of the sub-contractor or any of the sub-contractor's persons.

Clause 6.7.8 of the sub-contract conditions notes that, for the purposes of clause 6.7.2 only, materials and goods forming part of the sub-contract works shall be deemed to have been fully, finally and properly incorporated into the main contract works when in each case they have been completed by the sub-contractor to the extent indicated or referred to in the section of the sub-contract particulars that deals with the incorporation of the sub-contract works into the main contract works.

In the event that any loss or damage is occasioned to the sub-contract works or the sub-contract site materials, by whatever means, then, upon its occurrence or later discovery, the sub-contractor is to forthwith (i.e. immediately) give notice in writing to the contractor of its extent, nature and location. The contractor is then to issue his directions and if those directions are such, the sub-contractor is, with due diligence, to restore the lost or damaged sub-contract works, and/or replace or repair any lost or damaged sub-contract site materials, remove and dispose of any debris, and proceed with the carrying out of the sub-contract works.

Where, in line with the above provisions, the sub-contractor is not responsible for the cost of compliance, such compliance shall be treated as a variation.

It must be noted, however, that the occurrence of loss or damage affecting the sub-contract works occasioned by any of the Specified Perils (as defined under clause 6.1 of the sub-contract conditions) shall be disregarded in computing any amounts payable to the sub-contractor under the sub-contract, although it may be payable as part of an insurance claim under the joint names all risks policy noted above.

Where monies are to be paid out by the insurer under that joint names all risks policy, the sub-contractor shall not object to the payment of the monies to the employer (as clause 6.7.5 of the sub-contract conditions).

Clause 6.7.6 makes it clear that, except in the case of loss or damage caused by the negligence, breach of statutory duty, omission or default of the sub-contractor or any of the sub-contractor's persons, the sub-contractor shall not be responsible for loss or damage to the sub-contract works occurring after the terminal date.

Finally, clause 6.7.7 of the sub-contract conditions states that nothing in clause 6.7 shall in any way modify the sub-contractor's obligations in regard to defects in the sub-contract works.

MPSub

Clause 35.1 notes that the sub-contractor is liable for, and shall indemnify the contractor against, any expense, liability, loss, claim or proceedings arising under statute or at common law in respect of the loss, injury or damage to any property real or personal to the extent that such loss, etc., arises out of or in the course of carrying out of the sub-contract works, provided that the loss, etc., is not as a consequence of some act of neglect of the contractor or the employer (or persons that the contractor or the employer are responsible for). The sub-contractor's liability excludes any amount recoverable by the contractor or the employer under an applicable insurance policy under the main contract.

Clause 35.2 states that the contractor is liable for, and shall indemnify the sub-contractor against, any expense, liability, loss, claim or proceedings arising under statute or at common law in respect of the loss, injury or damage to any property real or personal to the extent that such loss, etc., arises out of or in the course of carrying out the project works as a consequence of some act of neglect of the contractor or the employer (or persons that the contractor or the employer are responsible for). The sub-contractor's liability excludes any amount recoverable by the sub-contractor under an applicable insurance policy under the main contract.

Where an insurance policy under the main contract includes the sub-contractor as an insured party, the sub-contractor can ask the contractor to provide documentary evidence that the policy has been taken out and remains in place. If the contractor does not provide the required documentary evidence within 14 days of being asked to, the sub-contractor may take out the insurance and the contractor will be liable to pay the sub-contractor the costs incurred in taking out and maintaining that insurance.

The sub-contractor is to advise the contractor if an event arises that may give rise to a claim under the insurance policy. Any such claim is to be disregarded in the computation of the amount due to the sub-contractor in accordance with the sub-contract. Any payment received by the sub-contractor or the contractor will be by way of the insurance policy. If a sub-contractor makes a claim against the insurance policy, then he is to pay the policy excess as stated in the sub-contract particulars to the contractor.

Where an insurance policy includes terrorism cover, but that terrorism cover is no longer available, then the risk of loss that would otherwise have been covered by that terrorism cover rests with the contractor. Any additional works necessary to complete the project as a consequence of a loss due to terrorism that would otherwise have been covered by the policy, shall be treated as a change (i.e. a variation).

ICSub

All as DBSub.

ICSub/D

All as DBSub.

ICSub/NAM

All as DBSub.

MWSub/D and ShortSub

This is not specifically dealt with under these sub-contracts, but it should be noted that clause 7.3 requires the sub-contractor to assume the liabilities of the main contractor to the extent that such obligations and liabilities relate to the sub-contract works.

SubSub

This is not specifically dealt with under the SubSub, but it should be noted that clause 7.3 requires the sub-subcontractor to assume the liabilities of the sub-contractor to the extent that such obligations and liabilities relate to the sub-subcontract works.

What are the employer's options where terrorism cover is not available?

DBSub

As noted in clause 6.8 of the sub-contract conditions, if the insurers named in the joint names policy referred to in clause 6.6 notify the contractor or the employer that terrorism cover will cease on a particular date (and will no longer be available) the contractor is to immediately inform the sub-contractor.

When the employer has notified the contractor of his decision (in the light of this lack of cover) either to continue with the main contract works or to terminate the contractor's employment under the main contract, the contractor is to notify the sub-contractor accordingly.

If the employer gives notice terminating the contractor's employment under the main contract, then upon and from the date stated by the employer in his notice to the contractor the sub-contractor's employment under the sub-contract shall terminate, in which case the provisions of clause 7.11 of the sub-contract conditions shall apply (but excluding for the recovery by the sub-contractor of any direct loss and/or damage caused to the sub-contractor by the termination).

If the employer does not give notice terminating the contractor's employment under the main contract, then upon and from the date that terrorism cover will cease:

- if the sub-contract work that has been executed and/or the sub-contract site materials suffer physical loss or damage caused by terrorism, the sub-contractor shall with due diligence restore lost or damaged work, repair or replace any lost or damaged site materials, remove and dispose of any debris, and proceed with the carrying out of the sub-contract works;
- the restoration, replacement or repair of such loss or damage and any removal and the disposal of debris shall be treated as a variation with no reduction in any amount payable to the sub-contractor by reason of any act or neglect of the sub-contractor or of any of his own sub-contractors (i.e. any sub-subcontractors) which may have contributed to the physical loss or damage; and
- where insurance option C applies, the requirement that the sub-contract works continue to be carried out shall not be affected by any loss or damage to the existing structures and/or their contents caused by terrorism.

MPSub

Not applicable.

ICSub

All as DBSub.

ICSub/D

All as DBSub.

ICSub/NAM

All as DBSub.

MWSub/D and ShortSub

This is not specifically dealt with under these sub-contracts, but it should be noted that clause 7.3 requires the sub-contractor to assume the liabilities of the main contractor to the extent that such obligations and liabilities relate to the sub-contract works.

SubSub

This is not specifically dealt with under the SubSub, but it should be noted that clause 7.3 requires the sub-subcontractor to assume the liabilities of the sub-contractor to the extent that such obligations and liabilities relate to the sub-subcontract works.

Is the contractor responsible for damage caused to sub-contractor's plant, etc.?

DBSub

The contractor is only responsible for the loss of or damage to temporary works, plant tools, equipment or other property belonging to or provided by the sub-contractor, or the sub-contractor's persons, or to any materials or goods of the sub-contractor which are not sub-contract site materials, when such loss or damage is due to any negligence, breach of statutory duty, omission or default of the contractor or any of the contractor's persons.

In all other instances, the sub-contractor is responsible, and the sub-contractor needs to ensure that his insurance arrangements with his own sub-contractors (i.e. the sub-subcontractors) are compatible with this liability.

MPSub

Clause 35.2 states that the contractor is liable for, and shall indemnify the sub-contractor against, any expense, liability, loss, claim or proceedings arising under statute or at common law in respect of the loss, injury or damage to any property real or personal to the extent that such loss, etc., arises out of or in the course of carrying out the project works as a consequence of some act of neglect of the contractor or the employer (or persons that the contractor or the employer are responsible for). The sub-contractor's liability excludes any amount recoverable by the sub-contractor under an applicable insurance policy under the main contract.

Where an insurance policy under the main contract includes the sub-contractor as an insured party, the sub-contractor can ask the contractor to provide documentary evidence that the policy has been taken out and remains in place. If the contractor does not provide the required documentary evidence within 14 days of being asked to, the sub-contractor may take out the insurance and the contractor will be liable to pay the sub-contractor the costs incurred in taking out and maintaining that insurance.

The sub-contractor is to advise the contractor if an event arises that may give rise to a claim under the insurance policy. Any such claim is to be disregarded in the computation of the amount due to the sub-contractor in accordance with the sub-contract. Any payment received by the sub-contractor or the contractor will be by way of the insurance policy. If a sub-contractor makes a claim against the insurance policy, then he is to pay the policy excess as stated in the sub-contract particulars to the contractor.

Where an insurance policy includes terrorism cover, but that terrorism cover is no longer available, then the risk of loss that would otherwise have been covered by that terrorism cover rests with the contractor. Any additional works necessary to complete the project as a consequence of a loss due to terrorism that would otherwise have been covered by the policy, shall be treated as a change (i.e. a variation).

ICSub

All as DBSub.

ICSub/D

All as DBSub.

ICSub/NAM

All as DBSub.

MWSub/D and ShortSub

This is not specifically dealt with under these sub-contracts, but it should be noted that clause 7.3 requires the sub-contractor to assume the liabilities of the main contractor to the extent that such obligations and liabilities relate to the sub-contract works.

SubSub

This is not specifically dealt with under the SubSub, but it should be noted that clause 7.3 requires the sub-subcontractor to assume the liabilities of the sub-contractor to the extent that such obligations and liabilities relate to the sub-subcontract works.

Is the sub-contractor required to take out professional indemnity insurance?

DBSub

Clause 6.10 of the sub-contract conditions places an obligation on the sub-contractor to insure in respect of PI insurance.

The sub-contractor is required to take out this insurance of the type and with the level of cover of at least that stated in item 14 of the sub-contract particulars as contained in the sub-contract agreement.

If no level of cover is stated there, then *no PI insurance is required*. This is an important point to note.

Clause 6.10.2 of DBSub/C notes that, provided that it remains available at commercially reasonable rates, the sub-contractor is to maintain the PI insurance until the expiry of the period stated under item 14 of the sub-contract particulars from the date of practical completion of the sub-contract works.

Normally, where the sub-contract is executed as a deed, the period will be twelve years, but if the sub-contract is executed under hand, the period should be six years.

Clause 6.10.3 of the sub-contract conditions notes that the sub-contractor is required to provide documentary evidence that the PI insurance is being maintained, as and when the contractor reasonably requires. If the PI insurance is not available at commercially reasonable rates, clause 6.11 requires the sub-contractor to notify the contractor immediately so that the contractor and sub-contractor can discuss the means of best protecting their respective positions in the absence of such PI insurance.

MPSub

Clause 37.1 states that the sub-contractor is only required to take out professional indemnity insurance if it is so stated in the sub-contract particulars.

The amount of cover for the professional indemnity insurance is to be for no less than the value stated in the sub-contract particulars, and provided that it remains generally available at commercially reasonable rates, it is to be maintained until 12 years after practical completion of the project. If the sub-contractor considers that the insurance is not available at commercially reasonable rates, he is to notify the contractor and is to cooperate with the contractor in seeking means by which the sub-contractor can be protected against professional liability claims arising out of the sub-contract works.

The contractor may request the sub-contractor to provide documentary evidence that the policy has been taken out and remains in place. If the sub-contractor does not provide the required documentary evidence within seven days of being asked to, the contractor may take out the insurance and the sub-contractor will be liable to pay the sub-contractor the costs incurred in taking out and maintaining that insurance.

ICSub

Not applicable.

ICSub/D

Clause 6.14 of the sub-contract conditions places an obligation on the sub-contractor to insure in respect of PI insurance.

The sub-contractor is required to take out this insurance of the type and with the level of cover of at least that stated in item 11 of the sub-contract particulars as contained in the sub-contract agreement.

If no level of cover is stated there, then *no PI insurance is required*. This is an important point to note.

Clause 6.14.2 of ICSub/D/C notes that, provided that it remains available at commercially reasonable rates, the sub-contractor is to maintain the PI insurance until the expiry of the period stated under item 11 of the sub-contract particulars from the date of practical completion of the sub-contract works.

Normally, where the sub-contract is executed as a deed, the period will be twelve years, but if the sub-contract is executed under hand, the period should be six years.

Clause 6.14.3 of the sub-contract conditions notes that the sub-contractor is required to provide documentary evidence that the PI insurance is being maintained, as and when the contractor reasonably requires. If the PI insurance is not available at commercially reasonable rates, clause 6.15 requires the sub-contractor to notify the contractor immediately so that the contractor and sub-contractor can discuss the means of best protecting their respective positions in the absence of such PI insurance.

ICSub/NAM, MWSub/D, ShortSub and SubSub

Not applicable.

When does the Joint Fire Code apply?

DBSub

Clause 6.12 of the sub-contract conditions notes that the Joint Fire Code only applies where the main contract particulars state that the Joint Fire Code applies, and clause 6.13 notes that where the Joint Fire Code applies, the parties shall comply with the Joint Fire Code; the contractor is to ensure compliance by all contractor's persons and the sub-contractor shall ensure such compliance by all sub-contractor's persons.

MPSub

There is no mention of the Joint Fire Code in MPSub.

ICSub

Clause 6.10 of the sub-contract conditions notes that the Joint Fire Code only applies where the main contract particulars state that the Joint Fire Code applies, and clause 6.11 notes that where the Joint Fire Code applies, the parties shall comply with the Joint Fire Code; the contractor is to ensure compliance by all contractor's persons and the sub-contractor shall ensure such compliance by all sub-contractor's persons.

ICSub/D

All as ICSub.

ICSub/NAM

All as ICSub.

MWSub/D and ShortSub

There is no mention of the Joint Fire Code in these sub-contracts, but it should be noted that clause 7.3 requires the sub-contractor to assume the liabilities of the contractor to the extent that such obligations and liabilities relate to the sub-contract works.

SubSub

There is no mention of the Joint Fire Code in the SubSub, but it should be noted that clause 7.3 requires the sub-subcontractor to assume the liabilities of the sub-contractor to the extent that such obligations and liabilities relate to the sub-subcontract works.

What happens if there is a breach of the Joint Fire Code?

DBSub

Clause 6.14 of the sub-contract conditions sets out the procedure to be followed in the event of a breach of the Joint Fire Code. This procedure being:

- If there is a breach of the Joint Fire Code, the insurers may require the contractor to carry out remedial measures.
- If the insurers do require the contractor to carry out remedial measures, the contractor is to send a copy of the notice requiring the remedial measures to the sub-contractor, forthwith.
- Upon receipt of the notice, the sub-contractor is to comply with any direction of the contractor that is reasonably necessary in respect of the sub-contract works to correct the breach of the Joint Fire Code.
- Except where the breach of the Joint Fire Code was as a result of an act, omission or default of the sub-contractor or of any of the sub-contractor's persons, then the compliance by the sub-contractor with any direction of the contractor, as noted above, will be treated as a variation.
- If the breach of the Joint Fire Code was as a result of (or was partly as a result of) an act, omission or default of the sub-contractor or of any of the sub-contractor's persons, then the sub-contractor shall be liable for the appropriate additional costs incurred by the contractor for the remedial measures and either an appropriate deduction shall be made in calculating the final sub-contract sum or it shall be recoverable by the contractor from the sub-contractor as a debt.
- If the sub-contractor, within four days of receipt of a direction as outlined above, does not begin to comply with it or thereafter fails without reasonable cause regularly and diligently to comply with it, then the provisions of clause 3.6 of the sub-contract conditions will apply.

Clause 3.6 makes it clear that if, within seven days after receipt of a written notice from the contractor which requires a sub-contractor to comply with a direction, the sub-contractor still does not comply with that direction, then the contractor may employ and pay other persons to execute any work whatsoever which may be necessary to give effect to that direction.

In such a situation, the sub-contractor would be liable for all additional costs incurred by the contractor in connection with such employment and an appropriate deduction would either be taken into account in the calculation of the final sub-contract sum, or would be recoverable by the contractor from the sub-contractor as a debt.

MPSub

Not applicable.

ICSub

All as DBSub, but clause 6.12 not clause 6.14.

ICSub/D

All as ICSub.

ICSub/NAM

All as ICSub.

MWSub/D and ShortSub

There is no mention of the Joint Fire Code in these sub-contracts, but it should be noted that clause 7.3 requires the sub-contractor to assume the liabilities of the contractor to the extent that such obligations and liabilities relate to the sub-contract works.

SubSub

There is no mention of the Joint Fire Code in the SubSub, but it should be noted that clause 7.3 requires the sub-subcontractor to assume the liabilities of the sub-contractor to the extent that such obligations and liabilities relate to the sub-subcontract works.

What are the consequences of amendments/revisions to the Joint Fire Code?

DBSub

Clause 6.15 of the sub-contract conditions makes it clear that if the Joint Fire Code is amended after the sub-contract base date, and if that amended code is to be applied in respect of the main contract works, the cost, if any, of compliance by the sub-contractor with any such amendment shall be added into the calculations of the final sub-contract sum.

MPSub

Not applicable.

ICSub

All as DBSub, but clause 6.13 not clause 6.15.

ICSub/D

All as ICSub.

ICSub/NAM

All as ICSub.

MWSub/D and ShortSub

There is no mention of the Joint Fire Code in these sub-contracts, but it should be noted that clause 7.3 requires the sub-contractor to assume the liabilities of the contractor to the extent that such obligations and liabilities relate to the sub-contract works.

SubSub

There is no mention of the Joint Fire Code in the SubSub, but it should be noted that clause 7.3 requires the sub-subcontractor to assume the liabilities of the sub-contractor to the extent that such obligations and liabilities relate to the sub-subcontract works.

Chapter 12
Termination of Sub-contract

What can cause a sub-contract to be terminated?

If there is a breach of an important condition of the sub-contract then a party has a right to terminate the sub-contract.

A breach of contract is committed when a party, without lawful excuse, fails or refuses to perform what is due from him under the contract, or performs defectively or incapacitates himself from performing what is due.

In common law, such a breach may be:

- a repudiatory breach, which is a breach whereby one party clearly indicates (by words or by conduct) that he no longer intends to be bound by the terms of the sub-contract, or perhaps where the default of a party has rendered himself unable to perform his outstanding contractual obligations (e.g. where a contractor needs to be NHBC registered but he is removed from the NHBC register); or
- an anticipatory breach, where one party states that he will not be carrying out his obligations under the sub-contract before the time for carrying out those obligations has actually arrived.

Once either a repudiatory breach or an anticipatory breach has occurred, then the innocent party may either affirm the sub-contract or may elect to accept the repudiatory breach or the anticipatory breach as appropriate, and then terminate the sub-contract.

It is important to note that the repudiatory breach or the anticipatory breach needs to be accepted, because, as noted in the *White & Carter* v. *McGregor*[1] case: 'Repudiation by one party standing alone does not terminate the contract. It takes two to end it, by repudiation on the one side, and acceptance of the repudiation on the other.'

Of course, there may always be arguments under common law about whether the condition that was breached was important enough to allow a party to terminate the contract in the first place.

In respect of contracts that contain a contractual termination clause this problem is overcome because those breaches that will allow contractual termination to take place are defined. Therefore, in a contractual termination it will not be necessary to show that the breach was of a sufficiently fundamental nature to warrant termination, all that will need to be shown is that the breach relied upon has been defined within the contract as being a breach that allows contractual termination to take place.

---

[1] *White & Carter (Councils) Ltd* v. *McGregor* [1962] AC 413, HL.

However, for a contractual termination to be successful it will be necessary for the contract procedures for notices, etc., to be followed meticulously, otherwise the party pursuing the contractual termination route may find itself accused of a repudiatory breach.

What is the effect of a sub-contract being terminated?

It is reasonably well established that when a sub-contract is terminated by a party because of a breach of an important condition by the other party the innocent party is released from all further performance of his obligations but he is entitled to recover full damages arising from the breach. Therefore, termination of the sub-contract in such a situation actually means that a party's employment under the sub-contract is terminated, rather than that the sub-contract is terminated per se. Although strictly speaking it is unnecessary to state this fact, it is normal practice to say that a party's 'employment under the sub-contract' has been terminated; and it is suggested that this practice should be followed for the avoidance of any possible doubt.

Are the effects of a sub-contract being terminated under common law or by contractual provisions the same?

Generally, the rights and remedies under a common law termination may well be different from those under a contractual termination.

For example, under a common law termination (unlike most contractual termination clauses) the innocent party may decide not to complete the project but may still claim damages.

On the other hand, under contractual termination, the innocent party is frequently entitled to seize materials to complete the works which would not be the case under a common law termination.

Also, under a contractual termination, the consequences arising from the termination are clearly set out, and this may include the payment to the innocent party of all sums due within a relatively short timescale.

In the case of common law termination, a high degree of freedom and flexibility of remedy is available, but the innocent party has the burden of establishing that the breach of sub-contract was sufficiently fundamental to justify the termination in law.

A fundamental point to note is that, unless specifically excluded, it is generally the case that the right to common law termination coexists with any contractual termination provisions. In other words, unless specifically excluded, a party's common law termination rights are preserved irrespective of any contractual termination provisions.

Are the party's common law termination rights preserved under the JCT sub-contracts?

DBSub

The party's common law rights are preserved under clause 7.3.1 of the sub-contract conditions.

MPSub

The party's common law rights are preserved under clause 39 of the sub-contract conditions.

ICSub

All as DBSub above.

ICSub/D

All as DBSub above.

ICSub/NAM

All as DBSub above.

MWSub/D, ShortSub and SubSub

No mention is made regarding the preservation of the parties' common law rights. However, given that a party's common law rights generally need to be specifically excluded, the fact that no such exclusion exists would imply that the parties' common law rights are preserved.

What reasons give the contractor a right to terminate the sub-contractor's employment (or a sub-contractor to terminate a sub-subcontractor's employment) under the JCT sub-contracts?

DBSub

The contractor may terminate the sub-contractor's employment under the sub-contract conditions for three principal reasons.

- Default by the sub-contractor.
- Insolvency of the sub-contractor.
- Corruption.

MPSub

Under clause 40.1 of the sub-contract conditions, the contractor may terminate the sub-contractor's employment under the sub-contract, if the sub-contractor commits a material breach and fails to remedy that material breach following notice being given.

A material breach by the sub-contractor is defined (under clause 1 of the sub-contract conditions) as being:

- failure to proceed regularly and diligently with the performance of his obligations under the sub-contract;

- failure to comply with clause 9.1[2] (i.e. the failure to comply with instructions);
- failure to provide documentary evidence of professional indemnity insurance, where required;
- suspension of the sub-contract works other than in accordance with the Housing Grants, Construction and Regeneration Act 1996, or in the circumstances described in clause 42.1;
- failure to make payment in the manner required by any payment advice issued under the sub-contract;
- any repudiatory breach of the sub-contract;
- a breach of the CDM Regulations; or
- a breach of the requirements of CIS (the Construction Industry Scheme).

Under clause 40.3 of the sub-contract conditions, in the event that a sub-contractor becomes insolvent the contractor may at any time by notice to the sub-contractor terminate the sub-contractor's employment under the sub-contract.

Under clause 42.1 of the sub-contract conditions, in the event that the sub-contract works or a substantial proportion of it is suspended for the period stated in the sub-contract as a result of an event listed under clause 42.1 the contractor may terminate the sub-contractor's employment under the sub-contract following notice being given.

Under clause 42.3 of the sub-contract conditions, in the event that terrorism cover under a required insurance policy is no longer available, the contractor may, by notice, terminate the sub-contractor's employment under this sub-contract.

Under clause 43.1 of the sub-contract conditions, the sub-contractor's employment under the sub-contract shall automatically terminate upon the termination of the contractor's employment under the main contract. The contractor is to immediately notify the sub-contractor of such termination.

In respect of clauses 40.1, 40.3, 42.1, 42.3 and 43.1, the notice must be in writing and must be given by actual delivery, registered post or recorded delivery (as clause 5.2 of the sub-contract conditions).

ICSub

As DBSub above.

ICSub/D

As DBSub above.

ICSub/NAM

As DBSub above.

MWSub/D and ShortSub

In line with clauses 14.1.1 to 14.1.3 of the sub-contract conditions, if the sub-contractor fails to remedy one of the following defaults following notice being given by the contractor:

---

[2] The original MPSub referred to clause 3.1, this was amended to clause 9.1 by way of Amendment No. 1.

- without reasonable cause, wholly or substantially suspends the carrying out of the sub-contract works;
- fails to proceed regularly and diligently with the sub-contract works; or
- fails to comply with clause 5.5 (i.e. non-assignment, or obtaining consent from contractor before sub-letting) of the sub-contract conditions;

the contractor may terminate the sub-contractor's employment under the sub-contract.

Also, the contractor may terminate the sub-contractor's employment, under clauses 14.2.1 to 14.2.5 of the sub-contract conditions, by serving a notice if the sub-contractor:

- enters into an arrangement, compromise or composition in satisfaction of his debts (except where a solvent company enters into such an arrangement for the purposes of amalgamation or reconstruction);
- without a declaration of solvency, passes a resolution or makes a determination that he be wound up;
- has a winding up order or bankruptcy order made against him; or
- has appointed to him an administrator or administrative receiver.

The manner in which notices should be served is detailed under clause 4 of the sub-contract conditions.

In line with clause 15.1 of the sub-contract conditions, if the contractor's employment under the main contract is terminated, the sub-contractor's employment under this sub-contract shall automatically be terminated.

SubSub

All as MWSub/D and ShortSub above, but amended to allow for a sub-contract/sub-subcontract relationship rather than a contractor/sub-contractor relationship.

What is deemed to be default by the sub-contractor (or a sub-subcontractor)?

DBSub

The specified defaults (*which to be defaults must be committed before practical completion of the sub-contract works*) are:

- The sub-contractor without reasonable cause wholly or substantially suspends the carrying out of the sub-contract works (as clause 7.4.1.1 of the sub-contract conditions).

  This ground for termination does not apply where the sub-contractor has reasonable cause for suspending the carrying out of the sub-contract works.

  Therefore, a suspension because of non-payment in line with clause 4.11 of the sub-contract conditions would be a reasonable cause for suspending the carrying out of the sub-contract works, as this would not be a specified default.

- The sub-contractor fails to proceed regularly and diligently with the sub-contract works (as clause 7.4.1.2 of the sub-contract conditions).

The word 'regularly' suggests a requirement to attend for work on a regular daily basis with sufficient men, materials and plant to have the physical capacity to progress the works substantially in accordance with the sub-contract obligations.

'Diligently' adds the concept of the need to apply the above physical capacity industriously and efficiently towards that same end.

Taken together, the obligation upon the sub-contractor is essentially to proceed continuously, industriously and efficiently with appropriate physical resources so as to progress the works steadily towards completion substantially in accordance with the contractual requirements as to time, sequence and quality of work.

- The sub-contractor refuses or neglects to comply with a written direction from the contractor requiring him to remove any work, materials or goods not in accordance with the sub-contract and by such refusal or neglect the main contract works are materially affected (as clause 7.4.1.3 of the sub-contract conditions).

- The sub-contractor fails to comply with clause 3.1 (i.e. non-assignment) or 3.2 (consent to be obtained before sub-letting) of the sub-contract conditions (as clause 7.4.1.4 of the sub-contract conditions).

- The sub-contractor fails to comply with the CDM Regulations (as clause 7.4.1.5 of the sub-contract conditions).

MPSub

A specified default is the same as a material breach under the MPSub, which has been dealt with above.

ICSub

As DBSub above.

ICSub/D

As DBSub above.

ICSub/NAM

As DBSub above.

MWSub/D and ShortSub

In line with clauses 14.1.1 to 14.1.3 of the sub-contract conditions, a specified default is if the sub-contractor:

- without reasonable cause, wholly or substantially suspends the carrying out of the sub-contract works;
- fails to proceed regularly and diligently with the sub-contract works; or

- fails to comply with clause 5.5 (i.e. non-assignment, or obtaining consent from contractor before sub-letting) of the sub-contract conditions.

SubSub

All as MWSub/D and ShortSub, but amended to allow for a sub-contract/sub-subcontract relationship rather than a contractor/sub-contractor relationship.

What happens when a sub-contractor (or a sub-subcontractor) defaults or commits a material breach?

DBSub

The basic process that is followed is:

- The sub-contractor defaults.
- The contractor gives a notice to the sub-contractor specifying the default.
- If the sub-contractor continues a specified default for ten calendar days (excluding any public holiday days) after receipt of the contractor's notice, then the contractor may on, or within ten calendar days (excluding any public holiday days) from the expiry of the initial notice period, issue a further notice terminating the sub-contractor's employment under the sub-contract.
- If the contractor does not give the further notice referred to above (for any reason) but the sub-contractor repeats a specified default (whether previously repeated or not), then, upon or within a reasonable time after such repetition, the contractor may by notice to the sub-contractor terminate the sub-contractor's employment under the sub-contract.
- The notices given above must be in writing, and must be given by actual, special or recorded delivery.
- Where the notice is given by special or recorded delivery it shall, subject to proof to the contrary, be deemed to have been received on the second business day after the date of posting (as clause 7.2.3 of the sub-contract conditions).

  The words 'subject to proof to the contrary' are important, because this means that (subject to proof) a notice could have been actually received prior to the second business day after the date of posting[3].
- Termination shall take effect on receipt of the relevant notice (as clause 7.2.2 of the sub-contract conditions).
- Notice of termination of the sub-contractor's employment shall not be given unreasonably or vexatiously (as clause 7.2.1 of the sub-contract condition).

MPSub

The basic process that is followed (in respect of the matters noted under clause 40.1) is:

- The sub-contractor commits a material breach.
- The contractor gives a notice to the sub-contractor identifying the material breach and states that the contractor may terminate the sub-contractor's employment if the sub-contractor fails to remedy the material breach within either:

---

[3] *Lafarge (aggregotes) Ltd* v. *Newham Borough Council* [2005] (unreported) QBD, 24 June 2005.

— eight days of the notice[4], where the material breach relates to a matter in respect of which the employer has itself served a notice on the contractor in accordance with clause 38.1 of the main contract; or

— fourteen days of the notice, in all other cases.

● If the sub-contractor does not remedy the material breach in the period stipulated in the notice, then the contractor may issue a further notice terminating the sub-contractor's employment under the sub-contract. Such further notice must be issued within 14 days from the expiry of the initial 8 or 14-day notice period (as applicable).

The basic process that is followed (in respect of matters noted under clause 42.1) is:

● If the carrying out of the sub-contract works or a substantial proportion of it is suspended for the period stated in the sub-contract particulars (which, if no period is specifically stated is, by default, 13 weeks) as a consequence of *force majeure*, the occurrence of any specified peril (as defined under clause 1 of the sub-contract conditions) or hostilities involving the United Kingdom, or the use or threat of terrorism, then either party may issue to the other party a notice specifying the circumstances of the suspension and stating that if the specified circumstances continues for a further 19 days they may terminate the sub-contractor's employment under the sub-contract.

● If the circumstances of the suspension continue for 19 days, then the party that issued the initial notice may issue a further notice at any time within the subsequent 14 days to terminate the sub-contractor's employment under the sub-contract.

It should be noted that in respect of a named specialist, the contractor is not to terminate the sub-contract with the named specialist without the prior written consent of the employer (such consent is not to be unreasonably delayed or withheld).

ICSub

As DBSub above.

ICSub/D

As DBSub above.

ICSub/NAM

As DBSub above.

MWSub/D and ShortSub

If, in line with clauses 14.1.1 to 14.1.3 of the sub-contract conditions, the sub-contractor:

---

[4]The MPSub guide notes that this notice period reflects a desire to ensure that actions under the sub-contract can be coordinated with similar remedies under the main contract. The analogous notice period under clause 38.1 of the main contract is 14 days, and the sub-contractor's default should therefore cease within the 14-day main contract notice period.

- without reasonable cause, wholly or substantially suspends the carrying out of the sub-contract works;
- fails to proceed regularly and diligently with the sub-contract works; or
- fails to comply with clause 5.5 (i.e. non-assignment, or obtaining consent from contractor before sub-letting) of the sub-contract conditions;

the contractor may give notice to the sub-contractor which specifies the default and requires it to be ended. If the sub-contractor does not end the default within seven days, the contractor may terminate the sub-contractor's employment by serving a further notice. The termination takes effect when the notice is served.

The manner in which notices should be served is detailed under clause 4 of the sub-contract conditions.

SubSub

All as MWSub/D and ShortSub but amended to allow for a sub-contract/sub-subcontract relationship rather that a contractor/sub-contractor relationship.

What does insolvency of the sub-contractor mean?

DBSub

In line with clause 7.1 of the sub-contract conditions, and *for the purposes of the sub-contract conditions only*, a sub-contractor is insolvent if:

.1 he enters into an arrangement, compromise or composition in satisfaction of his debts (excluding a scheme of arrangement as a solvent company for the purposes of amalgamation or reconstruction); or
.2 without a declaration of insolvency, he passes a resolution or makes a determination that he be wound up; or
.3 he has a winding up order or bankruptcy order made against him; or
.4 he has appointed to him an administrator or administrative receiver; or
.5 he is the subject of an analogous arrangement, event or proceedings in any other jurisdiction; or
.6 (additionally in the case of partnerships) each partner is the subject of an individual arrangement or any other event or proceedings referred to in clause 7.1.1 to 7.1.5.

Clause 7.5.2 of the sub-contract conditions requires the sub-contractor to immediately inform the contractor in writing if he makes any proposal, gives notice of any meeting or becomes the subject of any proceedings or appointment to any of the matters referred to above.

MPSub

Clause 1 of the sub-contract conditions defines 'Insolvent' as being when a party:

- makes a composition or arrangement with his creditors, or becomes bankrupt;
- makes a proposal for a voluntary arrangement for a composition of debts or scheme of arrangement to be approved in accordance with the Companies Act 1985 or Insolvency Act 1986, as the case may be;

- has a provisional liquidator appointed;
- has a winding up order made or bankruptcy order made against him;
- passes a resolution for voluntary winding up (except for the purposes of amalgamation or reconstruction); or
- under the Insolvency Act 1986 has an administrator or an administrative receiver appointed.

ICSub

As DBSub above.

ICSub/D

As DBSub above.

ICSub/NAM

As DBSub above.

MWSub/D and ShortSub

Under these sub-contracts insolvency is not defined. But the matters detailed under clauses 14.2.1 to 14.2.5 of the sub-contract conditions relate to the matter of insolvency.

SubSub

Under the SubSub, insolvency is not defined. But the matters detailed under clauses 14.2.1 to 14.2.5 of the sub-subcontract conditions relate to the matter of insolvency.

What happens when the sub-contractor becomes insolvent?

DBSub

Clause 7.5.1 of the sub-contract conditions states that if the sub-contractor is insolvent, the contractor may at any time by notice to the sub-contractor terminate the sub-contractor's employment under the sub-contract.

It is interesting to note that upon a sub-contractor becoming insolvent, his employment under the sub-contract is not automatically terminated and a notice from the contractor (which may be made at any time) is required to bring the termination of employment into effect.

However, the sub-contractor's obligations to carry out and complete the sub-contract works (including the design works, where applicable) are suspended.

The assumed reason why there is not automatic termination, but there is a period of suspension, is to allow a period following the happening of a specified insolvency event, during which the sub-contract arrangements will remain in force while the parties attempt to agree a way forward.

In addition, clause 7.5.3.2 of the sub-contract conditions states that as from when the sub-contractor becomes insolvent (which may be at a date earlier than when the sub-contractor's employment under the sub-contract has been terminated) the contractor may take reasonable measures to ensure that the sub-contract works and the sub-contract site materials are retained on site. The sub-contractor shall allow and shall not hinder or delay the taking of those measures.

MPSub

Clause 40.3 of the sub-contract conditions states that if the sub-contractor becomes insolvent, the contractor may at any time by notice to the sub-contractor terminate the sub-contractor's employment under the sub-contract.

If a sub-contractor's employment is terminated under clause 40 of the sub-contract conditions, then the sub-contractor is not to remove any materials, plant or equipment from site unless expressly permitted to do so by the contractor, and the sub-contractor is to provide to the contractor all design documents prepared in connection with the sub-contract.

The contractor may make such other arrangements as he considers appropriate to complete the sub-contract works, and is not obliged to make any further payment to the sub-contractor other than as set out below.

In line with clause 40.5 of the sub-contract conditions, when the project has been completed and the defects rectification provision of the main contract fulfilled the contractor is to issue a payment advice setting out:

- the additional (i.e. extra-over) cost incurred by the contractor in completing the sub-contract works;
- any loss and/or damage suffered by the contractor and for which the sub-contractor is liable, whether arising as a consequence of the termination or otherwise arising out of the sub-contract.

The contractor may issue to the sub-contractor interim payment advices in respect of amounts that are due to him as detailed above, as and when costs are incurred.

The payment due date of any such payment advice shall be the date of issue of same, and the final date for payment is 21 days thereafter.

ICSub

As DBSub above.

ICSub/D

As DBSub above.

ICSub/NAM

As DBSub above.

MWSub/D and ShortSub

If the matters listed under clauses 14.2.1 to 14.2.5 of the sub-contract conditions apply to a sub-contractor, the contractor may terminate the sub-contractor's employment by serving a notice on the sub-contractor.

The manner in which notices should be served is detailed under clause 4 of the sub-contract conditions.

SubSub

If the matters listed under clauses 14.2.1 to 14.2.5 of the sub-subcontract conditions apply to a sub-subcontractor, the sub-contractor may terminate the sub-subcontractor's employment by serving a notice on the sub-subcontractor.

The manner in which notices should be served is detailed under clause 4 of the sub-subcontract conditions.

What does corruption entail?

DBSub

Corruption is deemed only to be when the sub-contractor or any person employed by him or acting on his behalf has committed an offence under the Prevention of Corruption Acts 1889 to 1916, or, where the employer is a local authority, has given any fee or reward the receipt of which is an offence under sub-section (2) of section 117 of the Local Government Act 1972.

MPSub

Corruption is not expressly referred to within the MPSub.

ICSub

As DBSub above.

ICSub/D

As DBSub above.

ICSub/NAM

As DBSub above.

MWSub/D, ShortSub and SubSub

Corruption is not expressly referred to within these sub-contracts.

What are the consequences of the contractor terminating the sub-contractor's employment?

DBSub

If the sub-contractor's employment is terminated by the contractor for any of the reasons outlined above, the following consequences arise:

- The contractor may employ and pay other persons to carry out and complete the sub-contract works (including the design works, where applicable) and to make good any defects of the kind referred to in the sub-contract conditions.

- The contractor may enter upon and take possession of the sub-contract works and (subject to obtaining any necessary third party consents) may use all the sub-contractor's temporary buildings, plant, tools, equipment and sub-contract site materials for those purposes.

  It should be noted that there may be restrictions placed on the contractor's right to use site materials, etc., of a sub-contractor that has become insolvent, because this may prejudice other creditors of the sub-contractor and may not be compliant with the mandatory equal discharge of liabilities by a liquidator or receiver, something that cannot be excluded by contract[5].

- When required by the contractor in writing (but not beforehand) the sub-contractor shall remove or procure the removal from the site of all temporary buildings, plant, tools, equipment, goods and materials belonging to the sub-contractor or to the sub-contractor's persons (as clause 7.7.2.1 of the sub-contract conditions).

- Where applicable, and without charge, the sub-contractor is to provide the contractor with three copies of all the sub-contractor's design documents then prepared, whether or not previously provided.

- If required by the contractor, the sub-contractor shall within 14 days of the date of termination, assign to the contractor, without charge, the benefit of any agreement for the supply of materials or goods and/or for the execution of any work for the purposes of the sub-contract.

  Of course, the above will only be possible to the extent that the sub-contractor can legally carry out or is legally required to carry out such assignments (noting that assignment may not be legally possible in the case of the sub-contractor's insolvency), and the sub-contractor's own sub-contractor or supplier is prepared for the said agreement to be assigned.

- The contractor is not required to make any further payment or release any outstanding retention, other than as provided for under clause 7.7.4 of the sub-contract conditions (as clause 7.7.3 of the sub-contract conditions).

- Clause 7.7.4 of the sub-contract conditions sets out the sub-contractor's rights in respect of future payment, as follows. Upon the completion of the sub-contract works and the making good of defects (i.e. by the contractor or by a person appointed by the contractor), or upon the termination of the contractor's employment under the main contract, whichever occurs first, the sub-contractor may apply to the contractor for payment.

---

[5] *British Eagle International Air Lines* v. *Compagnie Nationale Air France* [1975] 1 WLR 758.

The contractor need take no action until an application for payment is received from the sub-contractor that is compliant with the above time restriction.

- When such an application for payment is received by the contractor, the contractor is to pay the sub-contractor the value of any work executed or goods and materials supplied by the sub-contractor to the extent not included in previous payments. Without prejudice to any other rights that the contractor may have, the contractor may deduct from the sum determined:

    — the amount of any direct loss and/or damage caused to the contractor as a result of the termination (e.g. the cost of completing the works, and other delay costs, etc.);
    — any other amount payable to the contractor under the sub-contractor.

    In that the amount due to the contractor exceeds the amount due to the sub-contractor the balance is recoverable from the sub-contractor as a debt.

MPSub

If a sub-contractor's employment is terminated under clause 40 of the sub-contract conditions or under clause 43.1 of the sub-contract conditions (but only where the termination under clause 38 of the main contract is as a result of a material breach by the sub-contractor in respect of which the contractor has previously issued a notice under clause 40.1 of the sub-contract conditions), then the sub-contractor is not to remove any materials, plant or equipment from site unless expressly permitted to do so by the contractor, and the sub-contractor is to provide to the contractor all design documents prepared in connection with the sub-contract.

The contractor may make such other arrangements as he considers appropriate to complete the sub-contract works, and is not obliged to make any further payment to the sub-contractor other than as set out below.

In line with clause 40.5 of the sub-contract conditions, when the project has been completed and the defects rectification provision of the main contract fulfilled the contractor is to issue a payment advice setting out:

- the additional (i.e. extra-over) cost incurred by the contractor in completing the sub-contract works;
- any loss and/or damage suffered by the contractor and for which the sub-contractor is liable, whether arising as a consequence of the termination or otherwise arising out of the sub-contract.

The contractor may issue to the sub-contractor interim payment advices in respect of amounts that are due to him as detailed above, as and when costs are incurred.

The payment due date of any such payment advice shall be the date of issue of same, and the final date for payment is 21 days thereafter.

If a sub-contractor's employment is terminated under clause 42 of the sub-contract conditions or under clause 43.3 of the sub-contract conditions (where the termination is under clause 40 of the main contract), then the sub-contractor is to remove all of his materials, plant or equipment from site without delay, and the sub-contractor is to provide to the contractor all design documents prepared in connection with the sub-contract.

The sub-contractor is to prepare an account setting out:

- his valuation of his entitlements under the sub-contract at the date of termination;
- his reasonable costs of removal from the site as a consequence of the termination.

The sub-contractor is to issue to the contractor a statement that compares the total amount included in the account prepared by the sub-contractor (as above) with the total previous payments received by the contractor to show the balance that is to be paid by one party to the other.

The payment due date of any such statement shall be the date of issue of same, and the final date for payment is 21 days thereafter.

If a sub-contractor's employment is terminated under clause 43.2 of the sub-contract conditions, then the sub-contractor is to remove all of his materials, plant or equipment from site without delay.

The sub-contractor is to prepare an account setting out:

- his valuation of his entitlements under the sub-contract at the date of termination;
- his reasonable costs of removal from the site as a consequence of the termination;
- any loss and/or damage suffered by the sub-contractor and for which the contractor is liable, whether arising as a consequence of the termination or otherwise arising out of the sub-contract.

The sub-contractor is to issue to the contractor a statement that compares the total amount included in the account prepared by the sub-contractor (as above) with the total previous payments received by the contractor to show the balance that is to be paid by one party to the other.

The payment due date of any such statement shall be the date of issue of same, and the final date for payment is 21 days thereafter.

In respect of a named specialist, either before or as soon as possible after the termination of the sub-contract, the contractor is to notify the employer of his proposed replacement named specialist, and that replacement named specialist shall be appointed by the contractor within seven days of the date of the notification, unless the employer raises a reasonable objection within seven days of the date of the notification. The contractor's liabilities and obligations under the main contract shall not be affected by the appointment of a replacement named specialist, and the contractor shall be entirely responsible for any delay or additional cost incurred as a consequence of the appointment of a replacement named specialist.

ICSub

As DBSub above.

ICSub/D

As DBSub above.

ICSub/NAM

As DBSub above.

In addition, and in respect of named sub-contractors only, where, under paragraph 10.2 of schedule 2 of the main contract, the contractor becomes obliged to seek to recover from a defaulting named sub-contractor the additional costs or losses incurred by the employer as a result of instructions issued following the named sub-contractor's termination, then, under clause 7.7.4 of the sub-contract (ICSub/NAM/C) there is an express obligation on the named sub-contractor to pay such sums to the contractor. The contractor in turn must then account to the employer.

To avoid a technical legal defence to the contractor's claim, the named sub-contractor undertakes not to contend that, by virtue of paragraph 8 or 9 of schedule 2 (of the main contract), the contractor has suffered no loss.

MWSub/D and ShortSub

In line with clause 14.3 of the sub-contract conditions, if the sub-contractor's employment is terminated (for reasons other than for the termination of the contractor's employment under the main contract), the sub-contractor shall immediately leave the site.

The contractor is not obliged to make any further payments to the sub-contractor until after completion of the sub-contract works and the making good of defects, but there will be due to the contractor from the sub-contractor the additional costs of completing the sub-contract works including any expenses and direct loss and/or damage incurred by the contractor as a result of the termination.

In line with clause 15 of the sub-contract conditions, if the sub-contractor's employment is terminated because the contractor's employment has been terminated under the main contract, the sub-contractor shall immediately leave the site.

Where the termination of the contractor's employment under the main contract is not as a result of a breach of the sub-contract by the sub-contractor, the sub-contractor is entitled (under clause 15.2.1 of the sub-contract conditions) to be paid the value of the sub-contract works properly carried out and the reasonable cost of removal from site (less sums already paid).

Where the termination of the contractor's employment under the main contract is not as a result of a breach of the sub-contract by the sub-contractor, but is because the contractor is in default or because the contractor is insolvent or has had an administrative receiver appointed or had a winding-up order or the like made against him, the sub-contractor is entitled (under clause 15.2.2 of the sub-contract conditions) to be paid the value of the sub-contract works properly carried out and the reasonable cost of removal from site (less sums already paid) plus any direct loss and/or damage caused to the sub-contractor as a result of the contractor's employment under the main contract being terminated.

SubSub

All as MWSub/D and ShortSub but amended to allow for a sub-contract/sub-subcontract relationship rather that a contractor/sub-contractor relationship.

What reasons give the sub-contractor (or a sub-subcontractor) the right to terminate his employment under the JCT sub-contracts?

DBSub

The sub-contractor may terminate his employment under the sub-contract conditions for three principal reasons:

- Default by the contractor.
- Termination of the main contractor's employment under the main contract (as clause 7.9 of the sub-contract conditions).
- Insolvency of the contractor.

MPSub

Under clause 41.1 of the sub-contract conditions, the sub-contractor may terminate his employment under the sub-contract if the contractor commits a material breach and fails to remedy that material breach following notice being given.

A material breach by the contractor is defined (under clause 1 of the sub-contract conditions) as being:

- failure to issue a payment advice in the manner required by the sub-contract;
- failure to insure as required by clause 36.3 of the sub-contract conditions;
- failure to make payment in the manner required by any payment advice issued under the sub-contract;
- any repudiatory breach of the sub-contract;
- a breach of the CDM Regulations;
- a breach of the requirements of CIS (the Construction Industry Scheme).

Under clause 41.3 of the sub-contract conditions, in the event that a contractor becomes insolvent the sub-contractor may at any time by notice to the contractor terminate his employment under the sub-contract.

Under clause 42.1 of the sub-contract conditions, in the event that the sub-contract works or a substantial proportion of it is suspended for the period stated in the sub-contract as a result of an event listed under clause 42.1, the sub-contractor may terminate his own employment under the sub-contract following notice being given.

ICSub

As DBSub above.

ICSub/D

As DBSub above.

ICSub/NAM

As DBSub above.

MWSub/D and ShortSub

The MWSub/D and the ShortSub do not give a sub-contractor the express right to terminate his own employment.

SubSub

The SubSub does not give a sub-subcontractor the express right to terminate his own employment.

What is deemed to be default by the contractor (or of a sub-contractor in the case of the SubSub)?

DBSub

The specified defaults are dealt with under clause 7.8.1 of the sub-contract conditions.

These specified defaults are:

- The contractor without reasonable cause wholly or substantially suspends the carrying out of the main contract works (as clause 7.8.1.1 of the sub-contract conditions).

  This ground for termination does not apply where the contractor has reasonable cause for suspending the carrying out of the main contract works.

- The contractor without reasonable cause fails to proceed regularly and diligently with the main contract works so that the reasonable progress of the sub-contract works is seriously affected (as clause 7.8.1.2 of the sub-contract conditions).

  Interestingly, it should be noted that a specified default for a sub-contractor is that he 'fails to proceed regularly and diligently with the Sub-contract Works, etc.', but, in respect of the contractor, the specified default is amended to read that 'without reasonable cause it fails to proceed regularly and diligently with the Main Contract Works, etc.' What actual effect the additional words 'without reasonable cause' have, is not fully clear.

- The contractor fails to make payment in accordance with the sub-contract (as clause 7.8.1.3 of the sub-contract conditions).

  This is an interesting and quite onerous clause, particularly given that the sub-contractor has the express right to interest on late payments (as clause 4.10.5 and clause 4.12.4 of the sub-contract conditions), and the right to suspend performance (as clause 4.11 of the sub-contract conditions) in the appropriate cases.

- The contractor fails to comply with the CDM Regulations (as clause 7.8.1.4 of the sub-contract conditions).

  It is submitted that minor or trifling failures would not be considered to be a default, given that a notice of default and/or termination of the sub-contractor's employment is not to be given unreasonably or vexatiously.

MPSub

A specified default is a material breach which is defined (under clause 1 of the sub-contract conditions) as being:

- failure to issue a payment advice in the manner required by the sub-contract;
- failure to insure as required by clause 36.3 of the sub-contract conditions;
- failure to make payment in the manner required by any payment advice issued under the sub-contract;
- any repudiatory breach of the sub-contract;
- a breach of the CDM Regulations;
- a breach of the requirements of CIS (the Construction Industry Scheme).

ICSub

As DBSub above.

ICSub/D

As DBSub above.

ICSub/NAM

As DBSub above.

MWSub/D and ShortSub

These sub-contracts do not specify any defaults by a contractor.

SubSub

The SubSub does not specify any defaults by a sub-contractor.

What happens when a contractor (or a sub-contractor in the case of the SubSub) defaults or an event occurs that entitles the sub-contractor to terminate his own employment under the sub-contract?

DBSub

The basic process that is followed is:

- The contractor defaults.

- The sub-contractor gives a notice to the contractor specifying the default.

- If the contractor continues a specified default for ten calendar days (excluding any public holidays) after receipt of the sub-contractor's notice, then the sub-contractor may on, or within ten calendar days (excluding any public holidays) from the expiry of the initial notice period, issue a further notice terminating the sub-contractor's employment under the sub-contract.

- If the sub-contractor does not give the further notice referred to above but the contractor repeats a specified default (whether previously repeated or not), then, upon or within a reasonable time after such repetition, the sub-contractor may by notice to the contractor terminate the sub-contractor's employment under the sub-contract.

- The notices given above must be in writing, and must be given by actual, special or recorded delivery.

- Where the notice is given by special or recorded delivery it shall, subject to proof to the contrary, be deemed to have been received on the second business day after the date of posting (as clause 7.2.3 of the sub-contract conditions).

  The words 'subject to proof to the contrary' are important, because this means that (subject to proof) a notice could have been actually received prior to the second business day after the date of posting[6].

- Termination shall take effect on receipt of the relevant notice (as clause 7.2.2 of the sub-contract conditions).

- Notice of termination of the sub-contractor's employment shall not be given unreasonably or vexatiously (as clause 7.2.1 of the sub-contract conditions).

MPSub

The basic process that is followed (in respect of matters noted under clause 41.1) is:

- The contractor commits a material breach.
- The sub-contractor gives a notice to the contractor specifying the material breach.
- If the contractor continues a material breach for 19 days after receipt of the sub-contractor's notice[7], then the sub-contractor may on, or within 14 days from the expiry of the initial notice period, issue a further notice terminating the sub-contractor's employment under the sub-contract.
- The notices given above must be in writing, and must be given by actual delivery, registered post or recorded delivery (as clause 5.2 of the sub-contract conditions).
- The notices take effect upon delivery.

It should be noted that there is no provision under the MPSub for the termination by the sub-contractor of his own employment by the simple issue of a 'termination' notice in the case where a sub-contractor does not give notice within 14 days but the contractor subsequently repeats a material breach. Should such a situation arise, the sub-contractor would need to issue a further notice and allow a period of 19 days to pass whilst the contractor does not remedy a material breach before a 'termination' notice can be issued.

The basic process that is followed (in respect of matters noted under clause 42.1) is:

- If the carrying out of the sub-contract works or a substantial proportion of it is suspended for the period stated in the sub-contract particulars (which, if no

---

[6] *Lafarge (aggregates) Ltd* v. *Newham Borough Council* [2005] (unreported) QBD, 24 June 2005.

[7] The MPSub guide notes that this notice period reflects a desire to ensure that actions under the sub-contract can be coordinated with similar remedies under the main contract. The period of 19 days allows sufficient time for the contractor to serve a notice on the employer in order that the employer's material breach might be remedied before the sub-contractor may terminate his employment under the sub-contract.

period is specifically stated, is, by default 13 weeks) as a consequence of *force majeure*, the occurrence of any specified peril (as defined under clause 1 of the sub-contract conditions) or hostilities involving the United Kingdom, or the use or threat of terrorism, then either party may issue to the other party a notice specifying the circumstances of the suspension and stating that if the specified circumstances continue for a further 19 days he may terminate the sub-contractor's employment under the sub-contract.

- If the circumstances of the suspension continue for 19 days, then the party that issued the initial notice may issue a further notice at any time within the subsequent 14 days to terminate the sub-contractor's employment under the sub-contract.

ICSub

As DBSub above.

ICSub/D

As DBSub above.

ICSub/NAM

As DBSub above.

MWSub/D and ShortSub

These sub-contracts do not give a sub-contractor the express right to terminate his own employment.

SubSub

The SubSub does not give a sub-subcontractor the express right to terminate his own employment.

What does insolvency of the contractor mean?

DBSub

Clause 7.10.1 of the sub-contract conditions states that if the contractor is insolvent or makes any proposal, gives notice of any meeting or becomes the subject of any proceedings relating to any of the matters referred to in clause 7.1 of the sub-contract conditions, the contractor shall immediately inform the sub-contractor.

Under clause 7.1 of the sub-contract conditions a contractor is insolvent if:

  .1 he enters into an arrangement, compromise or composition in satisfaction of his debts (excluding a scheme of arrangement as a solvent company for the purposes of amalgamation or reconstruction); or

  .2 without a declaration of insolvency, he passes a resolution or makes a determination that he be wound up; or

.3 he has a winding up order or bankruptcy order made against him; or

.4 he has appointed to him an administrator or administrative receiver; or

.5 he is the subject of an analogous arrangement, event or proceedings in any other jurisdiction; or

.6 (additionally in the case of partnerships) each partner is the subject of an individual arrangement or any other event or proceedings referred to in clause 7.1.1 to 7.1.5 of the sub-contract conditions.

MPSub

Clause 1 of the sub-contract conditions defines 'Insolvent' as being when a party:

- makes a composition or arrangement with his creditors, or becomes bankrupt;
- makes a proposal for a voluntary arrangement for a composition of debts or scheme of arrangement to be approved in accordance with the Companies Act 1985 or Insolvency Act 1986, as the case may be;
- has a provisional liquidator appointed;
- has a winding up order made or bankruptcy order made against him;
- passes a resolution for voluntary winding up (except for the purposes of amalgamation or reconstruction); or
- under the Insolvency Act 1986 has an administrator or an administrative receiver appointed.

ICSub

As DBSub above.

ICSub/D

As DBSub above.

ICSub/NAM

As DBSub above.

MWSub/D, ShortSub and SubSub

Insolvency is not defined in these sub-contracts.

What happens when the contractor becomes insolvent?

Clause 7.10.2 of the sub-contract conditions states that if the contractor is insolvent (as defined), then:

- The sub-contractor is entitled by notice to the contractor to terminate the sub-contractor's employment under the sub-contract.
- If, before the contractor became insolvent, the sub-contractor had already issued a notice of 'default' then the sub-contractor may issue the 'termination' notice immediately that the contractor becomes insolvent.

- If, however, before the contractor became insolvent the sub-contractor had *not* issued a notice of 'default' then, when the contractor becomes insolvent, the sub-contractor may either:

  — issue a notice of termination of the sub-contractor's employment (in line with clause 7.10.2.1 of the sub-contract conditions); or
  — issue a notice of default pursuant to clause 7.8.1 of the sub-contract conditions.

Whichever of these notices the sub-contractor decides to issue, clause 7.10.2.1 of the sub-contract conditions makes it clear that the sub-contractor is not to exercise his rights to actually terminate his employment under the sub-contract prior to the expiry of a period of three weeks (or such further periods as the parties may agree) from the date upon which the contractor became insolvent.

The reason for this three-week period is to allow a period following the happening of a specified insolvency event, during which the sub-contract arrangements will remain in force while the parties attempt to agree a way forward.

In line with clause 7.10.2.2 of the sub-contract conditions, the sub-contractor's obligations to carry out and complete the sub-contract works (including the sub-contractor's design works) are immediately suspended pending termination of the sub-contractor's employment; and such suspension shall, for the purposes of a relevant sub-contract event for extension of time and for the purposes of a relevant sub-contract matter, be deemed to be a default of the contractor.

MPSub

Clause 41.3 of the sub-contract conditions states that if the contractor becomes insolvent, the sub-contractor may at any time by notice to the contractor terminate the sub-contractor's employment under the sub-contract.

If a sub-contractor's employment is terminated under clause 41 of the sub-contract conditions, then the sub-contractor is to remove all of his materials, plant or equipment from site without delay.

The sub-contractor is to prepare an account setting out:

- his valuation of his entitlements under the sub-contract at the date of termination;
- his reasonable costs of removal from the site as a consequence of the termination.
- any loss and/or damage suffered by the sub-contractor and for which the contractor is liable, whether arising as a consequence of the termination or otherwise arising out of the sub-contract.

The sub-contractor is to issue to the contractor a statement that compares the total amount included in the account prepared by the sub-contractor (as above) with the total previous payments received by the contractor to show the balance that is to be paid by one party to the other.

The payment due date of any such statement shall be the date of issue of same, and the final date for payment is 21 days thereafter.

ICSub

As DBSub above.

ICSub/D

As DBSub above.

ICSub/NAM

As DBSub above.

MWSub/D, ShortSub and SubSub

These sub-contracts do not contain any express provisions for what happens when a contractor becomes insolvent.

What are the consequences of the sub-contractor terminating his own employment?

DBSub

If the sub-contractor's employment is terminated by the sub-contractor for any of the reasons outlined above, the following consequences arise:

- The contractor is not required to make any further payment or release any outstanding retention, other than as provided for under clause 7.11.1 of the sub-contract conditions.

- The sub-contractor shall with all reasonable dispatch remove or procure the removal from the site of any temporary buildings, plant, tools and equipment belonging to the sub-contractor and the sub-contractor's persons together with all goods and materials including site materials (except only for those site materials that will be valued in the sub-contractor's concluding account and which will be paid for by the contractor (refer to clause 7.11.5 of the sub-contract conditions)).

- Where applicable, and without charge, the sub-contractor is to provide the contractor with three copies of all the sub-contractor's design documents then prepared. This consequence is required irrespective of any request by the contractor.

- In line with clause 7.11.3 of the sub-contract conditions, the sub-contractor shall with reasonable dispatch prepare and submit to the contractor an account setting out:

  — The total value of the work properly executed at the date of termination ascertained and valued in line with the sub-contract conditions.
  — The total value of any other amounts due to the sub-contractor at the date of termination ascertained and valued in line with the sub-contract conditions, for example any applicable fluctuation payments.

  Also, given that the next item on the list only relates to any ascertained direct loss and/or expense, these 'other amounts' would presumably also include for other loss and/or expense items claimed by the sub-contractor, but not ascertained by the contractor.

— Any sums ascertained in respect of direct loss and/or expense under clause 4.19 of the sub-contract conditions.
— The reasonable cost of removal from the site of any temporary buildings, plant, tools and equipment belonging to the sub-contractor and the sub-contractor's persons together with all goods and materials including site materials, as appropriate.
— The cost of materials and goods (including sub-contract site materials) properly ordered for the sub-contract works for which the sub-contractor has already made payment, or for which the sub-contractor is legally liable to make payment.
— Any direct loss and/or damage caused to the sub-contractor by the termination of his employment. This could, of course, include, if applicable, the loss of future overheads' contribution and the loss of future profit.

In respect of the sub-contractor's recovery of direct loss and/or damage caused to the sub-contractor by the termination of his employment, this will *not* apply where:

— The termination of the sub-contractor's employment is due to the termination of the contractor's employment (by either the employer or the contractor) by reason of those events covered by clause 8.11.1 of the main contract conditions (refer to clause 7.11.4.1 of the sub-contract conditions).
— The termination of the sub-contractor's employment follows the employer's option in terminating the contractor's employment in the situation where there was a non-availability of insurance cover for terrorism (refer to clause 6.10.2.2 of the main contract conditions) (as clause 7.11.4.2 of the sub-contract conditions).
— The termination of the sub-contractor's employment is due to the termination of the contractor's employment (by either the employer or the contractor) because it is considered just and equitable to do so following loss or damage to the works occasioned by any of the risks covered by paragraph C.4.4 of schedule 3 to the main contract conditions) (as clause 7.11.4.2 of the sub-contract conditions).

● The contractor shall pay to the sub-contractor the amount properly due in respect of the account submitted by the sub-contractor (as detailed above), without any deduction for retention, but after taking into account amounts already paid, within *28 days* of its submission by the sub-contractor to the contractor.
Payment by the contractor for any materials and goods within the account is made on the basis that the said materials and goods become the property of the contractor upon payment being made (refer to clause 7.11.5 of the sub-contract conditions).

MPSub

If a sub-contractor's employment is terminated under clause 41 of the sub-contract conditions, then the sub-contractor is to remove all of his materials, plant or equipment from site without delay.

The sub-contractor is to prepare an account setting out:

- his valuation of his entitlements under the sub-contract at the date of termination;
- his reasonable costs of removal from the site as a consequence of the termination;
- any loss and/or damage suffered by the sub-contractor and for which the contractor is liable, whether arising as a consequence of the termination or otherwise arising out of the sub-contract.

The sub-contractor is to issue to the contractor a statement that compares the total amount included in the account prepared by the sub-contractor (as above) with the total previous payments received by the contractor to show the balance that is to be paid by one party to the other.

The payment due date of any such statement shall be the date of issue of same, and the final date for payment is 21 days thereafter.

If a sub-contractor's employment is terminated under clause 42 of the sub-contract conditions, then the sub-contractor is to remove all of his materials, plant or equipment from site without delay, and the sub-contractor is to provide to the contractor all design documents prepared in connection with the sub-contract.

The sub-contractor is to prepare an account setting out:

- his valuation of his entitlements under the sub-contract at the date of termination;
- his reasonable costs of removal from the site as a consequence of the termination.

The sub-contractor is to issue to the contractor a statement that compares the total amount included in the account prepared by the sub-contractor (as above) with the total previous payments received by the contractor to show the balance that is to be paid by one party to the other.

The payment due date of any such statement shall be the date of issue of same, and the final date for payment is 21 days thereafter.

ICSub

As DBSub above, except that where the termination of the sub-contractor's employment is due to the termination of the contractor's employment (by either the employer or the contractor) because it is considered just and equitable to do so following loss or damage to the works occasioned by any of the risks covered by paragraph C.4.4, that paragraph is contained in schedule 1 rather than schedule 3 to the main contract conditions.

ICSub/D

All as ICSub, above.

ICSub/NAM

All as ICSub, above.

MWSub/D and ShortSub

These sub-contracts do not give a sub-contractor the express right to terminate his own employment.

SubSub

The SubSub does not give a sub-subcontractor the express right to terminate his own employment.

If a sub-contractor's employment is terminated for any reason can it subsequently be reinstated?

DBSub

Clause 7.3.2 of the sub-contract conditions makes it clear that, irrespective of the grounds of termination, the sub-contractor's employment may at any time be reinstated on such terms as the parties may agree. It should be noted, however, that section 178 of the Insolvency Act 1986 gives a liquidator or a receiver the statutory right to disclaim unprofitable contracts, and this may close this avenue in certain circumstances.

MPSub

The MPSub makes no express provision for the reinstatement of a sub-contractor's employment after it has been terminated.

ICSub

As DBSub above.

ICSub/D

As DBSub above.

ICSub/NAM

As DBSub above.

MWSub/D and ShortSub

These sub-contracts make no express provision for the reinstatement of a sub-contractor's employment after it has been terminated.

SubSub

The SubSub makes no express provision for the reinstatement of a sub-subcontractor's employment after it has been terminated.

Chapter 13
Settlement of Disputes

How are disputes settled?

It is an unfortunate reality of life in the construction industry that disputes will arise.

Certain aspects of the construction process, and certain aspects of the construction industry culture, make it more likely that disputes will arise than perhaps in other industries.

Some of the more obvious factors that increase the probability of disputes arising are new or unfamiliar materials or construction techniques; multi-party involvement in projects; uncertainty regarding the physical and commercial environment within which the works will be performed; low margins; tender errors; incorrect allocation of risk; and incomplete design before construction, causing information and documentation to be issued late.

The vast majority of disputes within the construction industry are resolved (often on a daily basis) by negotiation. In those relatively few occasions where disputes cannot be resolved by negotiation, the dispute must be referred to a more formal dispute resolution process.

The reality that disputes may need to be resolved by some formal dispute resolution process is recognised within the sub-contracts being considered in this book, and these methods are mediation, adjudication, arbitration and legal proceedings.

What cannot be referred to a binding formal dispute resolution method?

DBSub

When the final payment notice has been issued, clause 1.9 of the sub-contract conditions notes that (except in the case of fraud) it is deemed to be conclusive evidence that the quality of any materials or goods or standard of an item of workmanship which had expressly been for the approval of the employer is to the reasonable satisfaction of the employer, and is also deemed to be conclusive evidence that the valuation of the sub-contract works, the sub-contractor's entitlement to extensions of time, the sub-contractor's entitlement to direct loss and expense, or the contractor's reimbursement entitlement have been taken fully into account within the final payment notice.

The effect of the above is that if it remains unchallenged the final payment notice cannot be opened up in any future adjudication, arbitration or litigation. There is only a very limited period (of ten days) for a challenge to be made. Therefore, if a sub-contractor disagrees with the final payment notice he should commence adjudication, arbitration or other proceedings within ten days of receipt of the notice if he wishes to avoid the notice being deemed conclusive evidence as detailed above.

It should be noted that if the proceedings only relate to one issue (for example, the sub-contractor's entitlement to loss and expense) then the final payment notice *will still be taken as being conclusive evidence of all other matters.*

In the situation where either party has already commenced proceedings before the final payment notice is issued, unless the notice is to be taken as being conclusive evidence of all matters outlined above, either or both parties will have to take a further step in such proceedings within 12 months from or after the date the final payment notice was given.

At the conclusion of such proceedings, the final payment notice will be taken as being conclusive evidence of the matters outlined above, subject only to the terms of any decision, award or judgment in or settlement of those proceedings.

If an adjudicator gives his decision on a dispute or difference after the date of the final payment notice and either party then wishes to have the dispute determined by arbitration or legal proceedings, such proceedings must be commenced within 28 days of the date on which the adjudicator gives his decision if the adjudicator's decision is not to be final in respect of the matters referred to him.

MPSub

When the final payment advice has been issued, clause 31.7 of the sub-contract conditions notes that it is final and binding on the parties in relation to the amounts due to the sub-contractor under or in connection with the sub-contract, including in respect of claims arising from breach of contract, breach of statutory duty, negligence or otherwise.

The effect of the above is that if it remains unchallenged the final payment advice cannot be opened up in any future adjudication or litigation. There is only a very limited period (of 28 days) for a challenge to be made. Therefore, if a sub-contractor disagrees with the final payment notice he should commence adjudication or litigation within 28 days of the final payment advice being issued.

ICSub

When the final payment notice has been issued, clause 1.9 of the sub-contract conditions notes that (except in the case of fraud) it is deemed to be conclusive evidence that the quality of any materials or goods or standard of an item of workmanship which had expressly been for the approval of the architect/contract administrator is to the reasonable satisfaction of the architect/contract administrator, and is also deemed to be conclusive evidence that the valuation of the sub-contract works, the sub-contractor's entitlement to extensions of time, the sub-contractor's entitlement to direct loss and expense, or the contractor's reimbursement entitlement have been taken fully into account within the final payment notice.

The effect of the above is that if it remains unchallenged, the final payment notice cannot be opened up in any future adjudication, arbitration or litigation. There is only a very limited period (of ten days) for a challenge to be made. Therefore, if a sub-contractor disagrees with the final payment notice he should commence adjudication, arbitration or other proceedings within ten days of receipt of the notice if he wishes to avoid the notice being deemed conclusive evidence as detailed above.

It should be noted that if the proceedings only relate to one issue (for example, the sub-contractor's entitlement to loss and expense) then the final payment notice *will still be taken as being conclusive evidence of all other matters.*

If an adjudicator gives his decision on a dispute or difference after the date of the final payment notice and either party then wishes to have the dispute determined by arbitration or legal proceedings, such proceedings must be commenced within 28 days of the date on which the adjudicator gives his decision if the adjudicator's decision is not to be final in respect of the matters referred to him.

ICSub/D

All as ICSub above.

ICSub/NAM

All as ICSub above.

MWSub/D and ShortSub

There is no express restriction in respect of commencing adjudication, arbitration or litigation in relation to a final payment notice. However, it could be argued that because of clause 7.3 of the sub-contract conditions, which says that the sub-contractor shall assume the liabilities of the contractor under the main contract, the effect of the final payment notice on a contractor may be 'passed down' to the sub-contractor through clause 7.3. Because of this possibility, the sub-contractor would be wise to assure that he commenced any formal dispute resolution action timeously.

SubSub

There is no express restriction in respect of commencing adjudication, arbitration or litigation in relation to a final payment notice. However, it could be argued that because of clause 7.3 of the sub-subcontract conditions, which says that the sub-subcontractor shall assume the liabilities of the sub-contractor under the sub-contract, the effect of the final payment notice on a sub-contractor may be 'passed down' to the sub-subcontractor through clause 7.3. Because of this possibility, the sub-subcontractor would be wise to assure that he commenced any formal dispute resolution action timeously.

Mediation

What is mediation?

Mediation is a method of settling disputes or differences in which a third party, an independent and impartial person called a mediator, assists both parties to reach an agreement which each party considers is acceptable.

In the UK, mediators are normally facilitative, which means that they use their skills to help the parties to reach their own mutually acceptable settlement, rather than being evaluative (i.e. evaluating the liability and/or the quantum of the dispute) and then effectively imposing a decision on the parties.

During the mediation process, mediators encourage parties to consider their interests and needs, and not merely to concentrate on their legal rights.

Any negotiated settlement between the parties in a mediation is not binding until the parties draw up a binding settlement agreement.

What is the mediation process?

Mediation is a very flexible process which is normally adapted to suit the particular circumstances, dispute or difference, and parties.

However, typically the stages of the mediation process might include:

- In advance of the mediation hearing
  Brief written summaries of the dispute or difference submitted to the mediator by each party.

- At the mediation hearing (which fairly typically lasts half of a day, or one day)

  — The signing of the mediation agreement in which the parties confirm their participation in the process and their understanding that information given in the mediation will be on a without prejudice basis and is to be kept confidential, unless otherwise agreed. The parties also confirm that they have the authority to settle the dispute or difference.
  — An initial joint meeting between the parties and the mediator with presentations made by the parties setting out the parties' respective positions.
  — Private sessions (sometimes referred to as caucus sessions) in which the mediator meets privately with each party in turn. The mediator goes between the parties gaining a better understanding of the issues and endeavours to identify potential settlement options. Any information given by the parties in these sessions remains confidential, unless the party in question agrees that the information given may be passed to the other party.
  — Further joint meetings are held as appropriate, during which face-to-face negotiations can take place between the parties, with the mediator acting as facilitator, and where agreements can be reached.

- The mediation settlement agreement
  Where an agreement is reached it is normal for a legally enforceable settlement agreement to be signed by the parties (not by the mediator). In fact, other than possibly for general advice, the mediator plays no part in producing the settlement agreement, and this document should be produced by the parties or by the parties' legal representatives.

What are the advantages of the mediation process?

The main advantages of mediation, as compared to other dispute resolution processes, are its speed, its flexibility and its relatively low cost.

However, another often overlooked advantage is that mediation allows for non-legal interests and needs (for example, the promise of future work) to be taken into account in any agreement reached between the parties.

Mediation uses a non-adversarial approach which is usually more conducive to a settlement being reached, and which allows relationships to be maintained or rebuilt.

How popular is the use of mediation?

The use of mediation is growing, and this is partly as a result of:

- parties seeking simpler and cheaper means of resolving disputes; and
- encouragement from the courts which now require parties to make every effort to resolve their dispute using ADR (Alternative Dispute Resolution) before it is referred to court.

How is mediation incorporated into the JCT sub-contracts?

DBSub

Clause 8.1 of DBSub/C states: 'the Parties may by agreement seek to resolve any dispute or difference arising under this Sub-contract through mediation.'

MPSub

Mediation is dealt with under clauses 44.1 and 45 of MPSub.

Clause 44.1 simply states that, where the parties agree to do so, a dispute or difference may be submitted to mediation in accordance with the provisions of clause 45.

Clause 45 provides for the following:

- Either party may identify to the other party a dispute or difference that he considers to be capable of resolution by mediation.
- The other party is to indicate, within seven days of the identification being made, whether or not he consents to the proposal to use mediation.
- The objective of mediation is to reach a binding agreement.
- The mediator or the selection method for the mediator is to be agreed between the parties.

ICSub

Clause 8.1 of ICSub/C says: 'the Parties may by agreement seek to resolve any dispute or difference arising under this Sub-contract through mediation.'

ICSub/D

All as ICSub above.

ICSub/NAM

All as ICSub above.

MWSub/D, ShortSub and SubSub

All as ICSub above, except that the applicable clause number is 16.1, not 8.1.

Under the JCT sub-contracts are the parties obliged to go to mediation to resolve their disputes or differences?

None of the JCT sub-contracts considered in this book makes it mandatory that parties use mediation in an attempt to resolve their disputes or differences.

Mediation remains as a voluntary process, although the parties need to be aware that there could be cost sanctions imposed by a court if a party proceeded to court without first trying to settle the dispute or difference by mediation or by some other ADR process. This possibility has been reinforced by virtue of a CPR practice direction issued in April 2006, which makes it mandatory upon a court to take into account, when considering the issue of costs, whether the parties have given proper consideration to the use of ADR (which may include discussion and negotiation, early neutral evaluation by an independent third party and mediation).

In respect of construction work, the impact of mediation is often lessened because of the CPR (Civil Procedure Rules) requirement that the pre-action protocol for construction and engineering disputes (as dealt with later in this chapter) is followed before a dispute is referred to court.

Adjudication

What is statutory adjudication?

Statutory adjudication is a statutory procedure by which any party to a construction contract has a right to have a dispute decided by an adjudicator. Adjudication is not arbitration. The adjudication process is a means by which an independent third party (i.e. the adjudicator) makes a (relatively) quick decision when the parties to a contract are in disagreement.

Why have adjudication provisions in JCT sub-contracts?

The Housing Grants, Construction and Regeneration Act 1996 (the Construction Act), Part II provides under s.108(1) that: 'A party to a construction contract has the right to refer a dispute arising under the contract for adjudication under a procedure complying with this section. For this purpose "dispute" includes any difference'.

The construction contract must be in writing.

Therefore, except for in certain circumstances (e.g. the construction contract is not in writing, the contract is with a residential occupier that has not received professional advice before entering into the contract), all parties to a construction contract, as defined under s.104 of the Construction Act, have the right to refer a dispute arising under the contract for adjudication under a procedure complying with s.108(1) of the Construction Act.

If there are not express terms in the written construction contract regarding adjudication then a term to that effect will be implied into the contract.

Obviously, parties generally prefer to have some control over the adjudication provisions that will apply to their contract, and rather than rely on an implied term prefer to incorporate their own conditions, which is why there are adjudication provisions in the JCT sub-contracts.

How are the adjudication provisions incorporated into the JCT sub-contracts?

DBSub

Article 4 of the sub-contract agreement states that if any dispute or difference arises under the sub-contract either party may refer it to adjudication in accordance with clause 8.2 of the sub-contract conditions.

Clause 8.2 of the sub-contract conditions states that the Scheme rules shall apply, subject to the following conditions:

(1) Clause 8.2.1. For the purposes of the Scheme the adjudicator shall be the person (if any) stated in item 15 of the sub-contract particulars, and the nominating body of the adjudicator shall be that stated in the same sub-contract particulars items.
(2) Clause 8.2.2. Where the dispute or difference is or includes a dispute or difference relating to clause 3.11.3 (i.e. instructions to open up for inspection or to test any further similar non-compliant works) and as to whether a direction issued under that clause is reasonable in all of the circumstances, then the adjudicator to decide such a dispute shall (where practicable) be an individual with appropriate expertise and experience in the specialist area or discipline relevant to the direction or issue in dispute. Where the adjudicator does not have the appropriate expertise and experience, the adjudicator shall appoint an independent expert with such expertise and experience to advise and report in writing on whether or not the direction under clause 3.11.3 is reasonable in all of the circumstances.

MPSub

Adjudication is incorporated into MPSub by way of clause 44.2 which confirms that the provisions of clause 46 apply. Clause 46.1 confirms that the Scheme rules apply, subject only to the condition (under clause 46.3) that where a dispute arises as to whether an instruction to open up for inspection or to test any (further) similar non-compliant works is reasonable, then in the event that the appointed adjudicator does not have the appropriate expertise and experience, the adjudicator shall appoint an independent expert with such expertise and experience to advise and report in writing on the matter. In such a situation, the adjudicator is to make available to the parties his written instructions to the expert, and the report received from the expert, as soon as practicable and, in any event, prior to the issue of the adjudicator's decision.

ICSub

Article 4 of the sub-contract agreement states that if any dispute or difference arises under the sub-contract either party may refer it to adjudication in accordance with clause 8.2 of the sub-contract conditions.

Clause 8.2 of the sub-contract conditions states that the Scheme rules shall apply, subject to the following conditions:

(1) Clause 8.2.1. For the purposes of the Scheme the adjudicator shall be the person (if any) stated in item 13 of the sub-contract particulars, and the nominating body of the adjudicator shall be that stated in the same sub-contract particulars items.

(2) Clause 8.2.2. Where the dispute or difference is or includes a dispute or difference relating to clause 3.10 (i.e. instructions to open up for inspection or to test non-compliant works) and as to whether a direction issued under that clause is reasonable in all of the circumstances, then the adjudicator to decide such a dispute shall (where practicable) be an individual with appropriate expertise and experience in the specialist area or discipline relevant to the direction or issue in dispute. Where the adjudicator does not have the appropriate expertise and experience, the adjudicator shall appoint an independent expert with such expertise and experience to advise and report in writing on whether or not the direction under clause 3.10 is reasonable in all of the circumstances.

ICSub/D

All as ICSub above.

ICSub/NAM

All as ICSub above, with the exception only that the adjudicator shall be the person (if any) stated under item IT16 of the invitation to tender, and the nominating body of the adjudicator shall be that stated under the same item number.

MWSub/D, ShortSub and SubSub

Article 4 of the sub-contract states that if any dispute or difference arises under the sub-contract either party may refer it to adjudication in accordance with clause 16.2 of the sub-contract conditions. Clause 16.2 of the sub-contract conditions states that the Scheme rules shall apply.

What is the Scheme?

The Scheme is Part 1 of the Schedule to the Scheme for Construction Contracts (England and Wales) Regulations 1998 (i.e. that part of the schedule that relates to adjudication).

Those regulations were made by the secretary of state in exercise of the powers conferred on him by sections 108(6), 114 and 146(1) and (2) of the Construction Act.

The Scheme, as defined above, has 26 paragraphs split under the following headings:

- Notice of intention to seek adjudication.
- Powers of the adjudicator.

- Adjudicator's decision.
- Effects of the decision.

The Scheme sets out the adjudication provisions that apply.

What can be referred to adjudication?

Generally, any dispute or difference arising under the sub-contract may be referred by either party to adjudication.

Although paragraph 1(1) of the Scheme refers to the singular word dispute, it is generally accepted that this can mean a range of matters within a single dispute. This position is strengthened by the fact that under paragraph 20 of the Scheme, the adjudicator is to decide 'the matters in dispute', and this position is also supported by case law[1].

In the *Fastrack* v. *Morrison*[2] case, Judge Thornton QC held that the dispute encompassed 'whatever claims, heads of claim, issues, contentions or causes of action that are then in dispute which the referring party has chosen to crystallise into an adjudication reference.'

What cannot be referred to adjudication?

Although in the past the words 'dispute or difference arising under the sub-contract' was taken to mean that tortious liability matters were excluded[3], there is now some doubt regarding this position following a court of appeal case[4] which stated that the words 'arising under' should be construed widely and would normally be taken to mean the same as 'arising under and out of'.

In addition, any dispute that has already been decided by an adjudicator following an earlier adjudication action cannot be referred to adjudication (i.e. there is a bar on re-adjudication of the same dispute[5]). In line with paragraph 9(2) of the Scheme, an adjudicator must resign where the dispute is the same or substantially the same as one which has previously been referred to adjudication, and a decision has been taken in that adjudication.

However, provided that an adjudicator is dealing with a different dispute, he is not bound by an earlier adjudication decision. Therefore, for example, a claim for variations which are required to be remeasured and revalued on the final account is not the same as a similar claim under a previous interim application.

Something that has not yet crystallised into becoming a dispute cannot be referred to adjudication.

---

[1] *Fastrack Contractors Ltd* v. *Morrison Construction Ltd and Impregilo UK Ltd* [2000] BLR 168; *Balfour Kilpatrick Ltd* v. *Glauser International SA* (2000) (unreported) QBD (TCC).
[2] *Fastrack Contractors Ltd* v. *Morrison Construction Ltd and Impregilo UK Ltd* [2000] BLR 168.
[3] *Shepherd Construction* v. *Mecright Ltd* [2000] BLR 489.
[4] *Fiona Trust & Holding Corporation and Others* v. *Yuri Privalov and Others* [2007] EWCA Civ 20.
[5] *Sherwood & Casson Ltd* v. *Mackenzie Engineering Ltd* [2000] CILL 1577; *VHE Construction plc* v. *RBSTB Trust Co. Ltd* [2000] CILL 1592.

How can you tell that a dispute has crystallised?

Simply making a claim and submitting it may not be enough.

For a dispute to exist a claim has to be notified and rejected; however, a rejection may stem from the other party's refusal to consider or answer a claim[6].

When can a dispute be referred?

A dispute can be referred at any time.

Even if a dispute is the subject of existing proceedings in court it can still be referred to an adjudicator and enforced pending a decision by the court. There is no provision in the Construction Act for a stay of execution of the adjudication action pending a decision by the court[7]. Similarly, if there is an arbitration clause in the contract a stay of the adjudication will not be granted pending the outcome of the arbitration[8].

How is the adjudication process commenced?

The adjudication process is commenced by issuing a notice of adjudication to all of the parties to the contract.

What should the notice of adjudication contain?

Paragraph 1(3) of the Scheme states that the notice of adjudication should set out briefly:

(1) the nature and a brief description of the dispute and the parties involved;
(2) details of where and when the dispute has arisen;
(3) the nature of the redress sought; and
(4) the names and addresses of the parties to the contract (including, where appropriate, the addresses which the parties have specified for the giving of notices).

It is very important to note that the notice of adjudication, which must be in writing[9], sets out the jurisdiction of the adjudicator[10] (i.e. sets out the boundaries of the adjudicator's authority).

When asking for a decision from the adjudicator for a sum of money it is normal to add after the request 'or such other sum as the adjudicator may decide'. Similarly, if a time period is requested it is normal to add the words 'or such other time

---

[6] *Amec Civil Engineering Ltd* v. *the Secretary of State for Transport* [2005] CILL 2189; *Fastrack Contractors Ltd* v. *Morrison Construction Ltd and Impregilo UK Ltd* [2000] BLR 168.
[7] *A & D Maintenance and Construction Ltd* v. *Pagehurst Construction Services Ltd* (1999) CILL 1518.
[8] *Absolute Rentals Ltd* v. *Gencor Enterprises Ltd* [2000] CILL 1637.
[9] *Strathmore Building Services Ltd* v. *Colin Scott Greig* (2001) 17 Const LJ 72.
[10] *Fastrack Contractors Ltd* v. *Morrison Construction Ltd and Impregilo UK Ltd* [2000] BLR 168.

period as the adjudicator may decide'. The purpose of these caveats is to extend the adjudicator's jurisdiction to make a decision as to the actual sum due or the actual time due, thereby preventing the adjudicator from deciding on an all-or-nothing basis.

How is an adjudicator appointed, and who will be the appointed adjudicator?

Paragraph 2(1) of the Scheme makes it clear that an adjudicator can only be appointed after a notice of adjudication has been issued.

Section 108(2)(b) of Part II of the Construction Act requires the procedure being followed to provide a timetable with the object of securing the appointment of an adjudicator and referral of the dispute to him within seven days of the date of the notice of adjudication.

After the notice of adjudication has been issued, the parties may agree between themselves who shall act as adjudicator (refer to paragraph 2(1) of the Scheme).

It is relatively unusual for the parties to attempt to agree the name of an adjudicator after a notice of adjudication has been issued, and the more normal way of appointing an adjudicator is to approach a nominating body.

It is considered that an adjudicator should not normally be named in both a sub-contract and the main contract, because of concerns regarding bias (actual or perceived) by the adjudicator when dealing with a sub-contractor/contractor dispute if he or she has already dealt with, or will deal with in future, a similar dispute between the contractor and the employer.

DBSub

The adjudicator will either be the person named (if one is so named) under sub-contract particulars item 15, or will be nominated by one of the nominating bodies listed under sub-contract particulars item 15 as selected by the parties.

If the dispute or difference relates to clause 3.11.3 of the sub-contract conditions, and as to whether a direction issued thereunder is reasonable in all the circumstances, the nominating body needs to be advised that the selected adjudicator needs to be an individual (where practicable) with appropriate expertise and experience in the specialist area or discipline relevant to the direction or issue in dispute.

The choice of nominator of adjudicator under sub-contract particulars item 15 of the sub-contract agreement is the president or a vice-president or chairman or a vice-chairman of:

- The Royal Institute of British Architects
- The Royal Institution of Chartered Surveyors
- The Construction Confederation
- The National Specialist Contractors Council
- The Chartered Institute of Arbitrators

When completing sub-contract particulars item 15 the parties should delete all but one of the five above listed nominating bodies.

In the event that an adjudicator has not been named and a nominating body has not been selected (or where an adjudicator has been named but has said that he or she is unwilling or unable to act, and a nominating body has not been selected), then the referring party may select any one of the five listed nominating bodies.

MPSub

The adjudicator will either be the person named (if one is so named) under item 46 of the sub-contract particulars, or will be nominated by one of the nominating bodies listed under that same sub-contract particulars item.

If the dispute or difference relates to clause 46.3 of the sub-contract conditions (i.e. where a dispute arises as to whether an instruction to open up for inspection or to test any (further) similar non-compliant works is reasonable), the nominating body needs to be advised that the selected adjudicator needs to be an individual (where practicable) with appropriate expertise and experience in the specialist area or discipline relevant to the direction or issue in dispute.

The choice of nominator of adjudicator under sub-contract particulars item 46 of the sub-contract agreement is the president or a vice-president or chairman or a vice-chairman of:

- The Royal Institute of British Architects
- The Royal Institution of Chartered Surveyors
- The Construction Confederation
- The National Specialist Contractors Council
- The Chartered Institute of Arbitrators

When completing sub-contract particulars item 46 the parties should delete all but one of the five above listed nominating bodies.

In the event that an adjudicator has not been named and a nominating body has not been selected (or where an adjudicator has been named but has said that he or she is unwilling or unable to act, and a nominating body has not been selected), then the referring party may select any one of the five listed nominating bodies.

ICSub

The adjudicator will either be the person named (if one is so named) under item 13 of the sub-contract particulars, or will be nominated by one of the nominating bodies listed under that same sub-contract particulars item.

If the dispute or difference relates to clause 3.10 of the sub-contract conditions (i.e. where a dispute arises in respect of instructions to open up for inspection or to test non-compliant works) and as to whether a direction issued under that clause is reasonable, the nominating body needs to be advised that the selected adjudicator needs to be an individual (where practicable) with appropriate expertise and experience in the specialist area or discipline relevant to the direction or issue in dispute.

The choice of nominator of adjudicator under sub-contract particulars item 13 of the sub-contract agreement is the president or a vice-president or chairman or a vice-chairman of:

- The Royal Institute of British Architects
- The Royal Institution of Chartered Surveyors

- The Construction Confederation
- The National Specialist Contractors Council
- The Chartered Institute of Arbitrators

When completing sub-contract particulars item 13 the parties should delete all but one of the five above listed nominating bodies.

In the event that an adjudicator has not been named and a nominating body has not been selected (or where an adjudicator has been named but has said that he or she is unwilling or unable to act, and a nominating body has not been selected), then the referring party may select any one of the five listed nominating bodies.

ICSub/D

All as ICSub, above.

ICSub/NAM

All as ICSub, above.

MWSub/D, ShortSub and SubSub

In line with clause 16.2 of the sub-contract conditions, the adjudicator will be the person nominated by one of the nominating bodies listed under that clause as selected by the referring party.

The choice of nominating bodies being:

- The Royal Institute of British Architects
- The Royal Institution of Chartered Surveyors
- The Construction Confederation
- The National Specialist Contractors Council
- The Chartered Institute of Arbitrators

What happens if either party objects to the appointment of a particular person as adjudicator?

Paragraph 10 of the Scheme makes it clear that if any party to a dispute objects to the appointment of a particular person as adjudicator, the objection shall not invalidate the adjudicator's appointment nor any decision he may eventually reach.

However, if an objection was made which the adjudicator considered was a justified complaint then the adjudicator may resign (refer to paragraph 9(1) of the Scheme), in which case the adjudication appointment procedure would need to recommence.

What happens after the adjudicator is appointed?

The referring party shall, not later than seven days from the date of the notice of adjudication, refer the dispute in writing (the referral notice) to the adjudicator (refer to paragraph 7(1) of the Scheme).

At the same time, the referring party shall send the referral notice (and all accompanying documents) to every other party to the dispute.

What happens if the referral notice is issued to the adjudicator more than seven days after the date of the notice of adjudication?

The latest case law[11] indicates that under the Scheme a referral notice issued more than seven days after the date of the notice of adjudication will be out of time and will therefore not have been effectively served.

What is the referral notice?

The referral notice is effectively the referring party's statement of case, and normally details the particulars of the dispute, details the contentions being relied upon and provides a statement of the relief or remedy sought.

Paragraph 7(2) of the Scheme makes it clear that the referral notice shall be accompanied by copies of, or relevant extracts from, the construction contract and such other documents as the referring party intends to rely upon.

What is the liability of the adjudicator?

Paragraph 26 of the Scheme makes it clear that neither the adjudicator, nor any employee or agent of the adjudicator, shall be liable for anything done or omitted in the discharge or purported discharge of his functions as adjudicator unless the act or omission is in bad faith.

What is the jurisdiction of the adjudicator?

The jurisdiction of an adjudicator may be considered as being the boundaries within which an adjudicator may act.

These boundaries are partly established by the terms of the sub-contract and partly by the wording of the notice of adjudication.

In respect of the jurisdictional boundaries established by the wording of the notice of adjudication, an adjudicator only has jurisdiction to deal with matters specifically raised in the notice of adjudication, although it has been found, in certain circumstances, that the notice of adjudication and the referral notice can be read in conjunction with one another.

Therefore, if an adjudicator was asked within a notice of adjudication to reach a decision that the referring party was entitled to a further payment of £500,000.00, it is considered that the adjudicator could not decide that the referring party was entitled to a further payment of £450,000.00 (although, if an adjudicator was asked within a notice of adjudication to reach a decision that the referring party was

---

[11] *Hart Investment* v. *Fidler and Others* [2006] EWHC 2857 (TCC).

entitled to a further payment of £500,000.00, *or such other sum as the adjudicator may decide*, then the adjudicator could decide that the referring party was entitled to a further payment of £450,000.00).

What happens if there is a query regarding the jurisdiction of the adjudicator?

An adjudicator has no power to rule on his own jurisdiction[12] although the parties can agree to give him that power[13].

Nevertheless, if an adjudicator's jurisdiction is challenged the adjudicator has to consider the challenge.

In considering the challenge, the adjudicator appears to have three options:

- he can ignore the challenge and proceed and leave it to a party to challenge his jurisdiction at enforcement proceedings;
- he can investigate the issue and decide that he has jurisdiction (but not rule that he has jurisdiction) in which case he will proceed; or
- he can decide he has no jurisdiction and decline to continue[14].

In considering a challenge, an adjudicator would normally ask both parties for their views, and would also have in mind that where a decision is subsequently challenged and the lack of jurisdiction is clearly arguable then the court may decline to enforce the decision pending its judgment on the jurisdiction issue[15].

Any points which are to be made as to whether the adjudicator has jurisdiction must be raised in a timely manner during the adjudication itself[16]. If it is not made in a timely manner the objection may be ignored by a court in any subsequent enforcement action.

When an adjudication has started and a court holds, before a decision is reached, that the adjudicator lacks jurisdiction, an injunction may be granted to restrain the referring party from taking any substantial step in the adjudication action or seeking to enforce or implement any award which may be made[17].

What are the powers of the adjudicator?

The adjudicator's powers relate solely to the contract under which he is appointed. The parties themselves give him such powers as he has.

The powers granted to the adjudicator are contractual not statutory.

In terms of JCT sub-contracts, the powers of the adjudicator are set out under paragraphs 12 to 19 inclusive of the Scheme.

Under Paragraph 12 of the Scheme the adjudicator is to:

---

[12] *The Project Consultancy Group* v. *The Trustees of the Gray Trust* [1999] BLR 377.
[13] *Nolan Davis Ltd* v. *Steven Catton* (2000) (unreported) QBD (TCC) No. 590.
[14] *Christiani & Nielsen Ltd* v. *The Lowry Centre Development Ltd* (2000) (unreported) TCC, 16 June 2000.
[15] *The Project Consultancy Group* v. *The Trustees of the Gray Trust* [1999] BLR 377.
[16] *Maymac Environmental Services Ltd* v. *Faraday Building Services Ltd* [2000] CILL 1686.
[17] *John Mowlem & Co. plc* v. *Hydra-Tight Ltd* (2000) CILL 1649.

(1) act impartially in carrying out his duties and shall do so in accordance with any relevant terms of the contract and shall reach his decision in accordance with the applicable law in relation to the contract; and

(2) avoid incurring unnecessary expense.

Paragraph 13 of the Scheme makes it clear that the adjudicator has fairly wide powers in respect of the procedure to be followed in the adjudication action, and paragraph 14 of the Scheme states that the parties shall comply with any request or direction of the adjudicator in relation to the adjudication.

What happens if a party does not comply with the adjudicator's direction?

Paragraph 14 of the Scheme states that the parties shall comply with any request or direction of the adjudicator in relation to the adjudication.

Paragraph 15 of the Scheme makes it clear that if, without showing sufficient cause, a party fails to comply with any request, direction or timetable of the adjudicator made in accordance with his powers, fails to produce any document or written statement requested by the adjudicator, or in any other way fails to comply with a requirement under these provisions relating to the adjudication, the adjudicator may:

(1) continue the adjudication in the absence of that party or of the document or written statement requested;

(2) draw such inferences from that failure to comply as circumstances may, in the adjudicator's opinion, be justified, and make a decision on the basis of the information before him attaching such weight (i.e. the degree of reliance that an adjudicator places upon evidence provided to him – the heavier the weight of evidence, the more reliance; and the lighter the weight of evidence, the less reliance) as he thinks fit to any evidence submitted to him outside any period he may have requested or directed.

When does the responding party submit its response to the referral notice?

Under the Scheme, the responding party has no right to issue a response.

However, the adjudicator has the power to request a response and to satisfy the requirements of natural justice (i.e. in this case that both parties must be given a fair opportunity to present their case) an adjudicator will almost invariably ask for a response to be issued.

As the Scheme does not specifically refer to a response from the responding party, no timetable is set for the issue of same. However, it is commonly the case that an adjudicator will direct that a response should be issued by no later than seven days after the issue of the referral notice.

Who can assist or represent the parties?

Unless agreed otherwise by the parties, paragraph 16(1) of the Scheme makes it clear that the parties can be assisted by or represented by such advisers or representatives (whether legally qualified or not) as they consider appropriate.

However, unless the adjudicator gives directions to the contrary, paragraph 16(2) of the Scheme makes clear that a party to a dispute may not be represented by more than one person where the adjudicator is considering oral evidence or representations.

What is the timetable for the adjudication action?

Under the Scheme there is no set timetable for the adjudication action other than that the adjudicator shall reach his decision not later than either 28 days after the date of the referral notice (as paragraph 19(1)(a) of the Scheme), or 42 days after the date of the referral notice if the referring party consents (as paragraph 19(1)(b) of the Scheme).

Any period in excess of 42 days after the date of the referral notice can only be agreed if both parties to the dispute consent (as paragraph 19(1)(c) of the Scheme).

Apart from the above long-stop dates, the Scheme does not set out any timetable for the submission of documents, the holding of meetings, etc., and any such time-table dates are set by the adjudicator to suit the circumstances of each individual case.

Any documents, etc., that the adjudicator directs should be issued to him by one party, must also be simultaneously issued by that party to the other party.

What can go wrong in an adjudication action?

First, an adjudicator may resign at any time on giving notice in writing to the parties to the dispute (as paragraph 9(1) of the Scheme).

If an adjudicator does simply resign then the referring party may serve a fresh notice of adjudication and may start the adjudication process anew (as paragraphs 9(3)(a) and 9(3)(b) of the Scheme).

Also, if an adjudicator dies or is otherwise unavailable, the parties may agree to replace the adjudicator or may apply for the nomination of an alternative adjudicator.

An adjudicator must resign because the dispute being referred to him is the same or substantially the same as one which has previously been referred to adjudication. Or an adjudicator may resign because the dispute varies significantly from the dispute referred to him in the notice of adjudication, and for that reason he is not competent to decide it. In such cases, the adjudicator will be entitled to payment of his reasonable fees and expenses, and will be entitled to apportion his fees as he sees fit (although the parties will remain jointly and severally liable for the adjudicator's fees and expenses) (refer to paragraph 9(4) of the Scheme).

Under paragraph 11(1) of the Scheme, the parties to a dispute (but not a single party unilaterally) can agree to revoke the appointment of the adjudicator – but if they do so they must pay the reasonable fees and expenses of the adjudicator other than if the revocation is because of some default or misconduct of the adjudicator (as paragraph 11(2) of the Scheme).

How does an adjudicator reach his decision?

An adjudicator is required to ascertain the facts and apply the law in reaching his decision.

It should be noted that a basic principle is that 'He who asserts must prove'. In other words, a party putting forward a case must prove that case. This is what is known as the 'burden of proof'.

As to evidence, an adjudicator must decide on which party's evidence is preferred on the 'balance of probabilities' principle.

If the evidence of the party with the burden of proof is no more convincing than the evidence of the other party, then the party with the burden fails.

Paragraph 17 of the Scheme requires the adjudicator to consider any *relevant* information submitted to him by any of the parties to the dispute and shall make available to them any information he has taken into account in reaching his decision.

The adjudicator is not obliged to give reasons for his decision, unless one of the parties requests that he does so (refer to paragraph 22 of the Scheme).

It is normally sensible to ask the adjudicator to give reasons because:

(1) if reasons are given, it is more likely that the parties will understand the adjudicator's decision, and therefore more likely that they will accept the adjudicator's decision as being the final resolution of the dispute; and

(2) a silent decision is more susceptible to attack in enforcement proceedings than a reasoned one[18].

What should the adjudicator's decision contain?

The adjudicator shall decide the matters in dispute, and he may take into account those matters which the parties to the dispute agree should be within the scope of the adjudication or are matters under the sub-contract which the adjudicator decides are necessarily connected with the dispute (refer to paragraph 20 of the Scheme).

What is the effect of an adjudicator's decision?

The decision of the adjudicator is binding on the parties, and the parties shall comply with it until the dispute is finally determined by legal proceedings, by arbitration (if the contract provides for arbitration or the parties otherwise agree to arbitration) or by agreement between the parties (as paragraph 23(2) of the Scheme).

It should be noted that when the dispute is finally determined by legal proceedings or by arbitration (as applicable) the adjudicator's decision is not open for review or appeal, the dispute is simply considered again with a replacement process as though the adjudicator's decision had not taken place.

---

[18] *Joinery Plus Ltd* (in administration) v. *Laing Ltd* [2003] BLR 184.

Because of the above, an adjudicator's decision is often referred to as being temporarily binding[19].

The papers used in an adjudication action are not generally privileged and they are open for all to see in a later arbitration or litigation action.

However, paragraph 18 of the Scheme states that the adjudicator and any party to the dispute shall not disclose to any other person any information or document provided to him in connection with the adjudication which the party supplying it has indicated is to be treated as confidential, except to the extent that it is necessary for the purposes of, or in connection with, the adjudication.

Obviously, if at some later date the parties finally decide to settle their dispute by arbitration or by litigation, all documents other than those that are privileged would be discoverable in any event.

Who pays the adjudicator's fees and expenses, and what level of fees and expenses is the adjudicator entitled to?

The adjudicator will normally inform the parties of his fees and expenses schedule at the time of his appointment.

The adjudicator is entitled to the payment of such reasonable amount as he may determine by way of fees and expenses reasonably incurred by him, and it is for the adjudicator to decide and direct how his fees and expenses should be apportioned between the parties (as paragraph 25 of the Scheme). If he does not apportion his fees and expenses then his fees and expenses fall equally between the parties.

Irrespective of the above, the parties to the adjudication are jointly and severally liable (which means that a party can be sued jointly with another party, or can be sued individually for the whole sum, leaving it to the party sued to recover the monies from the other party with whom he is jointly liable) for the adjudicator's fees and expenses (as paragraph 25 of the Scheme). Therefore, if one party defaults on any payment due to the adjudicator based upon the adjudicator's directions, then the other party must make payment, and must then sue the defaulting party for the amount paid on their behalf.

An adjudicator may attempt to hold a lien against his decision, in that he will not release his decision until the payment of his fees and expenses has been paid, but this approach is now rarely used following a case[20] where it was found that a decision issued late was effectively a nullity.

Who pays the party's own costs in connection with the adjudication action?

The general position is that each party pays for its own legal and other costs in connection with the adjudication action, irrespective of whether the party wins or loses in the adjudication action. Any application fee paid to an adjudicator

---

[19] *Macob Civil Engineering Ltd* v. *Morrison Construction Ltd* (1999) 1 BLR 93; *Herschel Engineering Ltd* v. *Bream Property Ltd* [2000] BLR 272.
[20] *Cubitt Building & Interiors Ltd* v. *Fleetglade Ltd* [2006] EWHC 3413 (TCC), 21 December 2006.

nominating body is normally considered to be part of the costs of the referring party in going to adjudication.

If both parties expressly give the adjudicator the jurisdiction to apportion the parties' own legal and other costs, then the adjudicator may (rather than must) do so.

How does the adjudicator issue his decision, and what happens if the adjudicator is late in issuing his decision?

The adjudicator must issue his decision to both parties (or their representatives) simultaneously, as soon as possible after he has reached his decision (as paragraph 19(3) of the Scheme). An adjudicator is bound to reach his decision within 28 days or any agreed extended period[21], and a decision which is not reached within 28 days or any agreed extended period is, under most circumstances, a nullity[22].

Paragraph 19(2)(a) of the Scheme notes that where the adjudicator fails, for any reason, to reach his decision in accordance with the stipulated time, any of the parties to the dispute may serve a fresh notice of adjudication and may request an alternative (properly appointed) adjudicator to act.

What happens if the adjudicator makes a mistake in his decision?

In the adjudication action between *Bouygues* v. *Dahl-Jensen*[23], when making calculations to answer the question of whether the payments made under the sub-contract represented an overpayment or an underpayment, the adjudicator overlooked the fact that that assessment should be based on the contract sum less retention, rather than on the gross contract sum. This was an error, but an error made within the jurisdiction of the adjudication. The question was could this error be rectified.

When this matter was reviewed by the Court of Appeal, it was held that, provided that an adjudicator acts within his jurisdiction, his decision will stand and be enforceable, even if a mistake is made.

In *C & B Scene* v. *Isobars*[24], the Court of Appeal found that even if an error on a matter of law was made, provided that the error was within the scope of the dispute between the parties, then the adjudicator's decision would be enforceable. In other words, the adjudicator had therefore answered the right question in the wrong way, and the referring party was entitled to enforce the decision.

However, these somewhat harsh effects may not apply if the adjudicator has made a simple slip or mathematical error.

In *Bloor* v. *Bowmer & Kirkland*[25], Judge Toulmin QC ruled that a term could be implied into adjudication agreements giving an adjudicator the power to correct decisions containing accidental errors or omissions or to clarify any ambiguity.

---

[21] *Barnes & Elliott* v. *Taylor Woodrow* [2003] EWHC 3100 (TCC).

[22] *Ritchie Brothers (PWC) Ltd* v. *David Philp (Commercials) Ltd* (2005) SLT 341.

[23] *Bouygues (UK) Ltd* v. *Dahl-Jensen (UK) Ltd* [2000] BLR 522, CA.

[24] *C & B Scene Concept Design Ltd* v. *Isobars Ltd* (2002) CILL 1829.

[25] *Bloor Construction* v. *Bowmer & Kirkland* (2000) CILL 1626.

This therefore introduced into adjudication a slip rule without which an adjudicator would not be able to correct decisions at all, since, without this slip rule, once he has made his decision an adjudicator would be *functus officio*, which means that having discharged his duty he had no further power to rescind his decision or re-try the case.

The slip rule was confirmed in a second case, *Edmund Nuttall v. Sevenoaks*[26], which provided that the power must be exercised within a reasonable time and must not cause prejudice (i.e. injury) to either party.

In applying the slip rule it is still left to the discretion of the adjudicator to acknowledge that he has made a simple mistake and to decide that he is prepared to correct that mistake.

How can you appeal against an adjudicator's decision?

You cannot appeal against an adjudicator's decision.

If either party is not prepared to accept the adjudicator's decision as being a final decision on the dispute or difference, it can refer the dispute to either legal proceedings or to arbitration, whichever is applicable.

When such a reference is made, it is not in the form of an appeal against the decision of the adjudicator but involves consideration of the dispute or difference as if no decision had been made by an adjudicator.

What happens if a party does not comply with the adjudicator's decision – how is an adjudicator's decision enforced?

In the vast majority of cases, parties do comply with the adjudicator's decision.

However, where this compliance does not occur, it must be remembered that adjudication is a contractual provision. Therefore, if an adjudicator's decision is not complied with, the non-compliance is a breach of contract and any action in respect of enforcement has to be under the contract.

First of all, the adjudicator's decision may not need to be enforced. For example, if an adjudicator decides that work that has been carried out complies with the contract, or if an adjudicator decides that a sub-contractor is entitled to an extension of time, then why do these decisions need to be enforced?

However, certain matters, normally those involving some form of payment, may need to be enforced if the party that is due to make payment (or to repay monies) refuses to make payment.

When a party is ordered by an adjudicator to make a payment and fails to do so and the other party starts proceedings for enforcement the party that was due to make payment will be liable for the other parties' reasonable costs in making the application even if the sum is paid in full prior to the court hearing[27].

The Scheme has its own mechanism for the enforcement of adjudicators' decisions through the use of an amended version of section 42 of the Arbitration Act 1996 (refer to paragraph 24 of the Scheme).

---

[26] *Edmund Nuttall Ltd v. Sevenoaks District Council* (2000) (unreported) QBD (TCC).
[27] *Outwing Construction Ltd v. H Randell & Son Ltd* [1999] BLR 156.

However, there are some drafting errors which exist when section 42 of the Arbitration Act 1996 is amended by paragraph 24 of the Scheme, and this makes the mechanism unwieldy and impractical. The problem with the resultant device was explored in *Macob* v. *Morrison*[28].

Consequently, and in essence, from *Macob* v. *Morrison* the application of (an amended) section 42 of the Arbitration Act 1996 is defunct, but a pattern for the enforcement of adjudicators' decisions in the English courts was set as being by way of a claim form followed by summary judgment.

This is now covered by Part 24 of the Civil Procedure Rules (CPR), which enables the court to decide a claim or a particular issue without trial.

What are the principal reasons why an adjudicator's decision may not be enforced?

The principal reasons why an adjudicator's decision may not be enforced are:

(1) Procedural unfairness or breach of natural justice[29].
(2) No clear identification as to the status of the decision or of the adjudicator himself.
(3) Unfitness of the adjudicator.
(4) Bias (actual or perceived).
(5) Dealing with matters which are not part of the adjudication, i.e. exceeding jurisdiction.
(6) Failing to deal with matters which are part of the adjudication.
(7) No clear decision, for example failing to state an identifiable sum, and therefore being incapable of performance.
(8) Illegality.

Arbitration

What is arbitration?

Arbitration may be defined as a private procedure for settling disputes whereby a dispute between parties is decided judicially by an impartial individual (or a panel of individuals) either selected by the parties or appointed for that purpose.

An individual so appointed is referred to as an arbitrator, whilst his decision is referred to as an award. An arbitrator's award is legally binding on all the parties to the arbitration proceedings to whom it is addressed.

What is the arbitration procedure?

The Arbitration Act 1996 and any 'rules' that are stipulated in a contract establish the procedure to be adopted in the arbitration procedure.

---

[28] *Macob Civil Engineering Ltd* v. *Morrison Construction Ltd* (1999) 1 BLR 93.
[29] *Discain Project Services* v. *Opecprime Developments Ltd* (No. 1) (2000) 8 BLR 402; *Woods Hardwick Ltd* v. *Chiltern Air Conditioning* [2001] BLR 23.

How is arbitration incorporated into the JCT sub-contracts?

DBSub

Arbitration is incorporated by way of article 5.

When the contract is executed, item 2 of the sub-contract particulars article 5 (relating to arbitration) should be marked either to apply or not to apply:

- If article 5 is marked to apply then arbitration will apply and legal proceedings will not apply.
- If article 5 is not marked either to apply or not to apply, article 5 will be deemed *not* to apply, and disputes and differences will be determined by legal proceedings.

Therefore, the default position is that disputes and differences under this sub-contract are to be finally determined by legal proceedings. If the parties wish arbitration to apply, they must mark article 5 so that arbitration applies.

MPSub

In the unamended MPSub form, there is no provision for arbitration at all.

ICSub

All as DBSub above.

ICSub/D

All as DBSub above.

ICSub/NAM

Arbitration is incorporated either by way of article 5 of IT16 (of the main contract information) or by agreement between the contractor and the sub-contractor as noted under item 2 of the sub-contract particulars:

- If either article 5 of IT16 or item 2 of the sub-contract particulars is marked to show that arbitration applies, then arbitration will apply and legal proceedings will not apply.
- However, if neither article 5 of IT16 nor item 2 of the sub-contract particulars is marked to show that arbitration either applies or does not apply, then arbitration will be deemed *not* to apply, and disputes and differences will be determined by legal proceedings.

Therefore, the default position is that disputes and differences under this sub-contract are to be finally determined by legal proceedings. If the parties wish arbitration to apply, they must ensure that either article 5 of IT16 or item 2 of the sub-contract particulars is marked so that arbitration applies.

MWSub/D, ShortSub and SubSub

Arbitration is incorporated by way of article 5.

When the contract is executed, article 5 (relating to arbitration) should be marked either to apply or not to apply:

- If article 5 is marked to apply then arbitration will apply and legal proceedings will not apply.
- If article 5 is not marked either to apply or not to apply, article 5 will be deemed *not* to apply, and disputes and differences will be determined by legal proceedings.

Therefore, the default position is that disputes and differences under this sub-contract are to be finally determined by legal proceedings. If the parties wish arbitration to apply, they must ensure that article 5 of the sub-contract particulars is marked so that arbitration applies.

Is arbitration automatically incorporated into the JCT sub-contracts?

In no case is arbitration automatically incorporated into the JCT sub-contracts. The default position in each case is that litigation applies. If the parties require arbitration to apply then they need to take a positive step to mark the contract accordingly.

If arbitration is incorporated into the JCT sub-contracts, can the parties ignore this and refer a dispute to court?

If all of the parties to the sub-contract agree not to refer a dispute to arbitration but agree to refer the dispute to litigation, then the dispute can be referred to court.

However, if one party attempts to refer a dispute to court in the case where an arbitration agreement exists, section 9(1) of the Arbitration Act 1996 makes it clear that the other party may apply to the court in which the proceedings have been brought to stay the proceedings in favour of arbitration.

Section 9(4) of the Arbitration Act 1996 states that upon such an application being made, 'the court shall grant a stay unless it is satisfied that the arbitration agreement is null and void, inoperative, or incapable of being performed'.

If a sub-contract ceases to exist, does the arbitration agreement also cease to exist?

It should be noted that although the parties may incorporate the arbitration agreement within the JCT sub-contracts, the arbitration agreement is separable from and survives the termination of the sub-contract. In this regard, section 7 of the Arbitration Act 1996 states that:

'Unless otherwise agreed by the parties, an arbitration agreement which forms or was intended to form part of another agreement (whether or not in writing) shall not be regarded as invalid, non-existent or ineffective because that other agreement is invalid, or did not come into existence or has become ineffective, and it shall for that purpose be treated as a distinct agreement'.

What can be referred to arbitration?

DBSub

Apart from the exceptions below, any dispute or difference between the parties of any kind whatsoever arising out of or in connection with the sub-contract (which could imply that matters such as misrepresentation and tort may be referred to arbitration), whether before, during the progress or after the completion of the abandonment of the sub-contract works or after the termination of the sub-contractor's employment, can be referred to arbitration.

The only exceptions to this being:

(1) any disputes or differences arising under or in respect of the Construction Industry Scheme or VAT, to the extent that legislation provides another method of resolving such disputes or differences; and
(2) any dispute or differences in connection with the enforcement of any decision of an adjudicator.

MPSub

In the unamended MPSub form, there is no provision for arbitration at all.

ICSub

All as DBSub above.

ICSub/D

All as DBSub above.

ICSub/NAM

All as DBSub above.

MWSub/D, ShortSub and SubSub

Apart from a matter that is in connection with the enforcement of any decision of an adjudicator, any dispute or difference arising under or in connection with the sub-contract can be referred to arbitration. This description of what can be referred to arbitration is probably no more restrictive than the description of what can be referred to arbitration under DBSub, as noted above.

What is the procedure that the arbitration must operate under?

DBSub

- Clauses 8.3 to 8.8 of the sub-contract conditions.
- The Construction Industry Model Arbitration Rules (CIMAR) current at the main contract base date (as noted in clause 8.3 of the sub-contract conditions). At the time that this book was written, the JCT 2005 edition of the CIMAR rules applied.

- The Arbitration Act 1996 (as noted in clause 8.8 of the sub-contract conditions).

If any amendments to the CIMAR rules have been issued by the JCT after the main contract base date, the parties may, by a joint notice in writing to the arbitrator, state that they wish the arbitration to be conducted in accordance with those amended rules.

MPSub

In the unamended MPSub form, there is no provision for arbitration at all.

ICSub

All as DBSub above.

ICSub/D

All as DBSub above.

ICSub/NAM

All as DBSub above.

MWSub/D, ShortSub and SubSub

- Clause 16.3 of the sub-contract conditions.
- The Construction Industry Model Arbitration Rules (CIMAR) applicable to the main contract.

What are the Construction Industry Model Arbitration Rules (CIMAR)?

CIMAR is a set of rules for the conduct of arbitrations that was originally initiated by the Society of Construction Arbitrators and which has now been adopted by most major construction institutions and bodies.

The Arbitration Act 1996 confers wide powers on the arbitrator unless the parties have agreed otherwise, but leaves detailed procedural matters to be agreed between the parties or, if not so agreed, to be decided by the arbitrator.

To avoid problems arising, it is advisable to agree as much as possible of the procedural matters in advance, and where arbitration applies in respect of the JCT sub-contracts, this is done by the incorporation of CIMAR, which are clearly written and self-explanatory.

What is the procedure to be followed under CIMAR?

CIMAR rule 5.1 makes it clear that the arbitrator has the right and duty to decide all procedural matters, subject to the parties' right to agree any matter.

Within 14 days of the arbitrator being appointed the parties must each send the arbitrator and each other a note indicating the nature of the dispute and amounts in issue, the estimated length for the hearing (if a hearing is necessary at all) and the procedures to be followed (in line with CIMAR rule 6.2).

Within 21 days of the arbitrator's acceptance of appointment the arbitrator is to hold a preliminary meeting with the parties to discuss the above matters (as CIMAR rule 6.3). A meeting does not need to be held if both parties agree, and if the arbitrator considers that a meeting is unnecessary (CIMAR rule 6.6).

The first decision for the arbitrator and the parties to make is whether:

- CIMAR rule 7 (a short hearing);
- CIMAR rule 8 (a documents only arbitration); or
- CIMAR rule 9 (the full procedure), is to apply.

The decision will, of course, depend on the scale and type of dispute.

Under all three rules referred to above, the parties exchange statements of claim and statements of defence, together with copies of documents and witness statements on which they intend to rely.

Under CIMAR rule 4, the arbitrator is given a wide range of powers, including:

- the power to obtain advice (CIMAR rule 4.2);
- the powers as set out in section 38 of the Arbitration Act 1996 (i.e. general powers exercisable by the tribunal) (CIMAR rule 4.3);
- the power to order the preservation of work, goods and materials even though they are part of work that is continuing (CIMAR rule 4.4);
- the power to request the parties to carry out tests (CIMAR rule 4.5); and
- the power to award costs (CIMAR rule 13).

Who will be the arbitrator?

DBSub

Clause 8.4.1 of this sub-contract notes that:

> 'the Arbitrator shall be an individual agreed by the Parties or, failing such agreement within 14 days (or any agreed extension to that period) after the notice of arbitration is served, appointed on the application of either Party in accordance with Rule 2.3 by the person named in the Sub-contract Particulars'.

The arbitrator selected by the parties should, of course, be somebody with the necessary expertise and experience to deal with the dispute or difference being referred to arbitration.

If the parties do not agree upon a particular arbitrator, an arbitrator appointing body should be used. An appointer cannot name an arbitrator until at least 14 days after the notice of arbitration is served.

The parties should agree upon one of a choice of three appointers under item 15 of the sub-contract particulars DBSub/A, namely president or a vice-president of:

- The Royal Institute of British Architects
- The Royal Institution of Chartered Surveyors
- The Chartered Institute of Arbitrators.

The parties should signify this agreement by deleting two of the listed bodies.

If the parties fail to select an appointer when executing the contract, then the default appointer will be the president or a vice-president of the Royal Institution of Chartered Surveyors.

MPSub

In the unamended MPSub form, there is no provision for arbitration at all.

ICSub

All as DBSub above, except that the three arbitrator appointers are included under item 13 of the sub-contract particulars ICSub/A (rather than under item 15 of the sub-contract particulars DBSub/A).

ICSub/D

All as DBSub above, except that the three arbitrator appointers are included under item 13 of the sub-contract particulars ICSub/D/A (rather than under item 15 of the sub-contract particulars DBSub/A).

ICSub/NAM

All as DBSub above, except that the three arbitrator appointers are included under item IT16 of the sub-contract information ICSub/NAM (rather than under item 15 of the sub-contract particulars DBSub/A).

MWSub/D

The arbitrator is to be an individual agreed by the parties within 14 days of service of the notice of arbitration.

If the parties are unable to reach agreement in the timescale stipulated, then the arbitrator is to be appointed by the same person as identified in the main contract particulars or, where no such person is identified, by the president or a vice-president of the Chartered Institute of Arbitrators (refer to clause 16.3.3).

ShortSub

The arbitrator is to be an individual agreed by the parties within 14 days of service of the notice of arbitration.

If the parties are unable to reach agreement in the timescale stipulated, then the arbitrator is to be appointed by the president or a vice-president of the Chartered Institute of Arbitrators (refer to clause 16.3.3).

SubSub

The arbitrator is to be an individual agreed by the parties within 14 days of service of the notice of arbitration.

If the parties are unable to reach agreement in the timescale stipulated, then the arbitrator is to be appointed by the president or a vice-president of the Chartered Institute of Arbitrators (refer to clause 16.3.3 of SubSub).

Is there any restriction on the appointment of an arbitrator?

Yes, there may be a restriction on the appointment of an arbitrator.

Clause 8.4.2 of DBSub/C, ICSub/C, ICSub/D/C and ICSub/NAM/C all note that where two or more related arbitral proceedings in respect of the main contract works or the sub-contract works fall under separate arbitration agreements, CIMAR rules 2.6, 2.7 and 2.8 shall apply.

CIMAR rules 2.6, 2.7 and 2.8 impose duties on persons having the function of appointing arbitrators to give consideration as to whether the same or a different arbitrator should be appointed where two or more related arbitral proceedings are commenced.

In such a situation, CIMAR rule 2.6 says that the same arbitrator should be appointed unless sufficient grounds are shown for not doing so; CIMAR rule 2.7 says that where an arbitrator has already been appointed in respect of one arbitral proceeding, due consideration should be given to the appointment of that same arbitrator in respect of a related arbitral proceedings; and CIMAR rule 2.8 says that two different appointers must also give due consideration for the appointment of the same arbitrator for related arbitral proceedings.

Although the MWSub/D, the ShortSub and the SubSub do not have a clause similar to clause 8.4.2 of DBSub/C, because the CIMAR rules apply to those sub-contracts, the above requirements may be implied in any event.

How is an arbitration action commenced?

An arbitration action is commenced by one party serving on the other party a written notice of arbitration. The date that the notice of arbitration is served will be taken as being the date that the arbitral proceedings commenced.

What is a notice of arbitration?

A notice of arbitration identifies (briefly) the dispute between the parties, and asks the other party to agree to the appointment of an arbitrator. It is quite common for a list of three names of prospective arbitrators to be included in the notice of arbitration.

What are the powers of the arbitrator?

The powers of the arbitrator are derived from the Arbitration Act 1996, and the Construction Industry Model Arbitration Rules (CIMAR).

Under clause 8.5 of DBSub/C, ICSub/C, ICSub/D/C and ICSub/NAM/C these powers include that the arbitrator may:

- rectify the sub-contract so that it accurately reflects the true agreement made by the parties;
- direct such measurements and/or valuations as may in his opinion be desirable in order to determine the rights of the parties;
- ascertain and award any sum which ought to have been the subject of or included in any certificate and to open up, review and revise any certificate, opinion, decision, requirement or notice;
- determine all matters in dispute which shall be submitted to him in the same manner as if no such certificate, opinion, decision, requirement or notice had been given.

Are there joinder provisions within the arbitration rules?

A joinder provision is where it is required that two or more related arbitral proceedings, in respect of the main contract works or the sub-contract works that fall under separate arbitration agreements, are heard under the same arbitration proceeding.

The JCT sub-contracts do allow for joinder provisions.

First, as noted above, clause 8.4.2 of DBSub/C, ICSub/C, ICSub/D/C and ICSub/NAM/C says that where two or more related arbitral proceedings in respect of the main contract works or the sub-contract works fall under separate arbitration agreements, CIMAR rule 2.6 applies. In such a situation, CIMAR rule 2.6 says that the same arbitrator should be appointed for related arbitral proceedings unless sufficient grounds are shown for not doing so.

Second, clause 8.4.3 of DBSub/C, ICSub/C, ICSub/D/C and ICSub/NAM/C says that after an arbitrator has been appointed either party may give a further notice of arbitration to the other party and to the arbitrator referring any other dispute which is to be decided in the arbitral proceedings, and CIMAR rule 3.3 shall apply. CIMAR rule 3.3 says that if the other party in the arbitral proceedings does not consent to the joinder provisions being operated in respect of a subsequent notice of arbitration, the arbitrator can decide whether or not the other dispute referred should be joined and consolidated within the same arbitral proceedings.

Although the MWSub/D, the ShortSub and the SubSub do not have clauses similar to clause 8.4.2 and clause 8.4.3 of DBSub/C, because the CIMAR rules apply to those sub-contracts, the above requirements may be implied in any event.

It should be noted that the courts will generally give a purposeful interpretation to joinder provisions that allow for the same arbitrator to be appointed in related disputes to avoid a multiplicity of proceedings, which might lead to excessive costs and inconsistent judgments[30].

---

[30] *City & General (Holborn) Ltd* v. *AYH plc* [2005] EWHC 2494 (TCC).

What is the effect of the arbitrator's award?

Other than for the exceptions listed below, the arbitrator's award shall be final and binding on the parties (refer to clause 8.6 of DBSub/C, ICSub/C, ICSub/D/C and ICSub/NAM/C).

The only exceptions being that the parties agree (pursuant to section 45(2)(a) and section 69(2)(a) of the Arbitration Act 1996) that either Party may (upon notice to the other party and to the arbitrator):

(1) apply to the courts to determine any question of law arising in the course of the reference; and
(2) appeal to the courts on any question of law arising out of an award made in an arbitration under the arbitration agreement.

Challenges to the award may also be made on the grounds of lack of substantive jurisdiction (section 67 of the Arbitration Act 1996), and on the grounds of serious irregularity (section 68 of the Arbitration Act 1996).

In respect of all of the challenges or appeals, section 70 of the Arbitration Act 1996 makes it clear that an application to the court cannot be made until any available arbitral process of appeal or review, or any available recourse under section 57 (correction or award or additional award) of the Arbitration Act 1996 have been exhausted.

If a party takes part, or continues to take part, in arbitral proceedings without making, either forthwith or within such time as is allowed by the arbitration agreement or the tribunal or by the relevant provision of the Arbitration Act 1996, an objection that the tribunal lacks substantive jurisdiction, that the proceedings have been improperly conducted, that there has been a failure to comply with the arbitration agreement, or that there has been any other irregularity affecting the tribunal or the proceedings, he may not raise that objection later, before the tribunal or the court, unless he shows that, at the time he took part in the proceedings, he did not know and could not with reasonable diligence have discovered the grounds for the objection (as section 73 of the Arbitration Act 1996).

Who pays for the costs of arbitration?

In respect of costs, the general principle is that costs should be borne by the losing party.

Subject to any agreement between the parties, the arbitrator has the widest discretion in awarding which party should bear what proportion of the costs of the arbitration.

In arbitration, costs may be awarded either on:

- an indemnity basis (i.e. reasonable costs reasonably incurred, with any doubt as to any costs that were reasonable or were reasonably incurred being resolved in favour of the receiving party); or on
- a standard basis (i.e. reasonable costs reasonably incurred proportionate to the matters in issue, with any doubt as to any costs that were reasonable or were reasonably incurred being resolved in favour of the paying party).

The standard basis is used in the majority of cases, and under the standard basis a successful party may (on average) only recover about 70 to 75% of his total costs incurred in the arbitration.

The 25 to 30% balance of the successful party's total costs is known as the non-recoverable costs, and this can be a significant factor when, on a large arbitration, a party's costs in the arbitration may run in to hundreds of thousands of pounds.

The question of costs in arbitration is normally greatly influenced by 'Calderbank' offers made by the defending party.

A Calderbank offer is the arbitral equivalent of making a CPR part 36 offer in the courts. When considering costs, the question for the arbitrator is whether the claimant has achieved more by rejecting the offer and going on with the arbitration than he would have achieved if he had accepted the offer. If the claimant fails to achieve more than he would have by accepting the offer, then he is likely to have an award of costs made against him.

The arbitrator will be entitled to charge fees and expenses and will apportion those fees between the parties as he sees fit.

Irrespective of how the arbitrator apportions his fees and expenses, the parties are jointly and severally liable to the arbitrator for the fees and expenses incurred.

Litigation

What is litigation?

Legal proceedings are often referred to as litigation and this involves taking a dispute for resolution through the civil courts.

What is the litigation procedure?

Procedure in the civil courts is governed by statutory Civil Procedure Rules (CPR).

The CPR are contained in separate parts which set out the specific procedural rules to be followed in specific situations (e.g. service of documents, statements of case, summary judgment, disclosure, evidence, costs, etc.).

The CPR implement recommendations of the Woolf report (*Access to Justice*, 1994), and seek to improve the speed, efficiency and accessibility of the civil court procedure.

CPR part 1 sets out the overriding objective of CPR and that is to enable the court to deal with cases justly (i.e. to ensure, so far as is practicable, that the parties are on an equal footing, to save expense, to deal with the case in a way that is proportionate, and ensure that the case is dealt with expeditiously and fairly).

What is the litigation process?

The basic steps involved in a civil action in the Queen's Bench Division (where most disputes in the construction industry would be dealt with) are:

- The action is begun by the claimant issuing and serving a claim form.
- The defendant must then serve a defence or an acknowledgement of service.
- There may be a counterclaim from the defendant and reply from the claimant, in which the cases are defined.
- There is then a procedure of disclosure of documents and inspection of documents.
- Witness statements are usually prepared.
- Eventually, a trial is held which culminates in a judgment.

The process leading up to the trial is known as the interlocutory proceedings.

At the commencement of any legal proceedings, the court will allocate the action to one of the three tracks on the following basis:

- Small claims track – appropriate for claims not exceeding £5,000. The small claims track is a largely documents only process.
- The fast track – appropriate for most cases where the amount claimed is over £5,000 but does not exceed £15,000. The fast track is used where the trial is not expected to exceed one day.
- The multi-track – appropriate for all other cases.

In certain situations, a claimant may apply to a court for judgment on his claim on the ground that there is no (or no sufficient) defence. This is known as summary judgment, and if the defendant is unable to satisfy the court that there is an issue which ought to be tried, the claimant will be entitled to immediate judgment (i.e. summary judgment) on the claim or the part of the claim in question. This application will only be successful where a defendant has no (or no sufficient) defence.

Who pays for the costs of litigation?

A successful party in legal proceedings is entitled to an order for payment of his legal costs by the loser (and the loser must also pay his own legal costs).

However, the award of costs is at the discretion of the courts, and because of this and because costs are normally awarded on a standard basis (i.e. any doubt as to whether costs were reasonably incurred or were reasonable and proportionate in amount to be resolved in favour of the paying party), a party that is entirely successful in its claim may only recover somewhere in the region of 70% of its actual legal costs.

Is there any protection against the costs of litigation?

A degree of protection against the costs of litigation may be obtained by a party making an offer of settlement.

CPR Part 36 allows a defendant to make an offer to settle, and (in simple terms) if the claimant does not exceed the amount of that offer at the end of the trial (even though the claimant may have won the action), then he is liable to pay both his own legal costs and the defendant's legal costs from the date of the notification of the CPR Part 36 offer.

Must a party use ADR (Alternative Dispute Resolution) before going to court?

There is no obligation on a party to use ADR before going to court; however, if a party does not use ADR, then the consequence could be that a cost sanction could be imposed, because in deciding what order (if any) to make about costs, the court must have regard to all of the circumstances, including the conduct of the parties. In respect of construction disputes, the impact of ADR is lessened by the pre-action protocol for construction and engineering disputes.

What is the pre-action protocol for construction and engineering disputes?

In respect of disputes in construction, the CPR incorporates a pre-action protocol for construction and engineering disputes.

The objectives of the protocol are:

(1) to encourage the exchange of early and full information about the prospective legal claim;
(2) to enable parties to avoid litigation by agreeing a settlement of the claim before commencement of proceedings; and
(3) to support the efficient management of proceedings where litigation cannot be avoided.

What is the procedure of the pre-action protocol for construction and engineering disputes?

The procedure of the pre-action protocol for construction and engineering disputes, is:

- The claimant issues to the defendant a letter of claim.
- The defendant must acknowledge the letter of claim within 14 days (if the defendant does not acknowledge the letter of claim within 14 days, the claimant is entitled to commence proceedings without further compliance with the protocol).
- The defendant must then issue a response and, if appropriate, a counterclaim within 28 days from the date of receipt of the letter of claim (the period of 28 days can be extended by agreement between the parties up to three months).
- The claimant is to issue a response to any counterclaim within the equivalent period allowed to the defendant to respond to the letter of claim.
- After the exchange of the above submissions, the parties should meet at a pre-action meeting in an attempt to narrow the issues in dispute, and to agree whether and what form of alternative dispute resolution procedure would be more suitable to settle the outstanding disputes, rather than litigation.

How is litigation (legal proceedings) incorporated into the JCT sub-contracts?

DBSub

Legal proceedings are incorporated by way of article 6.

When the contract is executed, item 2 of the sub-contract particulars article 5 (relating to arbitration) should be marked either to apply or not to apply:

- If article 5 is marked to apply then arbitration will apply and legal proceedings will not apply.
- If article 5 is not marked either to apply or not to apply, article 5 will be deemed *not* to apply, and disputes and differences will be determined by legal proceedings.

Therefore, the default position is that disputes and differences under this sub-contract are to be finally determined by legal proceedings.

If the parties wish the legal proceedings to be decided by a jurisdiction other than the English courts, then the appropriate amendment must be made to article 6.

MPSub

Legal proceedings are incorporated by way of clause 44.3.

In the unamended MPSub form there is no provision for arbitration at all.

ICSub

All as DBSub above.

ICSub/D

All as DBSub above.

ICSub/NAM

Legal proceedings are incorporated by way of article 6.

When the invitation to tender (ICSub/NAM/IT) is completed, item IT16 (of the main contract information) should be marked to show that article 5 (arbitration) either applies or does not apply. Alternatively, if the contractor and the sub-contractor agree that arbitration should apply, then this should be stated under item 2 of the sub-contract particulars:

- If either article 5 or item 2 of the sub-contract particulars is marked to show that arbitration applies, then arbitration will apply and legal proceedings will not apply.
- However, if neither article 5 nor item 2 of the sub-contract particulars is marked to show that arbitration either applies or does not apply, then arbitration will be deemed *not* to apply, and disputes and differences will be determined by legal proceedings.

Therefore, the default position is that disputes and differences under this sub-contract are to be finally determined by legal proceedings.

If the parties wish the legal proceedings to be decided by a jurisdiction other than the English courts, then the appropriate amendment must be made to article 6.

MWSub/D and ShortSub

Legal proceedings are incorporated by way of article 6.

When the contract is executed, article 5 (relating to arbitration) should be marked either to apply or not to apply:

- If article 5 is marked to apply then arbitration will apply and legal proceedings will not apply.
- If article 5 is not marked either to apply or not to apply, article 5 will be deemed *not* to apply, and disputes and differences will be determined by legal proceedings.

Therefore, the default position is that disputes and differences under this sub-contract are to be finally determined by legal proceedings.

Rather strangely this sub-contract does not specify that the English courts have jurisdiction, and, in the appropriate case, this is something that the parties may wish to clarify.

SubSub

All as MWSub/D and ShortSub above.

Table of Cases

Table of Statutes and Regulations

Subject Index